Lecture Notes in Computer Science   15530

Founding Editors

Gerhard Goos
Juris Hartmanis

Editorial Board Members

Elisa Bertino, USA
Wen Gao, China

Bernhard Steffen ⓘ, Germany
Moti Yung ⓘ, USA

Advanced Research in Computing and Software Science

Subline of Lecture Notes in Computer Science

Subline Series Editors

Giorgio Ausiello, *University of Rome 'La Sapienza', Italy*
Vladimiro Sassone, *University of Southampton, UK*

Subline Advisory Board

Susanne Albers, *TU Munich, Germany*
Benjamin C. Pierce, *University of Pennsylvania, USA*
Bernhard Steffen ⓘ, *University of Dortmund, Germany*
Deng Xiaotie, *Peking University, Beijing, China*
Jeannette M. Wing, *Microsoft Research, Redmond, WA, USA*

More information about this series at https://link.springer.com/bookseries/558

Krishna Shankaranarayanan ·
Sriram Sankaranarayanan · Ashutosh Trivedi
Editors

Verification, Model Checking, and Abstract Interpretation

26th International Conference, VMCAI 2025
Denver, CO, USA, January 20–21, 2025
Proceedings, Part II

Springer

Editors
Krishna Shankaranarayanan ⬤
Indian Institute of Technology Bombay
Mumbai, Maharashtra, India

Sriram Sankaranarayanan ⬤
University of Colorado Boulder
Boulder, CO, USA

Ashutosh Trivedi ⬤
University of Colorado Boulder
Boulder, CO, USA

ISSN 0302-9743   ISSN 1611-3349 (electronic)
Lecture Notes in Computer Science
ISBN 978-3-031-82702-0   ISBN 978-3-031-82703-7 (eBook)
https://doi.org/10.1007/978-3-031-82703-7

Preface

The 26th International Conference on Verification, Model Checking, and Abstract Interpretation (VMCAI 2025) was held in Denver, Colorado, USA from the 20th to the 21st of January 2025. VMCAI serves as a forum for research in the broad areas of Formal Verification, Model Checking, and Abstract Interpretation, facilitating interaction, cross-fertilization, and advancement of hybrid methods that combine these and related areas.

VMCAI 2025 received 48 submissions, of which 20 were accepted. The accepted papers included 2 accepted tool papers and 18 full-length regular and case-study papers. Each submission received at least three reviews from the program committee (PC) members. In some cases, PC members reached out to outside reviewers to provide expert opinions on the papers. The review period was followed by a discussion phase, wherein the PC members and chairs discussed each paper. We are pleased to report that VMCAI 2025 upheld the high standard for acceptance established in past editions. Each accepted paper either had a championing PC member and/or overall positive scores from the reviewers.

The VMCAI 2025 program featured two invited speakers:

- Jyotirmoy Deshmukh, University of Southern California, USA.
- Alexandra Silva, Cornell University, USA.

We extend our gratitude to the technical program committee and external reviewers for their hard work in reviewing the papers submitted to VMCAI 2025 and helping us put together a technical program consisting of high-quality papers. The program committee was made up of 50 members who represent the leading researchers in the core areas of VMCAI including abstract interpretation, programming languages, hardware and software model checking, cyber-physical systems, formal synthesis, formal methods in artificial intelligence, concurrency, and other areas. Following the tradition of previous years, VMCAI adopted a single-blind review process. After submission, the papers were reviewed by at least three PC members for a period of nearly 28 days. After submission of reviews, the online discussion period lasted ten days. During this process, PC members were asked to "champion" papers that they felt should be included in the VMCAI 2025 conference program. Every accepted paper had a "champion" who argued for the inclusion of the paper in the program, and overall positive review scores from all the reviewers. Where appropriate, one of the PC members undertook the task of summarizing the discussions as part of their review.

We also acknowledge the EasyChair conference management system for enabling smooth and efficient coordination of the submission and review processes.

VMCAI 2025 featured an artifact evaluation process, led by our artifact evaluation chairs:

- Ernst Moritz Hahn, University of Twente, The Netherlands.
- Andrea Turrini, Institute of Software, Chinese Academy of Sciences, China.

They assembled an artifact evaluation committee. Authors were encouraged to submit artifacts in addition to their submitted papers. The artifact submission was open for all papers. Authors of tool and case-study papers were strongly encouraged to submit artifacts. The artifact evaluation process was conducted in two stages: first, each artifact underwent a "smoke test" to allow authors to address minor technical issues. Then, each artifact was evaluated based on criteria for functionality, availability, and reusability by two or three members of the artifact evaluation committee. Overall, 17 artifacts were submitted, and 8 artifacts passed at least two out of the three evaluation criteria, earning appropriate reproducibility badges. The artifact evaluation committee concluded its work in a timely fashion allowing the artifact evaluation reviews to be considered during the main PC discussion phase. The artifact evaluations helped the PC make decisions on a few borderline papers. We thank the Artifact Evaluation Committee for their service to the community. We also express our gratitude to the VMCAI 2025 steering committee: Viktor Kuncak, Andreas Podelski, and Lenore Zuck for their valuable guidance. Special thanks go to POPL general chair Steve Zdancewic and Neringa Young, ACM SIGPLAN conference manager, for their indispensable assistance with the local organization. Finally, we are deeply grateful to Toyota Motors and Springer for offering valuable financial sponsorship for VMCAI 2025.

January 2025

Krishna Shankaranarayanan
Sriram Sankaranarayanan
Ashutosh Trivedi
Ernst Moritz Hahn
Andrea Turrini

Organization

Program Committee Chairs

S. Krishna   Indian Institute of Technology Bombay, India
Sriram Sankaranarayanan   University of Colorado Boulder, USA
Ashutosh Trivedi   University of Colorado Boulder, USA

Artifact Evaluation Committee Chairs

Ernst Moritz Hahn   University of Twente, The Netherlands
Andrea Turrini   Institute of Software, Chinese Academy of
                                Sciences, China

Program Committee

Parosh Aziz Abdulla   Uppsala University, Sweden
Elli Anastasiadi   Aalborg University, Denmark
Haniel Barbosa   Universidade Federal de Minas Gerais, Brazil
Ahmed Bouajjani   IRIF, Université Paris Cité, France
Soham Chakraborty   TU Delft, The Netherlands
Xin Chen   University of New Mexico, USA
Patrick Cousot   New York University, USA
Loris D'Antoni   University of California-San Diego, USA
Vijay D'Silva   Google, USA
Deepak D'Souza   Indian Institute of Science, India
Jyotirmoy Deshmukh   University of Southern California, USA
Roberto Giacobazzi   University of Arizona, USA
Priyanka Golia   Indian Institute of Technology Delhi, India
Eric Goubault   LIX, École Polytechnique, France
Shibashis Guha   Tata Institute of Fundamental Research, India
Ashutosh Gupta   Indian Institute of Technology Bombay, India
Franjo Ivancic   Google, USA
Marie-Christine Jakobs   Ludwig-Maximilians-Universität München,
                                Germany
Sumit Kumar Jha   Florida International University, USA

Amir Goharshady	Hong Kong University of Science and Technology, China
Katherine Kosaian	University of Iowa, USA
Jan Kretinsky	Masaryk University, Czechia
Viktor Kuncak	École Polytechnique Fédérale de Lausanne, Switzerland
Kaushik Mallik	Institute of Science and Technology Austria, Austria
Ravi Mangal	Colorado State University, USA
Eric Mercer	Brigham Young University, USA
Roland Meyer	TU Braunschweig, Germany
Antoine Miné	LIP6, UPMC, France
Sergio Mover	École Polytechnique, France
Koko Muroya	RIMS, Kyoto University, Japan
Ana Oliveira Da Costa	TU Wien, Austria
Andreas Pavlogiannis	Aarhus University, Denmark
Andreas Podelski	University of Freiburg, Germany
Vinayak Prabhu	Colorado State University, USA
Xavier Rival	Inria/ENS Paris, France
Philipp Rümmer	University of Regensburg, Germany
Sven Schewe	University of Liverpool, UK
Subodh Sharma	Indian Institute of Technology Delhi, India
Fabio Somenzi	University of Colorado Boulder, USA
Meera Sridhar	University of North Carolina Charlotte, USA
B. Srivathsan	Chennai Mathematical Institute, India
Aditya Thakur	University of California, Davis, USA
Saeid Tizpaz-Niari	University of Texas at El Paso, USA
Chao Wang	University of Southern California, USA
Piotr Wojciechowski	West Virginia University, USA
Dominik Wojtczak	University of Liverpool, UK
Zhen Zhang	Utah State University, USA
Đorđe Žikelić	Singapore Management University, Singapore

Artifact Evaluation Committee

Bruno Andreotti	Universidade Federal de Minas Gerais, Brazil
Julie Cailler	Loria, University of Lorraine, France
David Chocholatý	Brno University of Technology, Czechia
Ramiro Demasi	Universidad Nacional de Córdoba, Argentina
Antonio Di Stasio	City St George's, University of London, UK
Zafer Esen	Uppsala University, Sweden

Daniel Feldan	New York University, USA
Michal Hečko	Brno University of Technology, Czechia
Marian Lingsch-Rosenfeld	LMU Munich, Germany
Annabell Petri	University of Twente, The Netherlands
Edoardo Putti	University of Twente, The Netherlands
Qiyi Tang	University of Liverpool, UK
Zhe Tao	University of California, Davis, USA
Milla Valnet	Sorbonne Université, France
Zhonghan Wang	Institute of Software, Chinese Academy of Sciences, China
Emily Yu	Institute of Science and Technology Austria, Austria

Steering Committee

Viktor Kuncak	École Polytechnique Fédérale de Lausanne, Switzerland
Andreas Podelski	University of Freiburg, Germany
Lenore Zuck	University of Illinois Chicago, USA

Additional Reviewers

Agrawal, Prashant
Andreotti, Bruno
Azeem, Muqsit
Busatto-Gaston, Damien
Chalupa, Marek
Chassot, Samuel
Dong, Yifan
Dorfhuber, Florian
Gambhir, Sankalp
Jeppson, Joshua
Keskin, Eren

Kim, Jinwoo
Maseli, René
Nagy, Shaan
Paquet, Hugo
Qin, Xin
Rieder, Sabine
Tan, Yong Kiam
Taylor, Landon
Tepe, Jakob
Tunç, Hünkar Can
Williams, Sam

Contents – Part II

Contents – Part I

Abstract Interpretation

Abstract Interpretation

Abstract Local Completeness
A Local Form of Abstract Non-interference

Isabella Mastroeni$^{(\boxtimes)}$ (iD)

Computer Science Department, University of Verona, Verona, Italy
isabella.mastroeni@univr.it

Abstract. Abstract interpretation offers sound and decidable approximations for undecidable queries related to program behavior. The effectiveness of an abstract interpretation process relies entirely on the abstract domain itself, and the worst-case scenario is when the abstract interpreter responds with "don't know", meaning that anything could happen during runtime. The concept of completeness relates to the answer precision degree when performing computations within the abstract domain. However, completeness for a whole language is an ideal domain property, usually holding only on trivial situations [20]; for this reason, a local notion of completeness, holding on a *specific* program input, has been deeply investigated [5]. In this paper, we characterize an intermediate notion holding for sets of input *selected* by abstraction. In other words, completeness holds for a set of concrete inputs determined by one *abstract* input. In this sense, it is a form of local *abstract* completeness required locally on one specific *abstract* value. We provide a simple proof system for proving this weakening of completeness and several examples. Notably, this proof system is both language and domain-agnostic and can be readily incorporated to support static program analysis.

Keywords: Abstract interpretation · Abstract domain precision · Static analysis

1 Introduction

The accuracy of an abstract interpretation depends upon many factors [14]. (1) The quality of the abstract domain: In this case, the abstract domain has to represent in the most precise way all the intermediate invariants that hold at each program point along a computation trace [6,7,11,12,20]. (2) The precision of the fixpoint strategy: In this case, an appropriate fixpoint strategy can improve the precision of the analysis either by delaying widening/narrowing or dynamic trace partition [1,3,9,13,34]. (3) How the code is written: In this case, the non-compositional nature of the precision of abstract interpretation can be influenced by how the code is assembled [4,21]. All these factors imply that designing an optimal (namely, sound and complete) abstract interpretation for static program analysis is a very complex task. In this paper, we focus on the first issue. The

© The Author(s), under exclusive license to Springer Nature Switzerland AG 2025
K. Shankaranarayanan et al. (Eds.): VMCAI 2025, LNCS 15530, pp. 3–25, 2025.
https://doi.org/10.1007/978-3-031-82703-7_1

notion of completeness perfectly captures the structure of an abstract domain that will produce no false alarm for a given program, but unfortunately, in its original global formulation [18], it is too strong a requirement for a realistic analysis. For this reason, in the last decade, it has been weakened following different directions: a *local* formulation [5], holding only on a specific input, has been provided for characterizing a concretely verifying notion of precision; a *partial* formulation [8], where the precision is evaluated in a metric space built over the abstract domain, allows measured error in the precision requirement; a formulation focusing on the precision of data representation more than on precision of abstract computation, i.e., adequacy [23, 29], that proves domain precision w.r.t. a fixed loss threshold.

In this paper, we focus on completeness, following the same weakening direction followed by local completeness [5], but integrating it with the abstraction flexibility typical of abstract non-interference [24]. The idea we investigate is changing the perspective on completeness: We do not look at completeness as a *domain property* w.r.t. a function, but as a hyperproperty. In other words, we characterize it by studying how the function transforms *abstract* inputs into *abstract* outputs, exploiting the recent characterization of completeness as an instantiation of abstract non-interference [33]. In particular, we observe that by making abstract non-interference local on one of its inputs, we obtain a notion of completeness that lies in between global [18] and local completeness [5] requiring completeness to hold on a *set* of inputs. It is worth noting that, stated in this way, it may seem that this is nothing more than requiring local completeness on all the inputs in the set. Instead, the added value of the proposed notion is twofold: (1) We formally characterize the input *region* where completeness has to hold by abstraction, namely by fixing an *abstract* input; (2) We define a framework for weakening completeness, in which we can move from (global) completeness (corresponding to the top abstraction) to the local one (corresponding to the identity abstraction), passing through non-trivial abstractions.

To address this issue, in the context of program analysis, we also introduce a simple proof system based on structural induction on a simple but Turing-complete language (the same used in [5]) to verify whether completeness is preserved on specific *abstract* inputs. This is achieved by proving the validity of triples holding on a program r w.r.t. an input concrete assertion c and an output one d. Defining as $[\![r]\!]_A^\sharp$ the abstract semantics of r given by an abstract interpreter defined on the abstract domain A, namely $[\![r]\!]_A^\sharp : A \to A$, then if the triple $\vdash_A (c) \, r \, (d)$ is derived in the proof system, then we can say that A is complete on all the inputs c' sharing with c the same domain property A. The proof system we introduce is very simple and can be efficiently checked online using program analysis tools.

Paper Roadmap. Section 2 provides an overview of abstract interpretation and programming language semantics. We used regular commands as the general programming language to establish a language-agnostic framework. In Sect. 3, we recall abstract domain completeness and introduce the novel concept of local abstract non-interference (local ANI for short). Section 4 outlines the procedure for *adjusting* an abstract domain towards local ANI, finally, in Sect. 5.2, we present a sound proof system for local ANI, and Sect. 6 concludes the paper with closing remarks and potential future directions.

Related Works. Completeness in abstract interpretation [18] is one of the two notions on which the proposed framework is based. Completeness, in abstract interpretation, characterizes the accuracy of a computation on an abstract domain when compared with the concrete computation. The importance of this notion in the context of static analysis comes from the fact that if an abstract domain is complete, false alarms are avoided [12]. The proof system we propose here inherits the locality of the notion of *local completeness* introduced in [5] and [7], which also forms the basis for the refinement strategy in *abstract interpretation repair* (AIR) [6]. Consequently, the proof systems for completeness in [20] and local completeness in [7] can be seen somehow as instantiations of the one proposed here, simply by considering $A = \top$ when dealing with global completeness, and $A = id$ when dealing with the local one.

The other closest notion is abstract non-interference [15,22]. Non-interference is initially formalized in the context of language-based security for modeling confidentiality. ANI is proposed as a framework for weakening non-interference by characterizing attackers as abstract interpretations. More recently [16,22], ANI has been generalized as a hyperproperty [27,30,31], and in one of its general instantiations, it has been proved to be equivalent to completeness [33].

In [8], the concept of partial completeness is introduced to *measuring/quantifying* an abstract interpretation's accuracy. This is accomplished by incorporating the abstract domain into a (quasi) metric space, which relaxes the requirement for local completeness to hold up to a metric neighbor of the exact (complete) solution.

Another direction for changing perspective on the characterization of abstract interpretation precision is adequacy [23,29] This notion is not intended to provide any quantitative estimate of the quality of an abstraction but rather an answer of whether the resulting invariant is below (is approximated by) a given bound τ in the abstract domain. This guarantees that at least the information in τ will be included in the computed approximate invariant.

2   Background

If S is a set, $\wp(S)$ denotes the powerset of S. If $f : S \to T$, then we often abuse notation by calling f also its additive lifting $f : \wp(S) \to \wp(T)$ to sets of values: $f(X) \stackrel{\text{def}}{=} \{ f(x) \mid x \in X \subseteq S \}$. If $f : S \longrightarrow T$ and $g : T \to U$, we denote by $g \circ f$ (or simply gf) their composition. If $f : S \to S$, and $n \in \mathbb{N}$ we define $f^n : S \to S$ inductively as $f^0 \stackrel{\text{def}}{=} id_S$ (the identity on S), $f^{n+1} \stackrel{\text{def}}{=} f \circ f^n$. In a partial ordered structure C, we use $\leq_C$ to denote the partial order relation, $\vee_C$ for lub, $\wedge_C$ for glb, $\top_C$ and $\bot_C$ for respectively the greatest and the least elements (we avoid the annotation C when clear from the context). A function $f : S \to T$ between ordered structures is monotone if it preserves the order, i.e., $c \leq_S d \Rightarrow f(c) \leq_T f(d)$. It is additive if it preserves arbitrary lubs (co-additivity is dually defined).

2.1   Abstract Interpretation

Abstract interpretation [11,12], is a formal framework for approximating program semantics defined on a concrete domain C, by means of some abstraction A of C. Given complete lattices C and A, a pair of functions $\alpha : C \to A$ and $\gamma : A \to C$ forms a Galois connection (GC for short) if for any $c \in C$ and $a \in A$ we have $\alpha(c) \leq_A a \Leftrightarrow c \leq_C \gamma(a)$. In this case, α (resp. γ) is the abstraction/left adjoint (resp. concretization/right adjoint), and it is additive (resp. co-additive). Co-additive functions $g : A \to C$ admits left adjoint $g^- \overset{\text{def}}{=} \lambda c. \bigwedge_A \{ a \,|\, c \leq_C g(a) \}$. An *upper closure operator* (uco for short) $\rho : C \to C$ on a poset C is monotone, idempotent, and extensive, i.e., $\forall c \in C. \ c \leq_C \rho(c)$. If in a GC $\alpha \circ \gamma = id_A$ then it is a Galois insertion (GI) and $\gamma \circ \alpha$, simply written $\gamma\alpha$, is an uco. Let us denote by $Abs(C)$ the class of abstract domains (GI or uco) of C. It is well known that $Abs(C)$ is isomorphic to the lattice of all ucos on C, therefore when dealing with GI and clear from the context, we abuse notation by denoting as A both the domain of abstract objects ($A = \gamma\alpha(C)$) and the closure operator ($A = \gamma\alpha$). Given an abstract domain $A \in Abs(C)$ ($A = \gamma\alpha$) and a concrete function $f : C \to C$, an abstract function $f_A^\sharp : A \to A$ is a *sound* approximation of f when $\alpha \circ f \leq_A f_A^\sharp \circ \alpha^1$. The best correct approximation (BCA) of f in A is the function $f^A \overset{\text{def}}{=} \alpha \circ f \circ \gamma$; other abstractions are less precise. When dealing with the BCA, soundness becomes a domain property defined as $A \circ f \leq_C A \circ f \circ A$.

2.2   Regular Commands

Following [5,35] (see also [36]), we consider the language Reg_{Exp} of regular commands in the top of Fig. 1 (where $\oplus$ denotes non-deterministic choice and $*$ is the Kleene closure), parametric on a grammar of expressions Exp. This language is general enough to represent control-flow graphs of basic expressions, and therefore, it covers simple deterministic imperative languages.

The Concrete Semantics. Let Reg_{Exp} be a regular language. We assume the basic transfer expressions have a semantics $(\!| \cdot |\!) : \text{Exp} \to C \to C$ on a complete lattice C such that $(\!|e|\!)$ is an additive function. The concrete semantics [35] $[\![\cdot]\!] : \text{Reg}_{\text{Exp}} \to C \to C$ of regular commands is inductively defined as follows: Let $c \in C$

$$[\![e]\!]c \overset{\text{def}}{=} (\!|e|\!)c \qquad\qquad [\![r_1 \oplus r_2]\!]c \overset{\text{def}}{=} [\![r_1]\!]c \vee_c [\![r_2]\!]c$$
$$[\![r_1 ; r_2]\!]c \overset{\text{def}}{=} [\![r_2]\!]([\![r_1]\!]c) \quad [\![r^*]\!]c \overset{\text{def}}{=} \bigvee_c \{ [\![r]\!]^n c \,|\, n \in \mathbb{N} \}$$

The Abstract Semantics. Let $A \in Abs(C)$, the abstract semantics of regular commands $[\![\cdot]\!]_A^\sharp : \text{Reg}_{\text{Exp}} \to A \to A$ on the abstract domain A is defined by

---

[1] By $\leq_A$ we denote the partial order relation on A.

$$\text{Reg}_{\text{Exp}} \ni \mathbf{r} ::= \mathbf{e} \mid \mathbf{r}; \mathbf{r} \mid \mathbf{r} \oplus \mathbf{r} \mid \mathbf{r}^* \qquad \mathbf{e} \in \text{Exp}$$

$$\mathcal{L} = \text{Reg}_{\mathcal{L}\,\text{Exp}}$$
$$\mathcal{L}\,\text{Exp} \ni \mathbf{e} ::= \mathbf{skip} \mid x := \mathbf{a} \mid \mathbf{b}?$$
$$\text{AExp} \ni \mathbf{a} ::= x \mid n \mid \mathbf{a} + \mathbf{a} \mid \mathbf{a} - \mathbf{a} \mid \mathbf{a} * \mathbf{a}$$
$$\text{BExp} \ni \mathbf{b} ::= \mathbf{tt} \mid \mathbf{ff} \mid \mathbf{a} = \mathbf{a} \mid \mathbf{a} \leq \mathbf{a} \mid \mathbf{a} < \mathbf{a} \mid \mathbf{b} \wedge \mathbf{b} \mid \neg \mathbf{b}$$
$$\text{Var} \ni x \text{ (variables)}, \qquad n \in \mathbb{Z} \text{ (values)}$$

Fig. 1. The syntax of Reg_{Exp}, parametric on Exp, and of $\mathcal{L}$.

structural induction as follows:

$$\llbracket \mathbf{e} \rrbracket_A^\sharp \mathbf{a} \overset{\text{def}}{=} \llbracket \mathbf{e} \rrbracket^A \mathbf{a}$$
$$\llbracket \mathbf{r}_1; \mathbf{r}_2 \rrbracket_A^\sharp \mathbf{a} \overset{\text{def}}{=} \llbracket \mathbf{r}_2 \rrbracket_A^\sharp (\llbracket \mathbf{r}_1 \rrbracket_A^\sharp \mathbf{a})$$
$$\llbracket \mathbf{r}_1 \oplus \mathbf{r}_2 \rrbracket_A^\sharp \mathbf{a} \overset{\text{def}}{=} \llbracket \mathbf{r}_1 \rrbracket_A^\sharp \mathbf{a} \vee_A \llbracket \mathbf{r}_2 \rrbracket_A^\sharp \mathbf{a}$$
$$\llbracket \mathbf{r}^* \rrbracket_A^\sharp \mathbf{a} \overset{\text{def}}{=} \bigvee_A \left\{ (\llbracket \mathbf{r} \rrbracket_A^\sharp)^n \mathbf{a} \mid n \in \mathbb{N} \right\}$$

where we recall that $\llbracket \mathbf{e} \rrbracket^A$ is the BCA in A of $\llbracket \mathbf{e} \rrbracket$, i.e., $\llbracket \mathbf{e} \rrbracket^A = \alpha \circ \llbracket \mathbf{e} \rrbracket \circ \gamma^2$.

By structural induction, we can prove that this abstract semantics is monotonic and correct, i.e., $\alpha \circ \llbracket \mathbf{r} \rrbracket \leq_A \llbracket \mathbf{r} \rrbracket_A^\sharp \circ \alpha$ (or equivalently $\alpha \circ \llbracket \mathbf{r} \rrbracket \circ \gamma \leq_A \llbracket \mathbf{r} \rrbracket_A^\sharp$). Note that this semantics is computable only when the abstract domain has no infinite chains; indeed, to make it computable, we have to define a widening operator forcing convergence.

Programs. We consider standard basic transfer functions for expressions used in deterministic while programs: no-op instruction, assignments, and Boolean guards, i.e., we consider the regular language $\mathcal{L}$ defined in Fig. 1. Hence, we have to deal with integer variables and with stores. Let us denote $Var(\mathbf{r})$ the set of all the variables in $\mathbf{r} \in \text{Reg}_{\text{Exp}}$, and let $C \overset{\text{def}}{=} \wp(\mathbb{M})$ the concrete domain of sets of stores, where the store $\mathbb{m} \in \mathbb{M}$ is a function associating values to a set of variables, i.e., $\mathbb{m} : V \to \mathbb{Z}$, $V \subseteq_F \text{Var}$ (finite subset).

In particular, the basic transfer function semantics $(\!|\mathbf{e}|\!) : C \to C$ for the expressions of $\mathcal{L}$, is defined as: $M \in C$ (i.e., $M \subseteq \mathbb{M}$)

$$(\!|\mathbf{skip}|\!)M \overset{\text{def}}{=} M$$
$$(\!|x := \mathbf{a}|\!)M \overset{\text{def}}{=} \left\{ \mathbb{m}[x \mapsto (\!|\mathbf{a}|\!)\mathbb{m}] \mid \mathbb{m} \in M \right\}$$
$$(\!|\mathbf{b}?|\!)M \overset{\text{def}}{=} \left\{ \mathbb{m} \in M \mid (\!|\mathbf{b}|\!)\mathbb{m} = \mathbf{tt} \right\} = M \cap (\!|\mathbf{b}|\!)$$

Where $(\!|\mathbf{a}|\!) : \mathbb{M} \to \mathbb{Z}$ and $(\!|\mathbf{b}|\!) : \mathbb{M} \to \{\mathbf{tt}, \mathbf{ff}\}$ are the standard evaluation semantics for arithmetic and boolean expressions, respectively, and where we denote $(\!|\mathbf{b}|\!) \overset{\text{def}}{=} \left\{ \mathbb{m} \in \mathbb{M} \mid (\!|\mathbf{b}|\!)\mathbb{m} = \mathbf{tt} \right\}$ the truth semantics of $\mathbf{b}$.

---

[2] Note that, the abstract program semantics so far defined is, in general, an over-approximation of the BCA , i.e., $\forall \mathbf{r} \in \mathcal{L}, \forall \mathbf{a} \in A.\ \llbracket \mathbf{r} \rrbracket^A \mathbf{a} \leq_A \llbracket \mathbf{r} \rrbracket_A^\sharp \mathbf{a}$.

The concrete semantics of the regular language defined above instantiated to $\mathcal{L}$ corresponds precisely to the denotational semantics defined [10] starting from standard operational semantics of non-deterministic choice and iteration [36].

3   Localizing Abstract Non-interference

When speaking of abstract domain precision/completeness in static analysis, we usually pay attention only to the abstract domain, treating the function as a fixed parameter on which we build an analysis as *precise* as possible. Indeed, precision, modeled in terms of completeness [18], is usually presented as a domain property characterized w.r.t. a semantic function. In recent works, it has been realized that completeness, in its original, global formulation, is quite a strong domain property, hardly satisfied in real applications. For this reason, recent works weakened completeness mainly following two directions: By weakening the universal quantifier, namely requiring completeness to hold *only* on one input and not necessarily on all inputs (local completeness [5]); By weakening the equality, admitting a measured error, i.e., a distance instead of the equality (partial completeness [8]). Here, we follow the first direction and consider a middle choice, namely a "regional" completeness, where we require completeness only on a set of inputs. It is worth noting that local completeness can be checked for a set of inputs, but we would have to repeat the verification for each point in the set. Instead, the idea we propose here consists of choosing the set to verify the completeness property once for the whole set. In particular, we aim to use the input abstraction itself to fix the input set on which to check completeness.

We obtain this by looking at completeness from a different perspective, namely as a matter of *interference* between the abstraction process induced by the abstract domain and the semantic function, namely as a *local* hyperproperty. In other words, as a hyperproperty holding only for one set of computations. In fact, we observe that a local version of ANI allows us to precisely check the completeness of only a set of inputs, those sharing the same abstraction. In this sense, we can think of local ANI as an *abstract* local completeness, namely *local* on an abstract input, which is the abstraction of a concrete fixed input.

3.1   Completeness vs Abstract Non-interference

The standard way for characterizing precision in abstract interpretation is by means of completeness. Let $A \in Abs(C)$ be an abstraction of C and $f : C \rightarrow C$ a concrete computation on C, e.g., the program semantics, then the abstract function $f_A^\sharp$ is said to be a *complete/precise* [12,18] approximation of f on A if $\alpha \circ f = f_A^\sharp \circ \alpha$. Intuitively, it means that if we abstract $c \in C$ (the input of f), we apply the f approximation $f_A^\sharp$, we obtain the same abstract element that we would obtain by abstracting the result of f applied directly on c (without an initial abstraction). Note that if there exists a complete approximation $f_A^\sharp$ of f, then f^A (the BCA of f) is itself complete. Completeness of f^A intuitively means that f^A is the most precise approximation of f and, therefore, in this case, completeness can be characterized as a domain property: $A \in Abs(C)$ is said to be a complete abstraction for f if $A \circ f \circ A = A \circ f$. In a more general setting, we can provide the following notion of completeness [5,18] for a function potentially defined on different input and output domains[3].

Definition 1 (Completeness). *[11,18] Let $f : C_1 \rightarrow C_2$ be a function on complete lattices C_i (potentially different), and let $A_i \in Abs(C_i)$ (for $i \in \{1,2\}$) be abstractions, respectively, of input and output domains, with $A_i = \gamma_i \alpha_i$. An abstraction[4] $f^\sharp : A_1 \rightarrow A_2$ of f is* complete *for f if $\forall c \in C_1$ we have $\alpha_2 \circ f(c) = f^\sharp \circ \alpha_1(c)$. When the BCA of f is complete, then completeness of the pair of abstractions $\langle A_1, A_2 \rangle$ w.r.t. f is $\forall c \in C_1. A_2 \circ f \circ A_1(c) = A_2 \circ f(c)$.*

As observed before, completeness describes the problem of precision as a domain property, while we would like to model the problem in terms of how the computation of f *interferes* with the abstraction. In this sense, abstract non-interference[5] [22] comes in help since one of its forms has been proved to be equivalent to completeness [33].

Definition 2 (Abstract Non-Interference). *[22] Let $f : C_1 \rightarrow C_2$ be a function on complete lattices C_i (potentially different), and let $A_i \in Abs(C_i)$ (for $i \in \{1,2\}$) be abstractions, respectively, of input and output domains. We have abstract non-interference (ANI for short) of f on the pair of abstractions $\langle A_1, A_2 \rangle$ if $\forall c, d \in C_1. A_1(d) = A_1(c) \Rightarrow A_2 \circ f(d) = A_2 \circ f(c)$.*

In [33], it has been proved that the completeness domain property of Definition 1 is equivalent to Definition 2. The next result tells us that we can characterize ANI also by considering only the abstraction functions.

Proposition 1. *Let $f : C_1 \rightarrow C_2$ be a function on complete lattices C_i, and let $A_i \in Abs(C_i)$ (for $i \in \{1,2\}$) be abstractions, respectively, of input and output domains, with $A_i = \gamma_i \alpha_i$. We have ANI of f on the pair of abstractions $\langle A_1, A_2 \rangle$ iff $\forall c, d \in C_1. \alpha_1(d) = \alpha_1(c) \Rightarrow \alpha_2 \circ f(d) = \alpha_2 \circ f(c)$*

3.2   Is Local Completeness a Local Form of ANI ?

Giacobazzi et al. [20](Theorem 4.5) proved that completeness holds for all programs in a Turing complete programming language only for trivial abstract domains. This means that the only complete abstract domains for all programs are the straightforward ones: (1) The identical abstraction, making abstract and concrete semantics the same, and (2) The top abstraction, making all programs

---

[3] The choice of considering functions with potentially different input and output domains aims to underline the generality of the definition, independent from the field of application.

[4] In this case, for the sake of readability, we avoid annotating the abstract domains on which we define the abstraction.

[5] The name *(abstract) non-interference* is an inheritance from language-based security where the notion has been introduced for modeling when a (property of a) sensitive input was not interfering with (a property of) an observable output [15].

equivalent by abstract semantics. Giacobazzi et al. [20] observed that since skip is always trivially complete and composition, conditional, and loop statements all preserve the completeness of their subprograms, the only sources of incompleteness may arise from assignments and Boolean guards. At this point, since all interesting programs include Boolean guards, complete abstract domains refining a given domain may become very close to the concrete domain, limiting the effectiveness of this notion of completeness in program analysis [7]. In other words, the standard notion of (global) completeness for Boolean guards is too strong a requirement for abstract domains, often met in practice just by trivial guards or domains.

While completeness can be hard/impossible to achieve globally, i.e., for all possible sets of stores, it could well happen that completeness holds locally, i.e., just for some store properties [5]. For this reason, local completeness has been proposed, corresponding to considering completeness only along a fixed program execution [5].

Definition 3 (Local Completeness). *[5] Let $f : C_1 \to C_2$ be a function on complete lattices C_i, and let $A_i \in Abs(C_i)$ (for $i \in \{1,2\}$) be abstractions, respectively, of input and output domains, with $A_i = \gamma_i \alpha_i$. An abstraction $f^\sharp : A_1 \to A_2$ of f is local complete for f on $c \in C_1$ if $\alpha_1 \circ f(c) = f^\sharp \circ \alpha_2(c)$. If the BCA is local complete on c then we say that the pair of abstractions $\langle A_1, A_2 \rangle$ is local complete w.r.t. f on $c \in C_1$ if $A_2 \circ f \circ A_1(c) = A_2 \circ f(c)$.*

In the previous section, we showed how (global) completeness can be modeled as a matter of interaction of the function f with the abstraction of inputs and outputs. We can observe that the same relation does not hold when we focus on one specific input, namely when we deal with local completeness. In this case, local completeness can be shown to be weaker than a local version of abstract non-interference holding on a specific input. Starting from this observation, we look for a weaker form of ANI that is equivalent to local completeness. We call this form of ANI *relaxed abstract non-interference* (relaxed ANI for short).

Definition 4 (Relaxed ANI). *Let $f : C_1 \to C_2$ be a function on complete lattices C_i, and let $A_i \in Abs(C_i)$ (for $i \in \{1,2\}$) be abstractions, respectively, of input and output domains. We have relaxed ANI of f on the pair of abstractions $\langle A_1, A_2 \rangle$ for $c \in C_1$, if $\forall d \in C_1. A_1(d) = A_1(c) \Rightarrow A_2 \circ f(d) \leq A_2 \circ f(c)$.*

Intuitively, we have relaxed ANI on c if, for all the points sharing, with the fixed input, the same input property, the output property may be more concrete (or equal) but not necessarily equal, as ANI would require. Note that relaxed ANI only exists locally, since it is trivial to show that quantifying universally on c then we imply the equality of the abstract results, and therefore ANI (Definition 2).

Then relaxed ANI precisely models local completeness by looking at the *local* interaction between the function and the input and output observations and offers the possibility of providing a further characterization of local completeness.

Proposition 2. *Let $f : C_1 \to C_2$ be a function on complete lattices C_i, and let $A_i \in Abs(C_i)$ (for $i \in \{1,2\}$) be abstractions, respectively, of input and output domains. Let $c \in C_1$, and the following facts are equivalent:*

1. $\forall d \in C_1. A_1(d) = A_1(c) \Rightarrow A_2 \circ f(d) \leq A_2 \circ f(c)$ *(Relaxed* ANI *)*;
2. $A_2 \circ f \circ A_1(c) = A_2 \circ f(c)$ *(Local completeness)*;
3. $A_2(\bigvee \{ f(d) \mid A_1(d) = A_1(c) \}) = A_2 \circ f(c)$.

Relaxed ANI shows us that we can investigate an intermediate notion, namely a new *local* form of abstract non-interference, which now we know to be stronger than local completeness (Definition 3), but, being local, surely weaker than completeness (Definition 1).

3.3   Abstract Local Completeness as Local Abstract Non-Interference

Abstract non-interference is a hyperproperty [30,31]. Namely, it involves two computations starting from different inputs, and in fact, in standard (global) ANI , we have the universal quantifier for both inputs. Here, we decide to fix one of the inputs and let the other be universally quantified, which corresponds precisely to require the output equality for relaxed ANI . In this way, we are sure to obtain a notion that is stronger than relaxed ANI (and therefore of local completeness) but clearly weaker than completeness, which is equivalent to ANI .

Definition 5 (Local ANI).[6] *Let* $f : C_1 \rightarrow C_2$ *be a function on complete lattices* C_i, *and let* $A_i \in Abs(C_i)$ *(for* $i \in \{1, 2\}$*) be abstractions, respectively, of input and output domains. We have* local ANI *of* f *on the pair of abstractions* $\langle A_1, A_2 \rangle$ *for* $c \in C_1$ *if* $\forall d \in C_1. A_1(d) = A_1(c) \Rightarrow A_2 \circ f(d) = A_2 \circ f(c)$.

As with the previous notions of ANI , we can characterize local ANI equivalently by considering only the abstraction functions.

$$\forall d \in C_1. \alpha_1(d) = \alpha_1(c) \Rightarrow \alpha_2 \circ f(d) = \alpha_2 \circ f(c) \tag{1}$$

The next result tells us that when we move to single values, then (local) abstract non-interference is indeed stronger than (local) completeness, as expected.

Proposition 3. *Let* $f : C_1 \rightarrow C_2$ *be a function on complete lattices* C_i, *and let* $A_i \in Abs(C_i)$ *(for* $i \in \{1, 2\}$*) be abstractions, respectively, of input and output domains. Local* ANI *(Definition 5) implies local completeness (Definition 3), but the inverse implication does not hold.*

Proof. Suppose ANI holds for c, i.e., $\forall d \in C_1. A_1(d) = A_1(c) \Rightarrow A_2 \circ f(d) = A_2 \circ f(c)$. Then, consider $d = A_1(c) \in C_1$. Then we have that $A_1(A_1(c)) = A_1(c)$ by idempotence of A_1, and therefore, by definition of ANI on c, we have $A_2 \circ f(A_1(c)) = A_2 \circ f(c)$, which is completeness for c.

Let us prove that the inverse implication does not hold by showing a counterexample. Suppose $A_1 = Par = \{\top, odd, even, \bot\}$, $A_2 = \{\top, \mathbb{Z}_{\geq 0}, \mathbb{Z}_{=0}, \mathbb{Z}_{<0}, \bot\}$

---

[6] In the machine learning context, a similar notion recently introduced is abstract robustness [24].

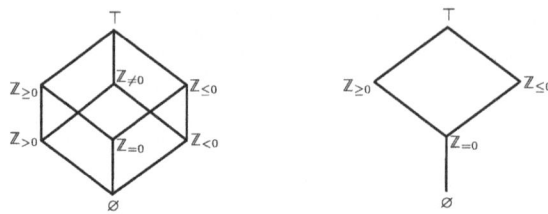

Fig. 2. Abstract domains for signs: *Sign* (left) and *Sign*$_1$ (right).

(an abstraction of *Sign*), and $f = \lambda z.(z+1)^2$. Then we have local completeness for $c = 3$, i.e., $A_2 \circ f \circ Par(3) = A_2 \circ f(odd) = \mathbb{Z}_{\geq 0}$ and $A_2 \circ f(3) = A_2(16) = \mathbb{Z}_{\geq 0}$. Consider $d = -1$, then $Par(3) = Par(-1) = odd$, but $A_2 \circ f(-1) = A_2(0) = \mathbb{Z}_{=0}$, hence local abstract non-interference does not hold.

We now wonder whether and how we can move backward from local completeness to local ANI . In other words, we aim to characterize the set of inputs on which local completeness has to hold to imply local ANI . The following result trivially derives from the proof of Prop. 3.

Corollary 1. *Let* $f : C_1 \rightarrow C_2$ *be a function on complete lattices* C_i, *and let* $A_i \in Abs(C_i)$ *(for* $i \in \{1,2\}$*) be abstractions, respectively, of input and output domains. Local* ANI *(Definition 5) holds for* c *iff* $\forall d \in \big\{ z \,\big|\, A_1(z) = A_1(c) \big\}$ *we have* local completeness *of the pair of abstractions* $\langle A_1, A_2 \rangle$ *w.r.t.* f *on* c *(Def. 3).*

We can summarize the proved relations in the following schema, making clear that local ANI for c corresponds to an intermediate, *regional* form of completeness, holding for a fixed set of inputs, determined by c, i.e., all the points sharing the same input property of c. We call this form of completeness *Abstract Local Completeness* since it corresponds to a local completeness w.r.t. an *abstract* input $A(c)$ instead of w.r.t. a concrete one c.

(Global) Completeness (Def. 1) $\equiv$ ANI (Def. 2)

$\Downarrow$

Abstract local completeness/Local ANI (Def. 5)

$\Downarrow$

Local Completeness (Def. 3) $\equiv$ Relaxed ANI (Def. 4)

4   Characterizing Locally Precise Abstract Domains

In this section, we exploit the framework of ANI [15,22], providing a different model for completeness, to investigate how it is possible to characterize domains satisfying the local ANI as a predicate holding on its elements.

In other words, we characterize what property of abstract elements makes the output domain satisfying local ANI for one computation, with respect to a given input observation, potentially equal to the output one. Afterward, we determine how to transform abstract domains towards local ANI .

4.1   Recognizing Domains Satisfying Abstract Local Completeness

Let $f : C_1 \to C_2$ be a function on complete lattices C_i, and let $A_i \in Abs(C_i)$ (for $i \in \{1,2\}$) be abstractions, respectively, of input and output domains, with $A_i = \gamma_i \alpha_i$. We can identify the elements that must share the same property in order to guarantee ANI , similarly to what has been done for standard ANI [22]: Let $a \in A_1$ ($a = A_1(c)$ for some $c \in C_1$, which must exist by surjectivity of α_1)

$$\Upsilon_{(f,A_1)}(a) \overset{\text{def}}{=} \{\ f(d) \,|\, A_1(d) = a \,\}$$

In the following, for the sake of readability, we avoid the subscript concerning the input abstract domain when it is clear from the context. By Prop. 2, this is the set of elements sharing the same abstraction in A_2 when we have local completeness. Hence, to guarantee local completeness on c, any element in A_2 must contain the set $\Upsilon_f(A_1(c))$ as a whole or has to be disjoint from it.

To formalize this relationship we define the predicate $Blind_f$ on elements $d \in A_2$, that, globally, considers any $c \in C_1$

$$\begin{aligned} Blind_{(f,A_1)}(d) &\Leftrightarrow \forall c \in C_1.\, (\exists d' \in \Upsilon_f(A_1(c)).\, d \geq d' \,\Rightarrow\, (d \geq \bigvee \Upsilon_f(A_1(c)))) \\ &\Leftrightarrow \forall a \in A_1.\, (\exists d' \in \Upsilon_f(a).\, d \geq d' \,\Rightarrow\, (d \geq \bigvee \Upsilon_f(a))) \end{aligned}$$

This predicate states that if d is greater than one element in $\Upsilon_f(A_1(c))$, then it must be greater than all the elements in $\Upsilon_f(A_1(c))$, since in this way we guarantee that all the elements in $\Upsilon_f(A_1(c))$ share the same abstraction.

Theorem 1. *Let* $f : C_1 \to C_2$ *be a function on complete lattices* C_i, *let* $A_i \in Abs(C_i)$ *(for* $i \in \{1,2\}$) *be abstractions, respectively, of input and output domains. Completeness holds for any* $c \subset C_1$ *(globally) iff* $\forall d \in A_2$ *we have* $Blind_{(f,A_1)}(d)$.

From this characterization, directly coming from the ANI framework [22], we can derive the one for local ANI , and therefore for the *regional* form of completeness, holding between global and local completeness.

As before, the set of elements on which we have to fix a property [22] is $\Upsilon_f(a)$, with $a \in A_1$ ($a = A_1(c)$ for some $c \in C_1$). Formally, we define a new predicate *Blind* parametric on abstract elements $a \in A_1$

$$Blind_{(f,a)}(d) \Leftrightarrow (\exists d' \in \Upsilon_f(a).\, d \geq d' \,\Rightarrow\, (d \geq \bigvee \Upsilon_f(a)))$$

Theorem 2. *Let* $f : C_1 \to C_2$ *be a function on complete lattices* C_i, *let* $A_i \in Abs(C_i)$ *(for* $i \in \{1,2\}$) *be abstractions, respectively, of input and output domains. Local ANI holds for* $c \in C_1$ *iff* $\forall d \in A_2$ *we have* $Blind_{(f,A_1(c))}(d)$.

This characterization of completeness allows us also to observe how local ANI , unlike local completeness, becomes independent of the specific choice of c inside the collection of concrete elements sharing the same abstract property $a = A_1(c)$; indeed, it only depends on a making it a *regional* property. Even if the blind predicate for (global) completeness has been defined for all concrete elements c, this trivially can be seen as a universal quantification on abstract elements a when we deal with Galois insertions since the predicate only looks at the abstract value. This dependency only on abstract elements becomes even clearer in the blind predicate for local ANI , where we directly consider only the abstraction of the fixed input. Things are different for local completeness, where the concrete element on which we look for precision is precisely fixed and, therefore, has to be kept.

5   Proving Abstract Local Completeness for $\mathcal{L}$ programs

In this section, we investigate local abstract completeness in the specific context of program analysis. It is well-known that the goal of a static analysis/verification tool is to soundly answer interesting questions on the dynamic (concrete) execution of programs [20]. In particular, the idea is to fix a level of observation of the program execution, to provide decidable answers possible on these abstract observations, and, finally, to derive, from them, answers about the concrete (undecidable) execution. In this context, completeness becomes a key aspect since completeness implies that the answers provided on the abstract observations precisely transfer to the concrete executions without loss of precision. For this reason, we focus now on programs (written in $\mathcal{L}$) and on the characterization of precision of a static analysis modeled by the abstraction A. Note that when f is a program semantics, then $C_1 = C_2 \overset{\text{def}}{=} C$, and analogously (since the abstract interpretation is performed inductively on the syntax) also, the input and the output observations collapse to the same abstraction, i.e., $A_1 = A_2 \overset{\text{def}}{=} A$ [5].

Formally, $f = [\![r]\!]$, for some $r \in \mathcal{L}$, where the concrete domain is $C = \wp(\mathbb{M})$, while the abstract observation is $A \in Abs(\wp(\mathbb{M}))$ (namely $\leq_C$ is $\subseteq$, while $\leq_A$ still depends on the abstract domain, and in particular on the abstraction α). When the abstraction A satisfies local ANI of r for $c \in C$ we write $\mathbb{L}_c^A(r)$.

In [20], the authors deeply investigate the role of each kind of statement on the precision/completeness of an abstract domain for program semantics. In particular, the authors prove that expressions Exp (guards and assignments) are the main sources of incompleteness (**skip** is trivially complete). Unfortunately, as the authors prove, the assignment can hardly be treated in a general way since a compositional rule for deriving the completeness of $x := e$ from the completeness of the expression e cannot be sound for a generic abstract domain A [20]. This means that assignments' completeness must be characterized/proved directly on the specific statement for the specific abstract domain. Instead, as far as the boolean guards are concerned, due to the intrinsic nature of their semantics, it is possible to provide a further formal characterization of the meaning of completeness for them.

5.1   Characterizing Abstract Local Complete Boolean Guards

In [5], the authors introduce a notion of expressibility, which strongly relates to completeness. Formally, b is expressible in A if its truth semantics is an element of A, i.e., $(\!|b|\!) \in A$. In particular, the authors show that if the semantics of b? is locally complete, then b and $\neg b$ are expressible, while the other implication does not always hold (Lemma III.2 [5]). The equivalence between global completeness and ANI allows us to exploit the abstract non-interference framework also for further characterizing completeness. In [5], it has been provided a characterization of abstract domains guaranteeing completeness for a boolean expression (and its negation) that here we can simplify by inheriting the ANI formalization, both for local and global completeness.

Proposition 4. *Let* $C = \wp(\mathbb{M})$ *and* $A \in Abs(C)$. *Let* $b \in BExp$ *and* $c \in C$, *then the following statements are equivalent:*

1. $A(A(c) \cap (\!|b|\!)) = A(c \cap (\!|b|\!))$ *(Local Completeness of* b? *semantics for* c *[5]);*
2. $\forall a \in A.\, \forall d \in C.\, A(d) = A(c)$ *we have* $a \supseteq d \cap (\!|b|\!) \Rightarrow a \supseteq A(c) \cap (\!|b|\!)$.

Corollary 2. *Let* $C = \wp(\mathbb{M})$ *and* $A \in Abs(\wp(\mathbb{M}))$. *Let* $b \in BExp$, *then the following statements are equivalent:*

1. $\forall c \in C.\, A(A(c) \cap (\!|b|\!)) = A(c \cap (\!|b|\!))$ *(Completeness of* b? *semantics);*
2. $\forall c \in C.\, \forall a \in A.\, \forall d \in C.\, A(d) = A(c)$ *we have* $a \supseteq d \cap (\!|b|\!) \Rightarrow a \supseteq A(c) \cap (\!|b|\!)$.

For instance, if we consider the abstract domain $A = Sign$, suppose $b \stackrel{\text{def}}{=} (x \geq 0)$, and consider $c = \{2, 4, 6\}$ then any d such that $Sign(d) = Sign(c) = \mathbb{Z}_{>0}$ is, by construction, such that $d \subseteq \mathbb{Z}_{>0} \subseteq (\!|b|\!)$, hence $d \cap (\!|b|\!) = d$ and $Sign(c) \cap (\!|b|\!) = Sign(c) = \mathbb{Z}_{>0}$. It is trivial to check that, in $Sign$, any element $a \supseteq d$ is such that $a \supseteq \mathbb{Z}_{>0}$, meaning that we have completeness. Consider $b \stackrel{\text{def}}{=} (x \bmod 2 = 1)$ and $c = \{0, -1, 2\}$, in this case any d such that $Sign(d) = Sign(c) = \mathbb{Z}$ is, by construction, such that d contains 0 and both positive and negative numbers, and $d \cap (\!|b|\!) = \{\, n \in d \mid n \bmod 2 = 1 \,\}$. On the other hand we have $Sign(c) \cap (\!|b|\!) = \{\, n \in \mathbb{Z} \mid n \bmod 2 = 1 \,\}$. Now consider $d = \{-3, 0, 2\}$, then $d \cap (\!|b|\!) = \{-3\}$ and, for instance, we have $\mathbb{Z}_{<0} \supseteq d \cap (\!|b|\!)$ with $\mathbb{Z}_{<0} \not\supseteq Sign(c) \cap (\!|b|\!)$ since this last set contains also positive values. Indeed, in this case, we don't have completeness.

Finally, the next theorem provides some relations between the local ANI of boolean guards and their expressibility.

Theorem 3. *Given* $b \in BExp$

1. $\mathbb{L}_b^A(\neg b?)$ *iff* b *expressible in* A *[$\mathbb{L}_{\neg b}^A(b?)$ iff $\neg b$ expressible in A];*
2. *If* b *expressible in* A *then* $\mathbb{L}_c^A(b?)$ *iff* $\forall d.\, A(c) = A(d).\, A(d \cap (\!|b|\!)) = A(c) \cap (\!|b|\!)$ *[If $\neg b$ expressible in A then $\mathbb{L}_c^A(\neg b?)$ iff $\forall d.\, A(c) = A(d).\, A(d \cap (\!|\neg b|\!)) = A(c) \cap (\!|\neg b|\!)$].*
3. *If* b *expressible in* A *and* $c \subseteq (\!|b|\!)$ *then* $\mathbb{L}_c^A(b?)$ *[If $\neg b$ expressible in A and $c \subseteq (\!|\neg b|\!)$ then $\mathbb{L}_c^A(\neg b?)$][7].*

In the examples above, we can observe that while $(x \geq 0)$ is expressible in $Sign$, $(x \bmod 2 = 1)$ instead is not expressible in it.

---

[7] Note that if the b is not expressible then we need $A(c) \subseteq (\!|b|\!)$ for implying local ANI , similarly for $\neg b$.

5.2   A Logic for Abstract Local Completeness

In this section, we define a proof system for program analysis of regular commands, parameterized by an abstraction $A = \gamma\alpha$ on the concrete domain of sets of stores $C = \wp(M)$. The provable triples of our logic are judgments of the form $\vdash_A (c)\, r\, (d)$, with $c, d \in C$, and $r \in \mathcal{L}$. $\vdash_A (c)\, r\, (d)$ guarantees that:

$$\forall c' \in C.\, \alpha(c') = \alpha(c) \text{ we have } [\![r]\!]^\sharp_A \alpha(c) = \alpha(d) = \alpha([\![r]\!](c')) \tag{2}$$

The next results tell us that the condition proved by the triple implies local ANI and implies that the $[\![r]\!]$ images of all the elements sharing the same abstraction of c also share the same abstraction of $[\![r]\!](c)$.

Lemma 1. *Let* $c, d \in C = \wp(M)$, $A \in Abs(C)$ $(A = \gamma\alpha)$. *If* $\forall c' \in C.\, \alpha(c') = \alpha(c)$ *we have* $[\![r]\!]^\sharp_A \alpha(c) = \alpha([\![r]\!](c'))$ *then*

1. $\mathbb{L}^A_c(r)$;
2. $\alpha \circ [\![r]\!](c') = \alpha \circ [\![r]\!](c)$

Proof. Consider $c' \in C.\, \alpha(c') = \alpha(c)$ then by hypothesis we have $[\![r]\!]^\sharp_A \alpha(c) = \alpha([\![r]\!](c'))$, note that, in particular we also have $[\![r]\!]^\sharp_A \alpha(c) = \alpha([\![r]\!](c))$. On the other hand, $A(c') = \gamma(\alpha(c')) = \gamma(\alpha(c)) = A(c)$ and

$$A([\![r]\!](c')) = \gamma \circ \alpha([\![r]\!](c')) = \gamma([\![r]\!]^\sharp_A(\alpha(c))) = \gamma \circ \alpha([\![r]\!](c)) = A([\![r]\!](c))$$

The other fact holds trivially by transitivity of equality and by the hypothesis $\alpha(c') = \alpha(c)$. ∎

Now, we can provide the rules of the proof system for local ANI in Fig. 3. In particular, the (**transfer**) rule requires the direct proof of local ANI on expressions; then, we can derive the triple for the expression semantics. Rule (**relax**) allows us to weaken the pre-condition and strengthen the post-condition of the triple (with the constraint of keeping the same abstractions). Rules (**seq**) and (**join**) are quite straightforward. While Rules (**iterate**) and (**rec**) handle the Kleene start statement.

The next result states that the proof system is sound w.r.t. local completeness.

Theorem 4. *Let* $c, d \in C = \wp(M)$, $A \in Abs(C)$ $(A = \gamma\alpha)$. *If* $\vdash_A (c)\, r\, (d)$ *then* $\forall c'.\, \alpha(c') = \alpha(c)$ *we have* $[\![r]\!]^\sharp_A \alpha(c) = \alpha(d) = \alpha([\![r]\!](c')$.

Proof. We prove the soundness of the rule system in Fig. 3 by structural induction on the derivation tree of $\vdash_A (c)\, r\, (d)$, by distinguishing the cases depending on the last rule applied.

(**transfer**) : Suppose $\mathbb{L}^A_c(e)$, this means by definition that $\forall c' \in C.\, A(c') = A(c)$ (or equivalently $\forall c'.\, \alpha(c') = \alpha(c)$) we have $A([\![e]\!](c')) = A([\![e]\!](c))$, which implies $\alpha \circ [\![e]\!](c) = \alpha \circ [\![e]\!] \circ \gamma \circ \alpha(c)$, being γ one-to-one. Then

$$[\![e]\!]^\sharp_A \circ \alpha(c) = [\![e]\!]^A \circ \alpha(c) = \alpha \circ [\![e]\!] \circ \gamma \circ \alpha(c) = \alpha(d) = \alpha \circ [\![e]\!](c)$$

$$\text{transfer:} \quad \frac{\mathbb{L}_c^A(e)}{\vdash_A (c)\, e\, ([\![e]\!]\gamma\alpha(c)))}$$

$$\text{relax:} \quad \frac{c' \le c \le \gamma\alpha(c') \quad \vdash_A (c')\, r\, (d') \quad d \le d' \le \gamma\alpha(d)}{\vdash_A (c)\, r\, (d)}$$

$$\text{seq:} \quad \frac{\vdash_A (c)\, r_1\, (d') \quad \vdash_A (d')\, r_2\, (d)}{\vdash_A (c)\, r_1; r_2\, (d)} \qquad \text{iterate:} \quad \frac{\vdash_A (c)\, r\, (d) \quad d \le \gamma\alpha(c)}{\vdash_A (c)\, r^*\, (c \lor d)}$$

$$\text{rec:} \quad \frac{\vdash_A (c)\, r\, (d') \quad \vdash_A (c \lor d')\, r^*\, (d) \quad \alpha(d) \le \alpha(d')}{\vdash_A (c)\, r^*\, (d)}$$

$$\text{join:} \quad \frac{\vdash_A (c)\, r_1\, (d_1) \quad \vdash_A (c)\, r_2\, (d_2)}{\vdash_A (c)\, r_1 \oplus r_2\, (d_1 \lor d_2)}$$

Fig. 3. A logic for local ANI .

(relax): By hypotheses for all $c'' \in C.\ \alpha(c'') = \alpha(c')$ we have $[\![r]\!]_A^\sharp(\alpha(c')) = \alpha(d') = \alpha \circ [\![r]\!](c'')$. Moreover, the hypotheses imply that $\alpha(d) = \alpha(d')$ and $\alpha(c) = \alpha(c') = \alpha(c'')$. Then

$$\begin{aligned}
[\![r]\!]_A^\sharp \circ \alpha(c) &= [\![r]\!]_A^\sharp \circ \alpha(c') && \text{(By hypothesis } \alpha(c) = \alpha(c')) \\
&= \alpha(d) = \alpha(d') && \text{(By hypothesis } \alpha(d) = \alpha(d')) \\
&= \alpha \circ [\![r]\!](c'')
\end{aligned}$$

(seq): By hypotheses, for all $c' \in C.\ \alpha(c) = \alpha(c')$ we have that $[\![r_1]\!]_A^\sharp(\alpha(c)) = \alpha([\![r_1]\!](c')) = \alpha(d') = \alpha([\![r_1]\!](c))$, and $\alpha([\![r_1]\!](c')) = \alpha(d')$, implies $\alpha \circ [\![r_2]\!]([\![r_1]\!](c')) = \alpha \circ [\![r_2]\!](d')$ (by the hypothesis on r_2). Then we have

$$\begin{aligned}
[\![r_1; r_2]\!]_A^\sharp \circ \alpha(c) &= [\![r_2]\!]_A^\sharp \circ [\![r_1]\!]_A^\sharp \circ \alpha(c) \\
&= [\![r_2]\!]_A^\sharp \circ \alpha \circ [\![r_1]\!](c) && \text{(By hypothesis on } r_1) \\
&= [\![r_2]\!]_A^\sharp \circ \alpha(d') && \text{(By hypothesis on } r_1)
\end{aligned}$$

$$\begin{aligned}
&= \alpha \circ [\![r_2]\!](d'') && \text{(By hypothesis on } r_2 \text{ and } \alpha(d') = \alpha(d'')) \\
&= \alpha(d) = \alpha \circ [\![r_2]\!](d') && \text{(By Lemma 1)} \\
&= \alpha \circ [\![r_2]\!]([\![r_1]\!](c')) = \alpha \circ [\![r_1; r_2]\!](c')
\end{aligned}$$

(iterate): Note that by hypothesis $\alpha(d) \le \alpha(c)$ and by [5](Lemma 2.1-c) we have that $[\![r]\!]_A^\sharp(\alpha(c)) = \alpha(d) \le \alpha(c)$ implies $[\![r^*]\!]_A^\sharp(\alpha(c)) = \alpha(c)$. Then $[\![r^*]\!]_A^\sharp(\alpha(c)) = \alpha(c) = \alpha(c) \lor \alpha(d) = \alpha(c \lor d)$. Moreover, $[\![r^*]\!]_A^\sharp(\alpha(c)) = \alpha(c) \le \alpha([\![r^*]\!](c))$ being $c = [\![r^0]\!](c) \le [\![r^*]\!](c)$, implying the equality.

(rec): Let $c' \in C$. $\alpha(c') = \alpha(c)$, first of all let us observe that $\alpha(d') = [\![r]\!]_A^\sharp(\alpha(c)) \leq [\![r^*]\!]_A^\sharp(\alpha(c)) \leq [\![r^*]\!]_A^\sharp(\alpha(c \vee d')) = \alpha(d)$, and let us observe that if c' is such that $\alpha(c') = \alpha(c)$ then $\alpha(c' \vee d') = \alpha(c') \vee \alpha(d') = \alpha(c) \vee \alpha(d') = \alpha(c \vee d')$. Hence, $\alpha \circ [\![r^*]\!](c' \vee d') = \alpha(d)$. Then we can prove for c' such that $\alpha(c') = \alpha(c)$

$$\alpha(d') = [\![r]\!]_A^\sharp \circ \alpha(c) = \alpha \circ [\![r]\!](c') \leq \alpha \circ [\![r^*]\!](c') \leq \alpha \circ [\![r^*]\!](c' \vee d')$$
$$= [\![r^*]\!]_A^\sharp \alpha(c' \vee d') = [\![r^*]\!]_A^\sharp \alpha(c \vee d') = \alpha(d) \leq \alpha(d')$$

Hence, they are all equalities and therefore $\alpha \circ [\![r^*]\!](c') = \alpha(d)$. Now, we have that

$$[\![r^*]\!]_A^\sharp \circ \alpha(c) \leq [\![r^*]\!]_A^\sharp \circ \alpha(c \vee d')$$
$$= \alpha(d) \text{ (By hypothesis)}$$
$$= \alpha \circ [\![r^*]\!](c') \text{ (For what we proved above)}$$

But since in general $[\![r^*]\!]_A^\sharp \circ \alpha(c) = [\![r^*]\!]_A^\sharp \circ \alpha(c') \geq \alpha \circ [\![r^*]\!](c')$, then we have the equality.

(join): By using the hypotheses and the additivity of α, we can show

$$[\![r_1 \oplus r_2]\!]_A^\sharp(\alpha(c)) = [\![r_1]\!]_A^\sharp(\alpha(c)) \vee [\![r_2]\!]_A^\sharp(\alpha(c))$$
$$= \alpha(d_1) \vee \alpha(d_2) = \alpha(d_1 \vee d_2)$$
$$= \alpha \circ [\![r_1]\!](c) \vee \alpha \circ [\![r_2]\!](c) = \alpha([\![r_1]\!](c) \vee [\![r_2]\!](c))$$
$$= \alpha([\![r_1 \oplus r_2]\!](c))$$

As a consequence, and analogously to what happens for local completeness [5], if a correctness specification π is expressible in A, namely, if $\pi = \gamma(a)$ for some abstract value $a \in A$, then any provable triple $\vdash_A (c) \, r \, (d)$ allows us to use d for witnessing either the correctness or the incorrectness of r for all the input conditions c' sharing the same input property A with c, as stated by the following result. The difference with local completeness is that in this case, we simultaneously prove completeness, and therefore π satisfaction, for a set of inputs and not for a single one.

Corollary 3. *If* $\vdash_A (c) r (d)$ *then* $\forall a \in A, \forall c'. A(c') = A(c)$ *we have* $[\![r]\!](c') \leq \gamma(a)$ *iff* $d \leq \gamma(a)$.

Proof. Suppose $a \in A$ and $d \leq \gamma(a)$, then $[\![r]\!](c') \leq \gamma \circ \alpha([\![r]\!](c')) = \gamma(\alpha([\![r]\!](c'))) = \gamma(\alpha(d))$ by the hypothesis that $\vdash_A (c) \, r \, (d)$, but then $\gamma \circ \alpha(d) \leq \gamma \circ \alpha(\gamma(a))$ (by hypothesis), and being $\alpha \circ \gamma = id$, we have $[\![r]\!](c') \leq \gamma(a)$.

Suppose now $[\![r]\!](c') \leq \gamma(a)$, again with $a \in A$. Then, by monotonicity of α, we have $\alpha([\![r]\!](c')) \leq \alpha(\gamma(a))$, and by hypothesis $\vdash_A (c) \, r \, (d)$ we have $\alpha(d) = \alpha([\![r]\!](c')) \leq \alpha(\gamma(a))$. But then we can prove that

$$d \leq \gamma \circ \alpha(d) \leq \gamma \circ \alpha(\gamma(a)) = \gamma(a)$$

by monotonicity of γ and being $\alpha \circ \gamma = id$.

Finally, we investigate the completeness of the rule system, observing that, as it happens for completeness (both global and local), also local ANI is intrinsically incomplete.

Theorem 5. *Let* $c, d \in C = \wp(M)$, $A \in Abs(C)$ $(A = \gamma\alpha)$, $r \in \mathfrak{L}$. (1) $d \leq A \circ [\![r]\!] \circ A(c)$ *and* (2) $\forall c'. \ \alpha(c') = \alpha(c)$ *we have* $[\![r]\!]_A^\sharp \alpha(c) = \alpha(d) = \alpha([\![r]\!](c'))$ *do not imply* $\vdash_A (c) \, r \, (d)$.

Proof (Sketch). We provide an example that cannot be deduced by using the rule system. Consider $r \stackrel{\text{def}}{=} (x := x - 1; x := x + 1)$ [5], let us denote by $r_1 \stackrel{\text{def}}{=} x := x - 1$ and by $r_2 \stackrel{\text{def}}{=} x := x + 1?$. Consider the sign domain $A \stackrel{\text{def}}{=} Sign = \gamma_S \alpha_S$ in Fig. 2. Let $c \stackrel{\text{def}}{=} \{2, 4, 6\}$ and $d \stackrel{\text{def}}{=} \{10\}$ Then we have that

$$[\![r]\!]_A^\sharp Sign(c) = [\![r]\!]_A^\sharp Sign(\{2, 4, 6\}) = [\![r]\!]_A^\sharp(\mathbb{Z}_{>0}) = \mathbb{Z}_{>0} = \alpha_S(d) = \alpha_S[\![r]\!](c) = \mathbb{Z}_{>0}$$

But, $\vdash_A (c) \, r_1 \, (d)$ cannot be proved, since $\alpha_S[\![r_1]\!](c) = \alpha_S(\{1, 3, 5\}) = \mathbb{Z}_{>0}$ but $[\![r_1]\!]_A^\sharp Sign(c) = [\![r_1]\!]_A^\sharp \mathbb{Z}_{>0} = \mathbb{Z}_{\geq 0}$, and therefore we cannot find $d' \in \wp(\mathbb{Z})$ such that $\vdash_A (c) \, r_1 \, (d')$, making not possible to apply rule seq.

Let us consider some simple examples of derivation. For the sake of simplicity, since the following programs have only one variable x, we abuse notation identifying the domain of stores M with the domain of values $\mathbb{Z}$. Hence, in the following examples, we will define A directly on $\wp(\mathbb{Z})$.

Example 1. Let us consider the regular command in Reg

$$r \stackrel{\text{def}}{=} (x \leq 0?; x := x * 2)^*; 0 \leq x?$$

Let us consider the abstract domain $A = Sign$ (let $A = \gamma_S \alpha_S$) on the left of Fig. 2. For the sake of readability, in the following, we identify each abstract element with its meaning, e.g., $\mathbb{Z}_{\leq 0}$ with $\{n \mid n \leq 0\}$. As observe previously, both the boolean expressions in r are expressible in *Sign*, since $x \leq 0$ is expressed by $\mathbb{Z}_{\leq 0}$ and $0 < x$ by $\mathbb{Z}_{>0}$. Let us consider an input property $c \stackrel{\text{def}}{=} \{-100, -10, 0\}$ $(Sign(c) = \mathbb{Z}_{\leq 0})$, by Th. 3 we have that both $\mathbb{L}_d^A(x \leq 0?)$ and $\mathbb{L}_d^A(0 < x?)$ hold for any d such that $Sign(d) = Sign(c) = \mathbb{Z}_{\leq 0}$. Indeed, if $Sign(d) = \mathbb{Z}_{\leq 0}$ it means that $d \subseteq Sign(d) = \mathbb{Z}_{\leq 0}$, hence $Sign(d \cap x \leq 0?) = Sign(d) = \mathbb{Z}_{\leq 0}$, on the other hand $Sign(d) \cap x \leq 0? = \mathbb{Z}_{\leq 0}$, therefore we have $\mathbb{L}_d^A(x \leq 0?)$ (analogous for $0 < x?$). Therefore by rule (transfer):

$$\frac{\mathbb{L}_c^A(x \leq 0?)}{\vdash_A (c) \, x \leq 0? \, (\mathbb{Z}_{\leq 0})}$$

being $[\![x \leq 0?]\!] Sign(c) = [\![x \leq 0?]\!] \mathbb{Z}_{\leq 0} = \mathbb{Z}_{\leq 0}$.

Now, note that $Sign [\![x := x * 2]\!](\mathbb{Z}_{\leq 0}) = \mathbb{Z}_{\leq 0} = [\![x := x * 2]\!]_A^\sharp(Sign(\mathbb{Z}_{\leq 0})) = [\![x := x * 2]\!]_A^\sharp(\mathbb{Z}_{\leq 0})$, hence $\mathbb{L}_{\mathbb{Z}_{\leq 0}}^A(x := x * 2)$, and therefore we can apply again rule (transfer)

$$\frac{\mathbb{L}_{\mathbb{Z}_{\leq 0}}^A(x := x * 2)}{\vdash_A (\mathbb{Z}_{\leq 0}) \, x := x * 2 \, (\mathbb{Z}_{\leq 0})}$$

Finally by rule (seq)

$$\frac{\vdash_A (c) \, x \leq 0? \, (\mathbb{Z}_{\leq 0}) \quad \vdash_A (\mathbb{Z}_{\leq 0}) \, x := x * 2 \, (\mathbb{Z}_{\leq 0})}{\vdash_A (c) \, x \leq 0?; x := x * 2 \, (\mathbb{Z}_{\leq 0})}$$

Now, we can apply rule (iterate) obtaining

$$\frac{\vdash_A (c)\, x \leq 0?; x := x * 2\ (\mathbb{Z}_{\leq 0})\quad \mathbb{Z}_{\leq 0} = Sign(c)}{\vdash_A (c)\, (x \leq 0?; x := x * 2)^*\ (c \vee \mathbb{Z}_{\leq 0})}$$

where $c \vee \mathbb{Z}_{\leq 0} = \mathbb{Z}_{\leq 0}$.

Finally, by Theorem 3 we have $\mathbb{L}^A_{\mathbb{Z}_{\leq 0}}(0 \leq x?)$, implying again by rule (transfer)

$$\frac{\mathbb{L}^A_{\mathbb{Z}_{\leq 0}}(0 \leq x?)}{\vdash_A (\mathbb{Z}_{\leq 0})\, 0 \leq x?\ (\mathbb{Z}_{=0})}$$

since $[\![0 \leq x]\!] Sign(\mathbb{Z}_{\leq 0}) = [\![0 \leq x]\!] \mathbb{Z}_{\leq 0} = \mathbb{Z}_{=0}$ In this way we can conclude, by rule (seq), that $\vdash_A (c)\, (x \leq 0?; x := x * 2)^*; 0 \leq x?\ (\mathbb{Z}_{-0})$, meaning that $\mathbb{L}^A_c(\mathbf{r})$.

Example 2. Let us consider the regular program $\mathbf{r} \stackrel{\text{def}}{=} \mathbf{r}_1 \oplus \mathbf{r}_2$, where

$$\mathbf{r}_1 \stackrel{\text{def}}{=} (0 < x?; x := x - 1)^*$$
$$\mathbf{r}_2 \stackrel{\text{def}}{=} (x < 90?; x := x + 1)^*$$

Let us consider as abstract domain $A = Int$ (let $A = \gamma_i \alpha_i$), and $c \stackrel{\text{def}}{=} \{0, 10, 100\}$ ($\alpha_i(c) = [0, 100]$). In the following, for the sake of readability, we identify $[n, m]$ with its meaning $\{i \mid n \leq i \leq m\}$. In the considered abstract domain, both the expressions are expressible. By Theorem 3, we have $\forall \mathbf{d} \in C.\ Int(\mathbf{d}) = Int(c) = [0, 100]$ that $\mathbb{L}^A_c(0 < x?)$, and that $\mathbb{L}^A_c(x < 90?)$.

Let us consider $\mathbf{r}_1$ first. By rule (transfer)

$$\frac{\mathbb{L}^A_c(0 < x?)_\tau}{\vdash_A (c)\, 0 < x?\ ([1, 100])}$$

Now we have that $Int([\![x := x - 1]\!][1, 100]) = [0, 99] = [\![x := x - 1]\!]^\sharp_A([1, 100])$, hence again by rule (transfer) we derive

$$\frac{\mathbb{L}^A_{[1, 100]}(x := x - 1)}{\vdash_A ([1, 100])\, x := x - 1\ ([0, 99])}$$

and therefore by rule (seq) we have $\vdash_A (c)\, 0 < x?; x := x - 1\ ([0, 99])$. Now, by rule (iterate), we prove

$$\frac{\vdash_A (c)\, 0 < x?; x := x - 1\ ([0, 99])\quad [0, 99] \subseteq Int(c) = [0, 100]}{\vdash_A (c)\, (0 < x?; x := x - 1)^*\ (c \vee [0, 99])}$$

where $c \vee [0, 99] = [0, 100]$.

Let us consider now $\mathbf{r}_2$ and $c' \stackrel{\text{def}}{=} \{0, 10, 50, 100\}$ ($Int(c') = [0, 100]$). Then by rule (transfer)

$$\frac{\mathbb{L}^A_{c'}(x < 90?)}{\vdash_A (c')\, x < 90?\ ([0, 89])}$$

then being $Int(\llbracket x := x + 1 \rrbracket [0, 89])) = [1, 90] = \llbracket x := x + 1 \rrbracket_A^{\sharp}([0, 89])$, we have by rule (transfer)

$$\frac{\mathbb{L}^A_{[0,89]}(x := x + 1)}{\vdash_A ([0, 89])\, x := x + 1\, ([1, 90])}$$

and therefore $\vdash_A (c')\, x < 90?; x := x + 1\, ([1, 90])$ by rule (seq). Now by rule (iterate)

$$\frac{\vdash_A (c')\, x < 90?; x := x + 1\, ([1, 90])\quad [1, 90] \leq Int(c) = [0, 100]}{\vdash_A (c')\, (x < 90?; x := x + 1)^*\, (c' \vee [1, 90])}$$

where $c' \vee [1, 90] = [0, 100]$. At this point, by rule (relax)

$$\frac{c \subseteq c' \subseteq Int(c) = [0, 100]\quad \vdash_A (c)\, (0 < x?; x := x - 1)^*\, ([0, 100])}{\vdash_A (c')\, (0 < x?; x := x - 1)^*\, ([0, 100])}$$

and therefore we can apply rule (join)

$$\frac{\vdash_A (c')\, (0 < x?; x := x - 1)^*\, ([0, 100])\quad \vdash_A (c')\, (x < 100?; x := x + 1)^*\, ([0, 100])}{\vdash_A (c')\, \mathbf{r}\, ([0, 100])}$$

meaning that $\mathbb{L}^A_c(\mathbf{r})$, being $Int(c') = Int(c)$.

Example 3. In this example, we show that we can derive limits for diverging loops. Consider $\mathbf{r} = (x \leq 0?; x := 2 * x)^*$, and $A = Int$ (let $A = \gamma_i \alpha_i$). In this case we can observe that if we start from a finite c then it is not possible to apply rule iterate since we enlarge at least one bound of the interval, while if we compute abstract local completeness on a point including the limit we can prove it.

Let c be such that $\alpha_i(c) = [n, 0]$, then $\vdash_A (c)\, x \leq 0?\, (c)$, while $\vdash_A (c)\, x := 2 * x\, ([2n, 0])$, hence $\alpha_i([2n, 0]]) \not\subseteq \alpha_i(c)$. It is trivial to observe that the only case in which we obtain a result contained in the starting point c is when $c = [-\infty, n]$ $(n \leq 0)$, since in this case we can derive, in the proof system, that $\vdash_A (c)\, x := 2 * x\, ([-\infty, 2n])$.

Let, for instance $n = -10$, then trivially $\vdash_A ([-\infty, -10])\, x \leq 0?\, ([-\infty, -10])$, $\vdash_A ([-\infty, -10])\, x := 2 * x\, ([-\infty, -20])$, hence, by using the composition rule, we prove $\vdash_A ([-\infty, -10])\, x \leq 0?; x := 2 * x\, ([-\infty, -20])$ and being $[-\infty, -20] \subseteq [-\infty, -10]$, we can derive $\vdash_A ([-\infty, -10])\,(x \leq 0?; x := 2*x)^*\, ([-\infty, -10])$. Hence, $\forall c \in \{\, [-\infty, n] \,|\, n < 0 \,\}$ we have that $\mathbb{L}^A_c(\mathbf{r})$.

6   Conclusions

This paper introduces a novel notion of abstract non-interference, made "local" on one of its abstract inputs. Classical ANI [22] is based on the comparison between two universally quantified executions. In the notion proposed here, we fix one execution (i.e., one input), and we let the other execution vary universally.

We obtain so far a *local* notion of ANI comparing pairs of executions such that their inputs share the same, *fixed*, abstract property.

We observe that such a notion is a form of completeness that lies in between completeness [18], holding for all possible inputs, and *local* completeness [5], holding precisely for one input. It requires completeness for all the inputs sharing the abstract property with a fixed input. The difference with requiring local completeness for all such points is that we can prove it for all such points simultaneously. Such a proof can be obtained by using the proposed logic, allowing us to check whether the abstract interpretation designed on a given abstract domain satisfies abstract local completeness (or, equivalently, local ANI). Our program logic is simple and can be checked online during program analysis.

The immediate application of this work is in static analysis, where we can exploit the proposed work to prove *areas* of completeness for a given program, where the satisfaction of a given specification could become verifiable. Moreover, whenever local abstract completeness is not satisfied, we also provide a way for transforming the abstract domain to enforce it, precisely as it happens in completeness [6,18] and in ANI [22]. To make Local ANI useful in practice, we should have, first to understand how it interacts with widening/narrowing in real abstract domains, e.g., intervals or octagons (in this paper we have only considered situations where the widening application was not necessary), or for which it does not exist a Galois connection [33], and second to implement a method for verifying it, either by using the provided proof system or by exploiting the hyperproperty characterization of ANI [27,30,31], potentially providing a higher-order static analysis for proving abstract local completeness.

Abstract local completeness can have applications beyond program analysis, such as language-based security and code protection. In language-based security, as specified by abstract non-interference [16,22,28], local ANI could guarantee that information is kept secret, at least on some sets of inputs, weakening the universal quantification. In this direction, it could be interesting to understand its relation with declassification, fixing by abstraction, and what should be kept secret [2,26]. In code protection, the notion of local ANI becomes stronger than completeness. The standard approach to protect code against program analysis is to make the abstract interpreter maximally imprecise with respect to the given program, which means incomplete [17,19,25]. In this case, we may replace (global) completeness with local abstract completeness and imagine code-protecting transformations, making an abstract interpreter incomplete [17,25,32] on specific sets of inputs for the transformed code. Finally, for program verification, it could be interesting to exploit the idea introduced for partial completeness [8], i.e., the use of a metric for weakening abstract local completeness.

Funding. This work was supported by the project SERICS (PE00000014) under the MUR National Recovery and Resilience Plan funded by the European Union - NextGenerationEU and by PRIN2022PNRR "RAP-ARA" (PE6) - codice MUR: P2022HXNSC.

References

1. Arceri, V., Mastroeni, I., Zaffanella, E.: Decoupling the ascending and descending phases in abstract interpretation. In: Sergey, I. (ed.) Programming Languages and Systems, pp. 25–44. Springer (2022)
2. Banerjee, A., Giacobazzi, R., Mastroeni, I.: What you lose is what you leak: information leakage in declassification policies. In: Proceedings of the 23th Internat. Symp. on Mathematical Foundations of Programming Semantics (MFPS '07). Electronic Notes in Theoretical Computer Science, vol. 1514. Elsevier (2007)
3. Bourdoncle, F.: Abstract interpretation by dynamic partitioning. J. Funct. Program. **2**(4), 407–435 (1992)
4. Bruni, R., Giacobazzi, R., Gori, R., Garcia-Contreras, I., Pavlovic, D.: Abstract extensionality: on the properties of incomplete abstract interpretations. Proc. ACM Program. Lang. **4**(POPL), 28:1–28:28 (2020). https://doi.org/10.1145/3371096
5. Bruni, R., Giacobazzi, R., Gori, R., Ranzato, F.: A logic for locally complete abstract interpretations. In: Symposium on Logic in Computer Science, LICS, pp. 1–13. IEEE (2021)
6. Bruni, R., Giacobazzi, R., Gori, R., Ranzato, F.: Abstract interpretation repair. In: Jhala, R., Dillig, I. (eds.) PLDI '22: 43rd ACM SIGPLAN International Conference on Programming Language Design and Implementation, San Diego, CA, USA, June 13 - 17, 2022, pp. 426–441. ACM (2022)
7. Bruni, R., Giacobazzi, R., Gori, R., Ranzato, F.: A correctness and incorrectness program logic. J. ACM **70**(2) (2023)
8. Campion, M., Preda, M.D., Giacobazzi, R.: Partial (in)completeness in abstract interpretation: limiting the imprecision in program analysis. Proc. ACM Program. Lang. **6**(POPL), 1–31 (2022). https://doi.org/10.1145/3498721
9. Cousot, P.: Asynchronous iterative methods for solving a fixed point system of monotone equations in a complete lattice. Research report R.R. 88, Laboratoire IMAG, Université scientifique et médicale de Grenoble, Grenoble, France (1977). 15 p
10. Cousot, P.: Constructive design of a hierarchy of semantics of a transition system by abstract interpretation. Theor. Comput. Sci. **277**(1–2), 47–103 (2002)
11. Cousot, P., Cousot, R.: Abstract interpretation: a unified lattice model for static analysis of programs by construction or approximation of fixpoints. In: Conference Record of the 4th ACM Symposium on Principles of Programming Languages (POPL '77), pp. 238–252. ACM Press (1977)
12. Cousot, P., Cousot, R.: Systematic design of program analysis frameworks. In: Conference Record of the 6th ACM Symposium on Principles of Programming Languages (POPL '79), pp. 269–282. ACM Press (1979)
13. Cousot, P., Cousot, R.: Comparing the Galois connection and widening/narrowing approaches to abstract interpretation. In: Bruynooghe, M., Wirsing, M. (eds.) PLILP 1992. LNCS, vol. 631, pp. 269–295. Springer, Heidelberg (1992). https://doi.org/10.1007/3-540-55844-6_142
14. Cousot, P.: Principles of Abstract Interpretation. MIT Press (2021)
15. Giacobazzi, R., Mastroeni, I.: Abstract non-interference: parameterizing non-interference by abstract interpretation. In: Proceedings of the 31st Annual ACM SIGPLAN-SIGACT Symposium on Principles of Programming Languages (POPL '04), pp. 186–197. ACM-Press (2004)
16. Giacobazzi, R., Mastroeni, I.: Adjoining classified and unclassified information by abstract interpretation. J. Comput. Secur. **18**(5), 751–797 (2010)

17. Giacobazzi, R., Mastroeni, I.: Making abstract interpretation incomplete - modeling the potency of obfuscation. In: Miné, A., Schmidt, D. (eds.) 19th International Static Analysis Symposium (SAS '12). Lecture Notes in Computer Science, vol. 7460, pp. 129–145 (2012)
18. Giacobazzi, R., Ranzato, F., Scozzari, F.: Making abstract interpretation complete. J. ACM **47**(2), 361–416 (2000)
19. Giacobazzi, R., Jones, N.D., Mastroeni, I.: Obfuscation by partial evaluation of distorted interpreters. In: Kiselyov, O., Thompson, S.J. (eds.) Proceedings of the ACM SIGPLAN 2012 Workshop on Partial Evaluation and Program Manipulation, PEPM 2012, Philadelphia, Pennsylvania, USA, January 23–24, 2012, pp. 63–72. ACM (2012)
20. Giacobazzi, R., Logozzo, F., Ranzato, F.: Analyzing program analyses. In: Rajamani, S.K., Walker, D. (eds.) Proceedings of the 42nd Annual ACM SIGPLAN-SIGACT Symposium on Principles of Programming Languages, POPL 2015, Mumbai, India, 15–17 January 2015, pp. 261–273. ACM (2015)
21. Giacobazzi, R., Mastroeni, I.: Making abstract models complete. Math. Struct. Comput. Sci. **26**(4), 658–701 (2016)
22. Giacobazzi, R., Mastroeni, I.: Abstract non-interference: a unifying framework for weakening information-flow. ACM Trans. Priv. Secur. **21**(2), 1–31 (2018)
23. Giacobazzi, R., Mastroeni, I., Perantoni, E.: How fitting is your abstract domain? In: Hermenegildo, M.V., Morales, J.F. (eds.) Static Analysis. pp. 286–309. Springer (2023)
24. Giacobazzi, R., Mastroeni, I., Perantoni, E.: Adversities in abstract interpretation - accommodating robustness by abstract interpretation. ACM Trans. Program. Lang. Syst. **46**(2) (2024). https://doi.org/10.1145/3649309
25. Giacobazzi, R., Mastroeni, I., Preda, M.D.: Maximal incompleteness as obfuscation potency. Formal Aspects Comput. **29**(1), 3–31 (2017)
26. Mastroeni, I., Banerjee, A.: Modelling declassification policies using abstract domain completeness. Math. Struct. Comput. Sci. **21**(6), 1253–1299 (2011)
27. Mastroeni, I., Pasqua, M.: Statically analyzing information flows: an abstract interpretation-based hyperanalysis for non-interference. In: Hung, C., Papadopoulos, G.A. (eds.) Proceedings of the 34th ACM/SIGAPP Symposium on Applied Computing, SAC 2019, pp. 2215–2223. ACM (2019). https://doi.org/10.1145/3297280.3297498
28. Mastroeni, I.: Abstract interpretation-based approaches to security - a survey on abstract non-interference and its challenging applications. In: Banerjee, A., Danvy, O., Doh, K., Hatcliff, J. (eds.) Semantics, Abstract Interpretation, and Reasoning about Programs: Essays Dedicated to David A. Schmidt on the Occasion of his Sixtieth Birthday, Manhattan, Kansas, USA, 19-20th September 2013. EPTCS, vol. 129, pp. 41–65 (2013)
29. Mastroeni, I.: Abstract domain adequacy: weakening completeness towards static analysis precision (2024). under submission
30. Mastroeni, I., Pasqua, M.: Hyperhierarchy of semantics - a formal framework for hyperproperties verification. In: Ranzato, F. (ed.) Static Analysis, pp. 232–252. Springer, Cham (2017)
31. Mastroeni, I., Pasqua, M.: Verifying bounded subset-closed hyperproperties. In: Podelski, A. (ed.) Static Analysis, pp. 263–283. Springer, Cham (2018)
32. Mastroeni, I., Pasqua, M.: Verifying opacity by abstract interpretation. In: Hong, J., Bures, M., Park, J.W., Cerný, T. (eds.) SAC '22: The 37th ACM/SIGAPP Symposium on Applied Computing, Virtual Event, April 25 - 29, 2022, pp. 1817–1826. ACM (2022). https://doi.org/10.1145/3477314.3507119

33. Mastroeni, I., Pasqua, M.: Domain precision in Galois connection-less abstract interpretation. In: Hermenegildo, M.V., Morales, J.F. (eds.) Static Analysis, pp. 434–459. Springer, Cham (2023)
34. Müller, M.N., Fischer, M., Staab, R., Vechev, M.: Abstract interpretation of fixpoint iterators with applications to neural networks. Proc. ACM Program. Lang. **7**(PLDI) (2023)
35. O'Hearn, P.W.: Incorrectness logic. In: Proceedings of the ACM on Programming Languages (POPL), vol. 4, no. 10 (2020)
36. Winskel, G.: The Formal Semantics of Programming Languages: An Introduction. MIT Press (1993)

An Abstract Domain for Heap Commutativity

Jared Pincus[1](✉) and Eric Koskinen[2]

[1] Boston University, Boston, MA 02215, USA
pincus@bu.edu
[2] Stevens Institute of Technology, Hoboken, NJ 07030, USA
eric.koskinen@stevens.edu

Abstract. Commutativity of program code (*i.e.* the equivalence of two code fragments composed in alternate orders) is of ongoing interest in many settings such as program verification, scalable concurrency, and security analysis. While some have explored static analysis for code commutativity, few have specifically catered to heap-manipulating programs. We introduce an abstract domain in which commutativity synthesis or verification techniques can safely be performed on abstract mathematical models and, from those results, one can directly obtain commutativity conditions for concrete heap programs. This approach offloads challenges of concrete heap reasoning into the simpler abstract space. We show this reasoning supports framing and composition, and conclude with commutativity analysis of programs operating on example heap data structures. Our work has been mechanized in Coq and is available in the supplement.

Keywords: Commutativity · Abstract Interpretation · Observational Equivalence · Separation Logic

1 Introduction

Commutativity describes circumstances in which two programs executed in sequence yield the same result regardless of their execution order. Fundamentally, commutativity offers a kind of independence that enables parallelization or simplifies program analysis by collapsing equivalent cases. As such, this property has long been of interest in many areas including databases [38], parallelizing compilers [35], scalable [8] and distributed systems [36], and smart contracts [12,32]. Commutativity is also used in verification, including efforts to simplify proofs of parameterized programs [18], and concurrent program properties [17] such as termination [27], dynamic race detection [19], and information flow [13].

Efforts in *commutativity analysis* seek to infer or verify *commutativity conditions*— the initial conditions (if any) under which code fragments commute. For example, Rinard and Kim [24] verify commutativity conditions by relating a data structure to its pre/postconditions, then showing commutativity of that abstraction. Later, methods were developed to automatically synthesize commutativity

© The Author(s), under exclusive license to Springer Nature Switzerland AG 2025
K. Shankaranarayanan et al. (Eds.): VMCAI 2025, LNCS 15530, pp. 26–49, 2025.
https://doi.org/10.1007/978-3-031-82703-7_2

conditions for specifications and imperative program fragments [2,3,6,7], which can be used to automatically parallelize programs [6,33,35].

Objectives. Considering its many applications, it would be appealing to extend commutativity analysis to the heap memory model. However, few works thus far have targeted heap-modifying programs. Pîrlea *et al.* [32] describe a commutativity analysis with some heap treatment, but with the limited scope of heap disjointness/ownership. Eilers *et al.* [13] employ, but do not verify, commutativity to reason about information flow security in concurrent separation logic. This paper thus aims to analyze commutativity of heap programs.

Challenges & Contributions. Commutativity reasoning can be challenging for heap-manipulating programs, but is more straightforward for functions over simple mathematical objects. With that in mind, this paper presents an abstract domain which is amenable to existing commutativity condition verification and synthesis techniques [2,3,7], and allows conditions found in said abstract space to immediately yield concrete separation-logic-style conditions for heap programs. For instance, given a simple mathematical list representation of a stack data structure, we can easily find that the functions *push* and *pop* operating on these lists commute when the first (i.e. top) list element equals the value being pushed. We may then thread this condition down to concrete push and pop methods. For this threading to be sound, we design the domain to handle various complications of heap programs, including observational equivalence of heap data structures, nondeterminism of address allocation, and improperly allocated heaps and resultant program failure.

In summary, the contributions of this paper include:

(Section 4) We introduce a simple abstract domain that is amenable to abstract commutativity reasoning, which can soundly entail concrete commutativity. The domain is designed to lift heap-allocated structures to mathematical objects (e.g. value lists to encode a stack) via separation-logic-style predicates, while cleanly handling ambiguous or improperly allocated heaps. We then describe the semantics of abstract programs, and establish a soundness relationship between abstract and concrete programs in the style of abstract interpretation [10], which enables the reduction of concrete to abstract commutativity.

(Section 5) We define concrete and abstract commutativity, and present our main soundness theorem: that a valid commutativity condition for two concrete programs may be derived immediately from commutativity reasoning performed (much more easily) on their corresponding abstract programs. These derived concrete conditions are also *framed*, i.e. if the programs commute under a given initial heap layout, they also commute in any larger layout.

(Section 6) We define a way to compose abstractions, which mirrors the separating conjunction on concrete heaps, and present compositional commutativity properties. Composition aids in settings with multiple allocated data structures, and working with arguments/return values.

(Section 7) We share four complete examples: a non-negative counter, two-place unordered set, linked-list stack, and the composition of a counter and stack.

(Supplement) Our work is implemented in Coq, on top of Separation Logic Foundations [5]. All theorems, lemmas and properties have been proved in Coq (except for the example programs written in the concrete semantics of Sect. 3.3).

Limitations. In this paper we establish the theory necessary for deriving heap commutativity from abstract commutativity reasoning. We leave implementation of this theory, including automation and synthesis, for future work.

2 Overview

We summarize our work through the example of a stack, implemented as a heap-allocated linked list, with typical **push** and **pop** methods (**pop** fails if the stack is empty). For simplicity, method arguments or return values are stored in method-specific pre-allocated heap cells (e.g. a).

$$
\begin{array}{ll}
\text{push}^a_\ell := \text{let } p = !\ell \text{ in} & \text{pop}^a_\ell := \text{let } p = !\ell \text{ in} \\
\quad \text{let } v = !a \text{ in} & \quad \text{if } p = \text{null then fail else} \\
\quad \text{let } x = (v, p) \text{ in} & \quad \text{let } x = !p \text{ in} \\
\quad \text{let } q = \text{ref } x & \quad \text{let } v = \text{fst } x \text{ in let } q = \text{snd } x \\
\quad \text{in } \ell \leftarrow q & \quad \text{in } \ell \leftarrow q \text{ ; free } p \text{ ; } a \leftarrow v
\end{array}
$$

Inconvenience of Concrete Heap Commutativity. We would like to know under what precondition P, if any, do **push** and **pop** *commute*. Verifying whether assume(P);push;pop is equivalent to assume(P);pop;push is not amenable to typical relational reasoning techniques [39], because the swapped orders of **push** and **pop** (or, in general, f and g) do not yield meaningful alignment points [1,28]. So, the problem amounts to performing four forward symbolic executions per method pair, meaning quadratically many executions as the number of methods grows. There are, however, efficient CEGAR/CEGIS methods for verifying and even synthesizing a condition P of logical method pre/postconditions [2,3,7]. We now discuss a simple abstraction of the Stack methods as mathematical functions, how our theory relates the concrete heap to this abstraction, and how this enables *concrete* commutativity conditions to be immediately derived from the results of simpler (automatable) abstract commutativity reasoning.

Abstract Commutativity. Let us, for the moment, set aside the concrete heap implementation, and consider a simple mathematical representation of a stack as a list of values, along with functions over the stack with an input/output "cell":

$$
push(s, v) := (v :: s, v) \qquad pop([], \_) := \bot \qquad pop(v :: s, \_) := (s, v)
$$

Here, *push* appends the input value to the stack, and *pop* removes and returns the topmost value, but fails (yields $\bot$) on an empty stack. Reasoning about commutativity of these mathematical objects is convenient, and synthesizing commutativity conditions can now be automated [2,3,7]. For this example, given an initial stack s, $push(u)$ and $push(v)$ commute when $u = v$; $push(u)$ and $pop()$ when $s = u :: \_$; and $pop()$ and $pop()$ when $s = x :: x :: \_$ for some x.

The question we ask in this paper is how we might abstract the concrete programs so that, from these abstract commutativity conditions, we can immediately obtain concrete conditions. We show that this is indeed possible; for example, in the case of push_ℓ^a and push_ℓ^b we obtain the (separation logic) precondition: $\exists u.\ \mathsf{stk}^+(\_) * a \mapsto u * b \mapsto u$, where $\mathsf{stk}^+(\_)$ asserts that the heap contains a properly allocated stack (we formally define stk^+ below). We now summarize how to achieve this using the Stack example.

Step 1: Projecting Heaps to Abstract Stacks. To soundly lift concrete heap structures into an abstract domain, we must accommodate various complications of the heap, such as improperly allocated heaps, observationally equivalent layouts, and allocation nondeterminism. We achieve this by choosing a set of abstract values X (such as value lists to represent stacks), and a projection function π which maps concrete heaps into X. Whether a heap is "well-structured" depends on the choice of π. For the Stack, π encodes that the only valid heaps for the abstraction are those containing a correctly allocated linked stack. We say that such heaps are "within the purview" of the abstraction at hand, meaning they will be amenable to abstract commutativity reasoning.

Of course, π must also account for heaps which are *not* properly allocated to be representable by a value in X. To this end, every abstract domain includes a special element ✗ ("cross"), to which π maps all inappropriately structured, or "out-of-purview", heaps. Commutativity w.r.t. such heaps cannot be reasoned about meaningfully in the abstract space.

For our Stack example, the abstract values X are lists of non-null concrete values, and the projection function π maps a heap to such a list by checking (via separation logic heap predicates [29]) whether it contains a valid linked list, and extracting the contents of said list. Precisely:

$$\pi(h) \quad := s \text{ if } \mathsf{stk}^+(s, h) \text{ for } s \in \mathrm{list}(\text{value}) \text{ else } ✗, \text{ where}$$

$$\mathsf{stk}(s) \quad := \exists p \in \text{address}.\ \ell \mapsto p * \psi(s, p)$$
$$\psi(v :: s, p) := \exists q \in \text{address}.\ p \neq \mathsf{null} * p \mapsto (v, q) * \psi(s, q)$$
$$\psi([], p) \quad := p = \mathsf{null}$$

By existentially quantifying the intermediate addresses of the linked stack and omitting them from the extracted abstract value, we render stacks observationally equivalent if and only if their contained values match.

Step 2: Abstract Domain for Commutativity. The next question is how this projection should induce an abstract domain, while soundly respecting concrete programs, and enabling commutativity reasoning. The key is to "wrap" the abstract values X into a domain that also includes elements which track if abstract commutativity still entails concrete commutativity. Namely, we add the abstract value ✗ for out-of-purview heaps, ✓ ("check") for an ambiguous collection of in-purview heaps, $\perp$ as a failure state, and $\top$ for a collection of heaps both in- and out-of-purview. Instrumenting π appropriately induces a Galois connection [11] between this abstract domain and the concrete heap domain.

For abstract commutativity to soundly entail concrete commutativity, we establish a soundness relation between concrete and abstract programs, in the style of abstract interpretation [10]. It demands, for instance, that *push* overapproximate push and that *push* fail when push fails on an in-purview heap.

Summary. From our user-provided Stack abstraction, and typical abstract interpretation techniques, we can immediately derive sound heap commutativity conditions from simple mathematical conditions:

Concrete Programs		Abstract Functions		Abstract Condition	Derived Concrete Condition
push	push	$push(u)$	$push(v)$	$u = v$	$\exists u.\, \mathsf{stk}^+(\_) * a \mapsto u * b \mapsto u$
push	pop	$push(u)$	$pop()$	$s = u :: \_$	$\exists u.\, \mathsf{stk}^+(u :: \_) * a \mapsto u * b \mapsto \_$
pop	pop	$pop()$	$pop()$	$s = x :: x :: \_$	$\exists x.\, \mathsf{stk}^+(x :: x :: \_) * a \mapsto \_ * b \mapsto \_$

In the remainder of this paper we formalize the above discussion, share properties of commutativity, and define composition on abstract domains. Then in Sect. 7 we elaborate on the Stack example (adding a peek operation) along with other examples. All definitions, lemmas, theorems, and examples have been mechanized in Coq (available in the supplement), except for those concerning the example concrete semantics in Sect. 3.3.

3 Preliminaries

Here we outline conventions and notations for heaps and heap predicates, as well as our choice of concrete heap semantics.

3.1 Heaps and Heap Predicates

A *heap* is a finite partial map from addresses ($\mathbb{L}$) to concrete values ($\mathbb{V}$). Denote the set of all heaps with $\mathbb{H}$, the set of non-null values with $\mathbb{V}^* = \mathbb{V} \setminus \{\mathsf{null}\}$, and the finite address domain of a heap h with $\mathsf{dom}(h)$.

Concrete values may be any typical datatypes (integers, booleans, pairs, etc.) which are generally fixed in size (e.g. an unbounded list would not typically be stored as one concrete value). Addresses themselves can be treated as concrete values. Assume unlimited addresses are available, namely $\forall h.\, \mathsf{dom}(h) \subsetneq \mathbb{L}$.

Heaps h_1 and h_2 are *disjoint* ($h_1 \perp h_2$) iff $\mathsf{dom}(h_1) \perp \mathsf{dom}(h_2)$. The *union* of disjoint heaps ($h_1 + h_2$) yields the combined partial map of h_1 and h_2. Dually, a heap may be partitioned into smaller disjoint heaps. h_1 is a *subheap* of h_2 ($h_1 \subseteq h_2$) iff $h_1 + k = h_2$ for some $k \perp h_1$. Heap equality is defined extensionally.

A *heap predicate* characterizes the addresses and values of heaps. Given heap predicate Ψ on $\mathbb{H}$, $\Psi(h)$ denotes that h satisfies Ψ, and $\mathbb{H}(\Psi)$ is the set of all Ψ-satisfying heaps. Sometimes we parameterize a heap predicate on some domain V. In this case, $\Psi(v, h)$ denotes that h satisfies Ψ with $v \in V$.

Aside from heap predicates containing pure logical expressions and quantifiers, we will describe heap contents directly with two standard predicates from separation logic [29,34]. The *points-to* predicate $p \mapsto v$, for $p \in \mathbb{L}$ and $v \in \mathbb{V}$, holds for h iff $\mathsf{dom}(h) = \{p\}$ and $h(p) = v$. The *separating conjunction* operator $\Psi * \Phi$, for predicates Ψ and Φ, holds for h iff $h = h_1 + h_2$ for some h_1 and h_2 s.t. $h_1 \perp h_2$ and $\Psi(h_1)$ and $\Phi(h_2)$. $*$ is commutative and associative.

We will also want to characterize when a *portion* of a heap satisfies a particular predicate. For this, we define heap predicate *extension*:

Definition 1. *For predicate* Ψ, h *satisfies* extension Ψ^+ *iff a subheap of* h *satisfies* Ψ. *Namely,* $\Psi^+(h) = (\exists \Phi. \Psi * \Phi)(h)$ *and* $\Psi^+(v, h) \equiv (\exists \Phi. \Psi(v, \cdot) * \Phi)(h)$.

3.2 Concrete Semantics

We choose to represent concrete programs as total functions over heaps, so that our semantics are highly general, and thus broadly applicable. We admit semantics which are nondeterministic, but must be constructive and must terminate. Nondeterminism mainly arises during address allocation, though this work is agnostic to the source of nondeterminism.

Definition 2. *A (concrete) program is a total function* $f : \mathbb{H} \to \wp(\mathbb{H})$, *subject to the following properties:*

1. *On any initial heap* h, f *terminates or fails in finite time. If* f *successfully terminates on* h, *then* $f(h)$ *yields the set of all possible final heaps in which* f *may terminate. If* f *fails on* h, *then* $f(h) = \varnothing$.
2. f *reads any input arguments from, and writes any outputs to, fixed locations on the heap. We explore this in Sect. 7.*
3. f *acts locally. Namely, if* $f(h) \neq \varnothing$, *then for any* $h' \perp h$: $f(h + h') \neq \varnothing$ *and* $\forall k \in f(h + h').\ h' \subseteq k$.

Local action enforces that f does not modify data outside its footprint, as a weak form of the typical frame rule of separation logic [34]. For any particular semantics that satisfies framing by construction, the local action constraint should immediately hold for all concrete programs. When working with compound abstractions, we will define a stronger framing property (see Definition 14).

For two programs to commute, they must terminate without failure in the first place. To that end, we classify the initial heaps upon which f can execute. *Sufficient* heaps contain the necessary addresses and appropriate values for f to run. In contrast, heaps in the *footprint* of f contain appropriate addresses, but may not contain appropriate values; thus a heap in f's footprint on which f fails cannot be "corrected" with further allocations. Precisely:

$$\mathsf{suff}(f) \triangleq \{h \mid f(h) \neq \varnothing\}$$
$$\mathsf{foot}(f) \triangleq \mathsf{suff}(f) \cup \{h \mid \forall h' \perp h.\ h + h' \notin \mathsf{suff}(f)\}$$

Our *concrete domain*, as the counterpart to our eventual abstract domain, will be $\mathscr{C} \triangleq \wp(\mathbb{H})$. *Concrete transformers* over this domain map over sets of heaps. We derive these transformers directly from programs:

Definition 3. *The* concrete transformer *constructed from f, denoted $\bar{f}$, is defined as $\bar{f}(C) = \bigcup_{h \in C} f(h)$ if $C \subseteq \mathsf{suff}(f)$, else $\varnothing$.*

$\bar{f}$ yields the set of all possible output heaps given a set of inputs, but fails entirely if f fails on any individual input. Relatedly, we define the composition/sequence of two programs so that the second program fails entirely if it fails on any one output of the first program, namely $f; g \triangleq \bar{g} \circ f$.

3.3   An Example Concrete Language

We share here a simple language called HIMP, which features allocation (`ref`), deallocation (`free`), writing ($\leftarrow$), reading (`!`), and branching. We will use HIMP in our example programs throughout the paper, as an instance of the general semantics of Sect. 3.2. (N.B. HIMP has not been formalized in Coq. However, all of our results regarding the general semantics have been proved in Coq.)

$$
\begin{aligned}
stmt &::= id \leftarrow val \mid \mathtt{free}\ id \mid stmt\ ;\ stmt \mid \mathtt{fail} \mid \mathtt{skip} \\
&\mid\ \mathtt{let}\ id := exp\ \mathtt{in}\ stmt \mid \mathtt{if}\ val\ \mathtt{then}\ stmt\ \mathtt{else}\ stmt \\
exp &::= \mathtt{ref}\ val \mid !(id \mid \mathbb{L}) \mid val + val \mid val = val \mid \cdots \\
val &::= id \mid \mathbb{V}
\end{aligned}
$$

HIMP has a nondeterministic allocation procedure: for any initial heap h, `ref` yields an address from a set $\mathsf{fresh}(h)$ s.t. $\mathsf{fresh}(h) \subseteq \mathbb{L} \setminus \mathsf{dom}(h)$. All other terms are deterministic. We elide the remaining details of HIMP's semantics, which are typical for an imperative heap language. When working with HIMP programs throughout the paper, we implicitly lift them to the concrete programs of Def. 2.

4   Abstract Domain

In this section we formalize our abstract domain and abstract programs operating within this domain, and establish a soundness relationship between concrete and abstract programs. This will let us perform commutativity reasoning easily with abstract objects, and obtain concrete commutativity results for free.

4.1   Constructing the Domain

We design our abstract domain to lift heap-allocated structures to mathematical objects. Each instance of an abstraction consists of a set of *abstract values* (e.g. integers, sets, sequences) and a mapping from heaps into said values. A heap with an allocated structure appropriate to be mapped to an abstract value is said to be *"within the purview"* of a given abstraction. On the other hand, poorly structured heaps are *"outside the purview"* of the abstraction, and get mapped to a special abstract value, ✗ ("cross").

When constructing an abstraction, we impose a *locality constraint* which demands that information extracted from heaps be finite in scope, so that any heap *containing* a valid structure is in-purview. Some results of this work actually do not depend on this constraint. However, it is a valuable property to strive for, and indeed it should hold for most typical data structure abstractions.

The abstract domain is defined formally as follows:

Definition 4. *Build an abstraction* $\mathsf{A} = \langle X, \pi \rangle$ *from a set of* abstract values $X = \{x_1, x_2, \dots\}$, *and a* projection function $\pi : \mathbb{H} \to X + ✗$ *subject to the* locality constraint *that for any disjoint h and h', $\pi(h) \in X \Rightarrow \pi(h + h') = \pi(h)$. The full domain is $\mathscr{A}_\mathsf{A} \triangleq X + \bot + \top + ✗ + ✓$, with partial ordering $\leq_\mathsf{A}$, where $\bot \leq_\mathsf{A} ✗ \leq_\mathsf{A} \top$ and $\forall x \in X. \bot \leq_\mathsf{A} x \leq_\mathsf{A} ✓ \leq_\mathsf{A} \top$ and $\forall x_1, x_2 \in X. x_1 \leq_\mathsf{A} x_2 \Leftrightarrow x_1 = x_2$.*

Denote all in-purview heaps of A with $\mathsf{Prv}(\mathsf{A}) \triangleq \{h \mid \pi(h) \in X\}$, and denote its abstract values and projection function with X_A and π_A.

$\mathscr{A}_\mathsf{A}$ is a complete lattice, and is thus equipped with a typical join ($\sqcup$) operation. Recalling from Sect. 3.2 that our concrete domain is $\mathscr{C} = \wp(\mathbb{H})$, we connect the concrete and abstract domains as follows:

Definition 5. *Define* abstraction *and* concretization *functions* $\alpha_\mathsf{A} : \mathscr{C} \to \mathscr{A}_\mathsf{A}$ *and* $\gamma_\mathsf{A} : \mathscr{A}_\mathsf{A} \to \mathscr{C}$, *where* $\alpha_\mathsf{A}(C) = \bigsqcup_{h \in C} \pi(h)$ *and* $\gamma_\mathsf{A}(x) = \{h \in \mathbb{H} \mid \pi(h) \leq_\mathsf{A} x\}$.

α_A and γ_A are each monotone, and together they form a Galois connection [11] which induces meanings for values in $\mathscr{A}_\mathsf{A}$: $\bot$ is the failure state, $x_i \in X$ abstracts sets of in-purview heaps which all map to x_i, ✓ abstracts any set of in-purview heaps, ✗ out-of-purview heaps, and $\top$ any heaps.

We now share three basic examples of abstract domains. In subsequent sections we will reason about concrete and abstract programs which operate on these defined structures. In Sect. 7 we explore each example in full.

Example 1 (Non-negative Counter).
A "hello world" data structure in commutativity is the non-negative counter (NNC)—an integer which may be incremented, decremented (but not below 0), and read. We abstract an NNC located at address p as $\mathsf{Ctr}_p := \langle \mathbb{N}, \pi \rangle$, where $\pi(h) := n$ if $(p \mapsto n)^+(h)$ for $n \in \mathbb{N}$, else ✗. The *extension* (Definition 1) on the predicate $(p \mapsto n)$ ensures that Ctr_p satisfies locality, allowing the domain to capture any heap *containing* a counter.

Example 2 (Two-Set).
Consider an unordered set containing at most two elements of type $T \subseteq \mathbb{V}^*$. These elements are stored at fixed addresses p and q, with null otherwise stored as a placeholder. One might construct X and π for this structure in a few equivalent ways, the key property being that concrete two-sets containing the same elements should map to the same abstract value, regardless of their order. In the following

approach, the abstract values are simply mathematical sets of size at most two. We define $\mathsf{Set}^T_{p,q} := \langle \{S \in \wp(T) \mid |S| \leq 2\}, \pi \rangle$ with:

$$\mathsf{set}_{p,q}\big((u,v),\cdot\big) := p \mapsto u * q \mapsto v$$

$$\pi(h) := \begin{cases} \{u,v\} & \mathsf{set}^+_{p,q}\big((u,v),h\big) \wedge u \neq v \text{ for } u,v \in T \\ \{u\} & \mathsf{set}^+_{p,q}\big((u,\mathsf{null}),h\big) \vee \mathsf{set}^+_{p,q}\big((\mathsf{null},u),h\big) \text{ for } u \in T \\ \varnothing & \mathsf{set}^+_{p,q}\big((\mathsf{null},\mathsf{null}),h\big) \\ \boldsymbol{\mathsf{X}} & \text{else} \end{cases}$$

Note how we use $\mathsf{set}^+_{p,q}$ throughout π to satisfy locality. One technicality is ensuring that π is well-defined, namely that $\pi(h)$ yields one unambiguous abstract value for each h. In Coq we prove that this π is indeed well-defined.

Example 3 (Linked Stack).

Reiterating our overview example (Sect. 2), consider a stack implemented on the heap as a linked list and accessed at fixed address ℓ. Constructing new cells when pushing to the stack involves nondeterministic address allocation; two stacks containing the same values but different intermediate addresses should be considered observationally equivalent. To achieve this, our recursive construction of π will existentially quantify intermediate addresses, and not lift them explicitly into the abstract domain. We define $\mathsf{Stk}_\ell := \langle \mathsf{list}(\mathbb{V}^*), \pi \rangle$ with

$$\pi(h) \quad := s \text{ if } \mathsf{stk}^+(s,h) \text{ for } s \in \mathsf{list}(\mathbb{V}^*) \text{ else } \boldsymbol{\mathsf{X}}, \text{ where}$$

$$\mathsf{stk}(s) \quad := \exists p \in \mathbb{L} + \mathsf{null}. \ \ell \mapsto p * \psi(s,p)$$
$$\psi(v :: s, p) := \exists q \in \mathbb{L} + \mathsf{null}. \ p \neq \mathsf{null} * p \mapsto (v,q) * \psi(s,q)$$
$$\psi([], p) \quad := p = \mathsf{null}$$

Again, we must show that each h satisfies $\mathsf{stk}^+(s,h)$ for at most one $s \in \mathsf{list}(\mathbb{V}^*)$.

4.2 Isomorphism of Abstract Domains

A notion of isomorphism between abstract domains will also prove useful:

Definition 6. *A bijection* $\varphi : X_\mathsf{A} \to X_\mathsf{B}$ *induces an isomorphism between* A *and* B, *denoted* $\mathsf{A} \cong_\varphi \mathsf{B}$, *if* $\forall h. \ \varphi(\pi_\mathsf{A}(h)) = \pi_\mathsf{B}(h)$. A *and* B *are isomorphic* $(\mathsf{A} \cong \mathsf{B})$ *if such a bijection exists.*

This definition is stronger than necessary for some of our results; future work may warrant distinguishing isomorphism from a weaker equivalence which relaxes the bijectivity requirement of φ. See the appendix of [31] for an example where we construct a two-set abstraction which is isomorphic to that of Example 2.

4.3 Abstract Semantics and Soundness

We design our abstract semantics so that concrete heap behavior (valid or invalid) can be threaded up to the abstract domain with enough precision that, later, abstract commutativity will soundly entail concrete commutativity (Sect. 5).

An *abstract program* in the context of abstraction $\mathsf{A} = \langle X, \pi \rangle$ is any function from X to $\mathscr{A}$, from which we derive an *abstract transformer* over $\mathscr{A}$ (recall from Definition 4 that $\mathscr{A} = X + \bot + \top + \boldsymbol{\mathsf{X}} + \boldsymbol{\checkmark}$). While these transformers could be constructed in various ways, for our purposes the following will suffice:

Definition 7. $\widehat{m}$ *denotes the abstract transformer derived from* $m : X \to \mathscr{A}$, *where* $\widehat{m}(x) = \bot$ *for* $x = \bot$, $m(x)$ *for* $x \in X$, *and* $\top$ *otherwise.*

To relate the behavior of an abstract program m defined in A to a concrete program f, we establish a notion of *soundness* (recall α from Definition 5):

Definition 8. m *soundly abstracts* f *in* A, *denoted* $m \rightsquigarrow_{\mathsf{A}} f$, *if*

1. $\forall C, x.\ \alpha_{\mathsf{A}}(C) \leq_{\mathsf{A}} x \implies \alpha_{\mathsf{A}}(\bar{f}(C)) \leq_{\mathsf{A}} \widehat{m}(x)$, *and*
2. $\forall C.\ \alpha_{\mathsf{A}}(C) \in X \wedge \alpha_{\mathsf{A}}(\bar{f}(C)) = \bot \implies \widehat{m}(\alpha_{\mathsf{A}}(C)) = \bot$.

The first condition is typical for abstract interpretation, demanding that m *over-approximate* f. This lets us be imprecise with the definition of abstract programs if desirable (e.g. to avoid complex behaviors), the trade-off being that concrete commutativity conditions eventually derived may yield false negatives (i.e. not be the weakest precondition). The second soundness condition demands that m fail (yield $\bot$) whenever f fails on an in-purview input. Because the failure sink state is at the bottom of the abstract lattice (Definition 4), without this condition there could be scenarios where two abstract programs successfully commute, but their corresponding concrete programs fail to execute.

We now return to the example of the non-negative counter and its abstraction Ctr_p (see Example 1), to define the corresponding concrete and abstract programs for increment and decrement. We cannot define *read* in Ctr_p, because *read* must write its output to an additional location on the heap. We similarly must wait to revisit our two-set and linked stack examples, whose methods all have inputs and/or outputs. We develop the machinery needed for inputs and outputs in Sect. 6, and reason fully about all three examples in Sect. 7.

Example 4 (Non-negative Counter).
  Consider an NNC allocated at address p. We begin with concrete implementations of increment and decrement, written in the HIMP language (Sect. 3.3), and implicitly lift them to concrete programs.

$$\begin{array}{ll} \mathtt{incr}_p := \mathtt{let}\ c = \ !p\ \mathtt{in} & \mathtt{decr}_p := \mathtt{let}\ c = \ !p\ \mathtt{in}\ \mathtt{let}\ i = c - 1\ \mathtt{in} \\ \qquad \mathtt{let}\ i = c + 1\ \mathtt{in}\ p \leftarrow i & \qquad \mathtt{if}\ i < 0\ \mathtt{then}\ \mathtt{skip}\ \mathtt{else}\ p \leftarrow i \end{array}$$

Next, defining abstract increment and decrement functions in Ctr_p is intuitive:

$$incr_p(n) := n + 1 \qquad\qquad decr_p(n) := \max(0, n - 1)$$

We can show that $incr_p \rightsquigarrow_{\mathsf{Ctr}_p} \mathtt{incr}_p$ and $decr_p \rightsquigarrow_{\mathsf{Ctr}_p} \mathtt{decr}_p$ with typical symbolic execution techniques. Had we defined **decr** to fail when the counter is 0, rather than skip, we would have to define $decr_p(0) = \bot$ to maintain soundness.

5 Sound Commutativity

With our concrete and abstract semantics established, we now define commutativity in the concrete and abstract spaces. We then relate the two with a soundness theorem that reduces concrete commutativity reasoning to simpler reasoning about abstract programs.

5.1 Defining Commutativity

We say concrete programs f and g *commute*, under a notion of observational equivalence on a subset of all heaps, when $f; g$ and $g; f$ both terminate successfully and yield observationally equivalent outcomes. We denote an obs. eq. relation with $[\Psi, \sim]$, for a predicate Ψ on $\mathbb{H}$ and an eq. relation $\sim$ on $\mathbb{H}(\Psi^+)$. Then concrete commutativity is defined precisely as:

Definition 9. *For commutativity condition P on $\mathbb{H}$, f and g commute under P w.r.t. $[\Psi, \sim]$, denoted $f \bowtie^P_{[\Psi, \sim]} g$, if $\forall h \in \mathbb{H}(P). (f; g)(h), (g; f)(h) \neq \varnothing$ and $(f; g)(h) \cup (g; f)(h) \subseteq [h']_\sim$ for some $h' \in \mathbb{H}(\Psi^+)$.*

The eq. relation we use for concrete commutativity is often very similar to the relation implicitly induced by an abstraction. We can make this explicit by deriving a concrete relation in terms of an abstraction, a technique we will use extensively in the remainder of the paper.

Definition 10. *For abstraction A, define concrete equivalence $[\mathsf{Prv}(\mathsf{A}), \sim_\mathsf{A}]$ with $h \sim_\mathsf{A} h' := \pi_\mathsf{A}(h) = \pi_\mathsf{A}(h')$. Use the shorthand $f \bowtie^P_\mathsf{A} g$ for $f \bowtie^P_{[\mathsf{Prv}(\mathsf{A}), \sim_\mathsf{A}]} g$.*

Naturally, a user is responsible for choosing an observational equivalence relation (or in light of Definition 10, an abstraction) which is sufficiently descriptive for their purposes. Properties of concrete commutativity include:

$$\frac{\mathbb{H}(P') \subseteq \mathbb{H}(P) \qquad f \bowtie^P_{[\Psi, \sim]} g}{f \bowtie^{P'}_{[\Psi, \sim]} g} \qquad \frac{\mathsf{A} \cong \mathsf{B} \qquad g \bowtie^P_\mathsf{A} f}{g \bowtie^P_\mathsf{B} f} \qquad \frac{f \bowtie^P_{[\Psi, \sim]} g}{g \bowtie^P_{[\Psi, \sim]} f} \qquad \frac{\forall h, h'. \, h \sim h' \implies h \sim' h' \qquad f \bowtie^P_{[\Psi, \sim]} g}{f \bowtie^P_{[\Psi', \sim']} g}$$

Next we define *abstract commutativity*, in a manner so that we can achieve a useful soundness guarantee in relation to concrete commutativity.

Definition 11. *Abstract programs m and n defined in A commute under predicate Q on X_A, denoted $m \hat{\bowtie}^Q_\mathsf{A} n$, if $\forall x \in X_\mathsf{A}. Q(x) \Rightarrow \hat{n}(m(x)) = \hat{m}(n(x)) \in X_\mathsf{A}$.*

Note how we only accept outcomes within the purview of A. We reject $\bot$, as non-failure is a prerequisite for commutativity. We also reject ✗, ✓, and ⊤, as such outcomes are not precise enough to guarantee meaningful commutativity.

5.2   Sound Commutativity Theorem

To relate abstract and concrete commutativity, we must first relate the relevant abstraction and observational equivalence relation:

Definition 12. A captures $[\Psi, \sim]$ *iff* $\mathbb{H}(\Psi^+) \subseteq \mathsf{Prv}(A)$ *and*
    $\forall h, h' \in \mathbb{H}(\Psi). \pi_A(h) = \pi_A(h') \in X_A \implies h \sim h'$.

This notion of *capture* is preserved by abstraction isomorphism. Additionally, A always captures $[\mathsf{Prv}(A), \sim_A]$, a fact we will frequently use implicitly.

   We now present the central result of our work, the *sound commutativity theorem*, which offloads the verification (via symbolic execution) of concrete commutativity conditions to much simpler reasoning about abstract programs:

Theorem 1. *If* $m \rightsquigarrow_A f$, $n \rightsquigarrow_A g$, $m \bowtie_A^Q n$, *and* A *captures* $[\Psi, \sim]$, *then*
    $f \bowtie_{[\Psi, \sim]}^{P+} g$, *where* $P(h) \equiv \pi_A(h) \in X_A \land Q(\pi_A(h))$.

Sound commutativity takes a valid abstract commutativity condition and transforms it automatically into a concrete condition. For instance, suppose π_A takes the fairly common form, for some Φ, of $\pi_A(h) = x$ if $\Phi(x, h)$ for $x \in X_A$, else ✗. In this case, the concrete condition we derive from abstract condition Q is $\exists x \in X. \Phi^+(x, h) * Q(x)$. Indeed, any condition we derive with Thm. 1 is *extended* (Definition 1). This provides *framing* for free—it guarantees if f and g commute in some heap context, then they also commute in any larger context.

   With sound commutativity established, we can summarize the overall pattern of reasoning: (i) construct an abstract domain, (ii) define abstract programs for each concrete program and prove soundness between them using symbolic execution, (iii) verify or synthesize commutativity conditions for abstract program pairs, and (iv) immediately derive concrete conditions.

Example 5 (Non-negative Counter).
   Returning to the NNC programs defined in Example 4, we can perform abstract commutativity reasoning easily. For instance, to verify that $incr_p \bowtie_{\mathsf{Ctr}_p}^Q decr_p$ for $Q(n) \equiv n > 0$, we simply evaluate:

$$\widehat{decr_p}(\widehat{incr_p}(n)) = \widehat{decr_p}(n + 1) = n \qquad \widehat{incr_p}(\widehat{decr_p}(n)) = \widehat{incr_p}(n - 1) = n$$

With proofs of soundness (Definition 8) of *incr* and *decr* w.r.t. incr and decr, we immediately derive via Thm. 1 that $f \bowtie_{\mathsf{Ctr}_p}^{P+} g$ where $P \equiv \exists n \in \mathbb{N}^+. p \mapsto n$. Note how we use the equivalence imposed by Ctr_p as our concrete observational eq. relation (Definition 10). We might *synthesize* an abstract condition interactively (we leave automation to future work), by trying to prove that *incr* and *decr* commute under ⊤, and seeing where the proof gets stuck. In this case, we will get stuck showing that $1 = 0$ under an initial abstract value of 0. By constraining that $n > 0$, the proof can be completed; thus $n > 0$ is a valid precondition.

6 Abstract Domain Composition

In this section, we expand upon our established tools for abstracting data structures and performing commutativity analysis, by defining composition of abstract domains. This enables compositional reasoning in settings with multiple data structures, and provides machinery needed to reason about concrete programs with heap-allocated inputs and outputs. Various compositional operators over abstract domains have been used in abstract interpretation, such as Cartesian product, reduced product, and reduced power [9,21–23]. While such an existing operator is likely suitable, we elect to design our own simple operator.

A composition of abstractions should yield a new abstraction, and should respect the disjointness of the concrete structures which are individually abstracted. We achieve this with the *abstract conjunction* operator, denoted $A*B$, which is defined to mirror the concrete separating conjunction. The purview of $A * B$ will contain heaps which can be partitioned into disjoint halves that are respectively in the purviews of A and B.

Definition 13. $A * B \triangleq \langle X_A \times X_B, \pi \rangle$, *where* $\pi(h) = (\pi_A(h_1), \pi_B(h_2))$
if $h = h_1 + h_2$ *for some* $h_1 \perp h_2$ *s.t.* $h_1 \in \mathsf{Prv}(A)$ *and* $h_2 \in \mathsf{Prv}(B)$, *else* ✗.

A technicality of Definition 13 is ensuring that $\pi(h)$ maps to a *unique* (a, b) pair (or to ✗) for every h. We show in Coq that this uniqueness does hold, due to the locality constraint of Definition 4. Additionally, when A and B satisfy locality, so too does $A * B$. Abstract conjunction is compatible with, and is commutative and associative up to, abstraction isomorphism; we use these facts implicitly throughout Sect. 7. Much like the heap separating conjunction, $A * B$ may be *unsatisfiable*, i.e. π_{A*B} maps all heaps to ✗. Commutativity reasoning w.r.t. an unsatisfiable domain is still valid, albeit meaningless, as there are no in-purview heaps. Applying the concrete observational equivalence construction of Definition 10 to composition, we get $[\mathsf{Prv}(A * B), \sim_{A*B}]$ where $h \sim_{A*B} h'$ if $h = h_a + h_b$ and $h' = h'_a + h'_b$ s.t. $h_a \sim_A h'_a$ and $h_b \sim_B h'_b$.

6.1 Concrete Program Capture

Effective compositional reasoning will often require that behavior of the concrete programs of interest are "captured" by the abstract domains in use. For instance, incr has well-defined behavior w.r.t. Ctr_p, but not Stk_ℓ. If we prove properties about incr w.r.t. Ctr_p, we should be able to derive how incr behaves w.r.t. $\mathsf{Ctr}_p * \mathsf{Stk}_\ell$ (namely it leaves the stack untouched). However, if we reason about incr w.r.t. Stk_ℓ, we will learn nothing about how incr behaves w.r.t. $\mathsf{Stk}_\ell * \mathsf{Ctr}_p$.

To this end, we say that an abstraction A *captures* program f if (1) heaps in the purview of A are in the footprint of f, (2) f maps in-purview heaps to in-purview heaps, and (3) f maintains its specific in-purview mappings when any disjoint heaps are joined onto the initial heap (thus strengthening *local action* in an abstraction-specific manner). Precisely:

Definition 14. A captures *concrete program* f *if:*

1. $\mathsf{Prv}(\mathsf{A}) \subseteq \mathsf{foot}(f)$, *and*
2. $\forall h \in \mathsf{Prv}(\mathsf{A}) \cap \mathsf{suff}(f). \ \alpha_\mathsf{A}(f(h)) \in X_\mathsf{A}$, *and*
3. $\forall h \in \mathsf{Prv}(\mathsf{A}) \cap \mathsf{foot}(f), h' \perp h. \ \alpha_\mathsf{A}(\{k - h' \mid k \in f(h + h')\}) = \alpha_\mathsf{A}(f(h))$,
   where $k - h'$ is the subheap of k with domain $\mathsf{dom}(k) \setminus \mathsf{dom}(h')$.

Proving capture can be less burdensome than it appears—if we have shown that $m \leadsto_\mathsf{A} f$ for m s.t. $\mathsf{image}(m) \subseteq X_\mathsf{A} + \perp$, then conditions (1) and (2) immediately hold for f. We also expect that (3) should hold automatically for all abstractions and all HIMP programs (and other typical heap languages). Capture is preserved by abstraction isomorphism; and, if A captures f then $\mathsf{A} * \mathsf{B}$ captures f.

Capture yields a common form of concrete commutativity: *commutativity from noninterference*. If two programs operate on disjoint structures, then they automatically commute whenever they individually terminate without failure:

Theorem 2. *If* A *and* B *capture* f *and* g *respectively, then* $f \bowtie_{\mathsf{A}*\mathsf{B}}^{P^+} g$ *where* $P(h) \equiv h \in \mathsf{suff}(f) \cap \mathsf{suff}(g) \cap \mathsf{Prv}(\mathsf{A} * \mathsf{B})$.

Finally, we will often discuss program capture and soundness in the same breath, so we provide a convenient shorthand: $m \overset{\mathrm{CAP}}{\leadsto}_\mathsf{A} f$ denotes that $m \leadsto_\mathsf{A} f$ and A captures f.

6.2 Compositional Commutativity

In service of sound commutativity reasoning in compound abstract domains, we define a way to compose abstract programs:

Definition 15. *Given* m *and* n *defined in* A *and* B *respectively, their* conjunction *is* $m * n$ *in* $\mathsf{A} * \mathsf{B}$, *where for* $a \in X_\mathsf{A}$ *and* $b \in X_\mathsf{B}$, $(m * n)(a, b) = \perp$ *if* $m(a) = \perp \vee n(b) = \perp$, $(m(a), n(b))$ *if* $m(a) \in X_\mathsf{A} \wedge n(b) \in X_\mathsf{B}$, *else* $m(a) \sqcup n(b)$.

The *compound program soundness theorem* then states that the conjunction of two abstract programs soundly abstracts the *sequence* of two concrete programs (in either order), so long as these concrete programs operate on disjoint structures (via *capture*). We also derive an *abstract frame rule* of sorts.

Theorem 3. *If* $m \overset{\mathrm{CAP}}{\leadsto}_\mathsf{A} f$ *and* $n \overset{\mathrm{CAP}}{\leadsto}_\mathsf{B} g$, *then* $m * n \overset{\mathrm{CAP}}{\leadsto}_{\mathsf{A}*\mathsf{B}} f; g$ *and* $m * n \overset{\mathrm{CAP}}{\leadsto}_{\mathsf{A}*\mathsf{B}} g; f$.

Corollary 1. *If* $m \overset{\mathrm{CAP}}{\leadsto}_\mathsf{A} f$, *then* $m * \mathsf{id} \overset{\mathrm{CAP}}{\leadsto}_{\mathsf{A}*\mathsf{B}} f$.

With composition of abstract domains and their programs established, we can demonstrate new compositional commutativity soundness properties, beginning with *compound abstract commutativity*:

Lemma 1. *If* $m \hat{\bowtie}_\mathsf{A}^Q n$ *and* $m' \hat{\bowtie}_{\mathsf{A}'}^{Q'} n'$, *then* $m * m' \hat{\bowtie}_{\mathsf{A}*\mathsf{A}'}^P n * n'$
   where $P(a, b) \equiv Q(a) \wedge Q'(b)$.

From this, along with sound commutativity (Theorem 1) and compound program soundness (Theorem 3), we yield *compound sound commutativity*, which allows us to compose individual commutativity results about pairs of programs operating on disjoint structures:

Theorem 4. *If* $m \bowtie_A^Q m'$, $n \bowtie_B^{Q'} n'$, $m \overset{\mathsf{CAP}}{\leadsto}_A f$, $m' \overset{\mathsf{CAP}}{\leadsto}_A f'$, $n \overset{\mathsf{CAP}}{\leadsto}_B g$, *and* $n' \overset{\mathsf{CAP}}{\leadsto}_B g'$, *then* $f; g \bowtie_{A*B}^{P^+} f'; g'$ *where* $P(h) \equiv \pi_{A*B}(h) = (a, b) \in X_{A*B} \wedge Q(a) \wedge Q'(b)$.

Note how this result still holds even if $A * B$ is unsatisfiable, because the concrete commutativity condition requires that the initial heap maps to a valid abstract value. From this theorem we derive *framed sound commutativity*:

Corollary 2. *If* $m \overset{\mathsf{CAP}}{\leadsto}_A f$ *and* $n \overset{\mathsf{CAP}}{\leadsto}_A g$ *and* $m \bowtie_A^Q n$, *then* $f \bowtie_{A*B}^{P^+} g$, *where* $P(h) \equiv h \in \mathsf{Prv}(A * B) \wedge Q(\mathrm{fst}(\pi_{A*B}(h)))$.

Let us contrast this result with our prior sound commutativity result (Theorem 1), which also includes a form of framing. Theorem 1 says that, *within* the parameters of a chosen abstraction and concrete observational eq. relation, the derived concrete commutativity condition is framed. On the other hand, Corollary 2 guarantees—under the stronger premise of *capture*—that the programs commute under a framed concrete condition within A, as well as within the purview and observational equivalence imposed by any compound abstraction *containing* A.

7   Examples in Full

In this section we reiterate and expand upon our working examples: a non-negative counter, a two-set, a linked-list stack, and the composition of a stack and counter. For each example, we (i) define concrete programs; (ii) construct an abstract domain and abstract programs; (iii) perform abstract commutativity reasoning; (iv) prove soundness between the concrete and abstract programs; and (v) derive concrete commutativity conditions. As (v) follows directly from (iii) and (iv), we sometimes omit it for brevity.

Throughout these examples, we will need to abstract the arguments and return values of methods. To that end we define the "address domain" Addr, which simply captures the value stored at a given heap address. For address p and values $V \subseteq \mathbb{V}$ considered valid, construct $\mathsf{Addr}_p^V := \langle V, \pi \rangle$, where $\pi(h) := v$ if $(p \mapsto v)^+(h)$ for $v \in V$, else $\boldsymbol{X}$.

We will also use some shorthands. We may omit abstract program compositions rendered trivial by (Corollary 1), e.g. if m is defined in A and we are reasoning in $A * B$, we may write m rather than $m * id$. Furthermore, we will leverage the associativity and commutativity of abstract composition up to isomorphism to reason about abstract programs which have been defined in a similar but "out-of-order" compound abstraction. Namely, we use the property that if $m \leadsto_A f$ and $A \cong_\varphi B$, then $\varphi \circ m \circ \varphi^{-1} \leadsto_B f$. The addresses on which an abstract program is parameterized will serve as unique identifiers for how the input and output tuples of said program should be implicitly permuted.

7.1   Non-negative Counter

Consider a non-negative counter structure allocated at p, featuring the following concrete operations written in HIMP:

$$
\begin{array}{lll}
\text{incr}_p \coloneqq & \text{decr}_p \coloneqq \text{let } c = !p \text{ in} & \text{read}_p^r \coloneqq \\
\quad \text{let } c = !p \text{ in} & \quad \text{let } i = c - 1 \text{ in} & \quad \text{let } c = !p \text{ in} \\
\quad \text{let } i = c + 1 \text{ in} & \quad \text{if } i < 0 \text{ then skip} & \quad r \leftarrow c \\
\quad p \leftarrow i & \quad \text{else } p \leftarrow i &
\end{array}
$$

Note how read writes its output to a location r. Define $\mathsf{Ctr}_p \coloneqq \langle \mathbb{N}, \pi \rangle$, where $\pi(h) \coloneqq n$ if $(p \mapsto n)^+(h)$ for $n \in \mathbb{N}$, else $\mathbf{✗}$. Next construct the abstract $incr$ and $decr$ programs in Ctr_p, and $read$ in $\mathsf{A} = \mathsf{Ctr}_p * \mathsf{Addr}_r^{\mathbb{N}}$ to accommodate its output:

$$
incr_p(n) \coloneqq n + 1 \qquad decr_p(n) \coloneqq \max(0, n - 1) \qquad read_p^r(n, \_) \coloneqq (n, n)
$$

We verify commutativity conditions for $incr$ and $decr$ w.r.t. an initial value $n \in X_{\mathsf{Ctr}_p}$. By evaluating the abstract programs (as in Example 5) we find that $incr_p$ and $incr_p$ commute under $\top$, $incr_p$ and $decr_p$ under $n > 0$, and $decr_p$ and $decr_p$ under $\top$. We cannot compare $incr$ or $decr$ with $read$ directly, as they are defined in different domains. So we consider $incr_p * id$ and $decr_p * id$ in $\mathsf{Ctr}_p * \mathsf{Addr}_r^{\mathbb{N}}$, while preserving soundness w.r.t. the concrete programs via Corollary 1. We can verify that, for an initial value $(n, \_) \in X_{\mathsf{Ctr}_p * \mathsf{Addr}_r^{\mathbb{N}}}$, $incr_p * id$ and $read_p^r$ never commute, and $decr_p * id$ and $read_p^r$ commute under $n = 0$.

Lastly, we pair $read$ with itself, recognizing that the two $read$ executions should write their outputs to different locations (this idea returns in the subsequent examples). So with addresses r and s, we consider $read_p^r * id$ and $read_p^s * id$ defined in $\mathsf{Ctr}_p * \mathsf{Addr}_r^{\mathbb{N}} * \mathsf{Addr}_s^{\mathbb{N}}$, and find that they commute under $\top$.

To derive concrete commutativity from these abstract results, we first find with symbolic execution the soundness and capture facts that $\text{incr}_p \overset{\text{CAP}}{\rightsquigarrow}_{\mathsf{Ctr}_p} incr_p$, $\text{decr}_p \overset{\text{CAP}}{\rightsquigarrow}_{\mathsf{Ctr}_p} decr_p$, and $\text{read}_p^r \overset{\text{CAP}}{\rightsquigarrow}_{\mathsf{Ctr}_p * \mathsf{Addr}_r^{\mathbb{N}}} read_p^r$ (recall the shorthand in Sect. 6.1). Now we may immediately derive *framed* concrete commutativity conditions via Theorem 1. For each program pair, we characterize $f \bowtie_{\mathsf{A}}^{P+} g$ (recall shorthand 10):

f	g	A	P
incr_p	incr_p	Ctr_p	$\exists n \in \mathbb{N}.\ p \mapsto n$
incr_p	decr_p	Ctr_p	$\exists n \in \mathbb{N}^+.\ p \mapsto n$
decr_p	decr_p	Ctr_p	$\exists n \in \mathbb{N}.\ p \mapsto n$
incr_p	read_p^r	$\mathsf{Ctr}_p * \mathsf{Addr}_r^{\mathbb{N}}$	$\bot$
decr_p	read_p^r	$\mathsf{Ctr}_p * \mathsf{Addr}_r^{\mathbb{N}}$	$\exists x \in \mathbb{N}.\ p \mapsto 0 * r \mapsto x$
read_p^r	read_p^s	$\mathsf{Ctr}_p * \mathsf{Addr}_r^{\mathbb{N}} * \mathsf{Addr}_s^{\mathbb{N}}$	$\exists n, x, y \in \mathbb{N}.\ p \mapsto n * r \mapsto x * s \mapsto y$

7.2   Two-Set

Consider an unordered set containing at most two elements of type $T \subseteq \mathbb{V}^*$. We define concrete add, remove, and member-test methods, each parameterized on the set locations p and q (we omit p and q subscripts for brevity), as well as input/output addresses a and r.

```
add_a := let u = !p in          rem_a := let u = !p in          mem_a^r := let u = !p in
    let v = !q in                   let v = !q in                   let v = !q in
    let x = !a in                   let x = !a in                   let x = !a in
    if u = x ∨ v = x then            if u = x then                   if u = x then
        skip else                       p ← null else                   r ← true else
    if u = null then                 if v = x then                   if v = x then
        p ← x else                      q ← null else                   r ← true else
    if v = null then                 skip                            r ← false
        q ← x else
    skip
```

Define abstraction $\mathsf{Set}^T_{p,q} := \langle \{S \in \wp(T) \mid |S| \leq 2\}, \pi \rangle$ with:

$$\mathsf{set}_{p,q}((u,v), \cdot) := p \mapsto u * q \mapsto v$$

$$\pi(h) := \begin{cases} \{u,v\} & \mathsf{set}^+_{p,q}((u,v),h) \wedge u \neq v \text{ for } u,v \in T \\ \{u\} & \mathsf{set}^+_{p,q}((u,\mathsf{null}),h) \vee \mathsf{set}^+_{p,q}((\mathsf{null},u),h) \text{ for } u \in T \\ \varnothing & \mathsf{set}^+_{p,q}((\mathsf{null},\mathsf{null}),h) \\ \text{✗} & \text{else} \end{cases}$$

Next we construct abstract programs add and rem in $\mathsf{B} = \mathsf{Set}^T_{p,q} * \mathsf{Addr}^T_a$, and mem in $\mathsf{C} = \mathsf{Set}^T_{p,q} * \mathsf{Addr}^T_a * \mathsf{Addr}^{\mathbb{B}}_r$:

$$add_a(S,v) := (S', v) \text{ where } S' = S \cup \{v\} \text{ if } |S \cup \{v\}| \leq 2 \text{ else } S$$
$$rem_a(S,v) := (S \setminus \{v\}, v)$$
$$mem_a^r(S,v,\_) := (S, v, \mathsf{true} \text{ if } v \in S \text{ else } \mathsf{false})$$

We analyze commutativity w.r.t. $\mathsf{A} = \mathsf{Set}^T_{p,q} * \mathsf{Addr}^T_a * \mathsf{Addr}^T_r * \mathsf{Addr}^T_b * \mathsf{Addr}^T_s$ (though only the mem/mem pair utilizes all four Addr domains).

Programs		Condition on $(S, u, \_, v, \_) \in X_\mathsf{A}$
add_a	add_b	$S = \varnothing \vee |S| = 2 \vee u \in S \vee v \in S \vee u = v$
add_a	rem_b	$|S| < 2 \wedge u \neq v \vee$ $|S| = 2 \wedge (u,v \notin S \vee u \in S \wedge u \neq v)$
rem_a	rem_b	$\top$
mem_a^r	add_b	$u \neq v \vee v \in S \vee |S| = 2$
mem_a^r	rem_b	$u \neq v \vee v \notin S$
mem_a^r	mem_b^s	$\top$

After proving that $add_a \overset{\mathsf{CAP}}{\leadsto}_\mathsf{B} add_a$, $rem_a \overset{\mathsf{CAP}}{\leadsto}_\mathsf{B} rem_a$, and $mem_a^r \overset{\mathsf{CAP}}{\leadsto}_\mathsf{C} mem_a^r$, we again can derive concrete commutativity conditions for add, rem, and mem. In particular, we use the observational equivalence induced by A as our concrete eq. relation, which ensures that concrete two-sets with the same values in different orders are considered equivalent. We omit for brevity the concrete conditions, which each amount to "the program pair commutes w.r.t. $[\mathsf{Prv}(\mathsf{A}), \sim_\mathsf{A}]$ on h if $\pi(h)$ satisfies the corresponding abstract condition."

7.3 Linked Stack

We now expand upon the example from the overview (Sect. 2) of a stack of non-null values, implemented as a linked list accessible at address ℓ. We consider its push, pop, and peek methods, where pop fails on an empty stack. To account for the lack of an `is_empty` operation, peek outputs null on an empty stack.

$$
\begin{array}{lll}
push^a_\ell := \mathtt{let}\ p = !\ell\ \mathtt{in} & pop^a_\ell := \mathtt{let}\ p = !\ell\ \mathtt{in} & peek^a_\ell := \mathtt{let}\ p = !\ell\ \mathtt{in} \\
\quad \mathtt{let}\ v = !a\ \mathtt{in} & \quad \mathtt{if}\ p = \mathtt{null}\ \mathtt{then} & \quad \mathtt{if}\ p = \mathtt{null}\ \mathtt{then} \\
\quad \mathtt{let}\ x = (v, p)\ \mathtt{in} & \quad\quad \mathtt{fail}\ \mathtt{else} & \quad\quad a \leftarrow \mathtt{null}\ \mathtt{else} \\
\quad \mathtt{let}\ q = \mathtt{ref}\ x\ \mathtt{in} & \quad \mathtt{let}\ x = !p\ \mathtt{in} & \quad \mathtt{let}\ x = !p\ \mathtt{in} \\
\quad \ell \leftarrow q & \quad \mathtt{let}\ v = \mathtt{fst}\ x\ \mathtt{in} & \quad \mathtt{let}\ v = \mathtt{fst}\ x\ \mathtt{in} \\
& \quad \mathtt{let}\ q = \mathtt{snd}\ x\ \mathtt{in} & \quad a \leftarrow v \\
& \quad \ell \leftarrow q\ ;\ \mathtt{free}\ p\ ; \\
& \quad a \leftarrow v
\end{array}
$$

Once again, we abstraction define $\mathsf{Stk}_\ell := \langle \mathrm{list}(\mathbb{V}^*), \pi \rangle$ where

$$\pi(h) := s\ \mathrm{if}\ \mathsf{stk}^+(s, h)\ \mathrm{for}\ s \in \mathrm{list}(\mathbb{V}^*)\ \mathrm{else}\ \textbf{\textsf{X}}$$

$$\mathsf{stk}(s) := \exists p \in \mathbb{L} + \mathsf{null}.\ \ell \mapsto p * \psi(s, p)$$

$$\psi(v :: s, p) := \exists q \in \mathbb{L} + \mathsf{null}.\ p \neq \mathsf{null} * p \mapsto (v, q) * \psi(s, q)$$

$$\psi([], p) := p = \mathsf{null}$$

Define abstract $push/pop$ in $\mathsf{B} := \mathsf{Stk}_\ell * \mathsf{Addr}^{\mathbb{V}^*}_a$, and $peek$ in $\mathsf{C} := \mathsf{Stk}_\ell * \mathsf{Addr}^{\mathbb{V}}_a$:

$$
\begin{array}{lll}
push^a_\ell(s, v) := (v :: s, v) & pop^a_\ell([], \_) := \bot & peek^a_\ell(v :: s, \_) := (v :: s, v) \\
& pop^a_\ell(v :: s, \_) := (s, v) & peek^a_\ell([], \_) := ([], \mathsf{null})
\end{array}
$$

Next we verify commutativity conditions for each abstract program pair w.r.t. $\mathsf{A} = \mathsf{Stk}_\ell * \mathsf{Addr}^U_a * \mathsf{Addr}^V_b$, where U and V depend on the pair.

Programs		U	V	Cond. on $(s, u, v) \in X_\mathsf{A}$
$push^a_\ell$	$push^b_\ell$	$\mathbb{V}^*$	$\mathbb{V}^*$	$u = v$
$push^a_\ell$	pop^b_ℓ	$\mathbb{V}^*$	$\mathbb{V}^*$	$s = u :: \_$
$push^a_\ell$	$peek^b_\ell$	$\mathbb{V}^*$	$\mathbb{V}$	$s = u :: \_$
pop^a_ℓ	pop^b_ℓ	$\mathbb{V}^*$	$\mathbb{V}^*$	$\exists x.\ s = x :: x :: \_$
pop^a_ℓ	$peek^b_\ell$	$\mathbb{V}^*$	$\mathbb{V}$	$\exists x.\ s = x :: x :: \_$
$peek^a_\ell$	$peek^b_\ell$	$\mathbb{V}$	$\mathbb{V}$	$\top$

We then find through symbolic execution that $push_\ell^a \overset{\text{CAP}}{\leadsto}_B push_\ell^a$, $pop_\ell^a \overset{\text{CAP}}{\leadsto}_B pop_\ell^a$, and $peek_\ell^a \overset{\text{CAP}}{\leadsto}_C peek_\ell^a$. Finally, we freely derive commutativity conditions for each concrete pair w.r.t. $[\mathsf{Prv}(A), \sim_A]$, which we again omit for brevity.

7.4   Composition of Stack and Counter

Suppose we augment our stack data structure with a counter to track the stack's size. The concrete operations for this structure would be

$$\mathtt{cpush}_{\ell,p}^a := \mathtt{push}_\ell^a\,; \mathtt{incr}_p \qquad\qquad \mathtt{cpeek}_\ell^a := \mathtt{peek}_\ell^a$$

$$\mathtt{cpop}_{\ell,p}^a := \mathtt{pop}_\ell^a\,; \mathtt{decr}_p \qquad\qquad \mathtt{csize}_p^r := \mathtt{read}_p^r$$

We can similarly define abstract programs with composition. Namely, the following are defined in $\mathsf{Stk}_\ell * \mathsf{Ctr}_p$ (with Addr domains joined on as appropriate):

$$cpush_{\ell,p}^a := push_\ell^a * incr_p \qquad\qquad cpeek_\ell^a := peek_\ell^a * id$$

$$cpop_{\ell,p}^a := pop_\ell^a * decr_p \qquad\qquad csize_p^r := id * read_p^r$$

We analyze commutativity of these abstract programs in $A = \mathsf{Stk}_\ell * \mathsf{Ctr}_p * \mathsf{Addr}_a^U * \mathsf{Addr}_b^V$. For each pair, we leverage Lemma 1 to derive a condition for free from our prior analyses (ℓ and p subscripts are omitted):

Programs		U	V	Cond. on $(s, n, u, v) \in X_A$
$cpush^a$	$cpush^b$	V^*	V^*	$u = v$
$cpush^a$	$cpop^b$	V^*	V^*	$s = u :: \_ \,\wedge n > 0$
$cpush^a$	$cpeek^b$	V^*	V	$s = u :: \_$
$cpush^a$	$csize^b$	V^*	N	$\bot$
$cpop^a$	$cpop^b$	V^*	V^*	$\exists x.\ s = x :: x :: \_$
$cpop^a$	$cpeek^b$	V^*	V	$\exists x.\ s = x :: x :: \_$
$cpop^a$	$csize^b$	V^*	N	$s = \_ :: \_ \,\wedge n = 0$
$cpeek^a$	$cpeek^b$	V	V	$\top$
$cpeek^a$	$csize^b$	V	N	$\top$
$csize^a$	$csize^b$	N	N	$\top$

For our requisite soundness and capture facts, we use compound soundness (Theorem 3) to derive for free that $cpush_{\ell,p}^a \overset{\text{CAP}}{\leadsto}_A \mathtt{cpush}_{\ell,p}^a$, $cpop_{\ell,p}^a \overset{\text{CAP}}{\leadsto}_A \mathtt{cpop}_{\ell,p}^a$, $cpeek_\ell^a \overset{\text{CAP}}{\leadsto}_A \mathtt{cpeek}_\ell^a$, and $csize_p^a \overset{\text{CAP}}{\leadsto}_A \mathtt{csize}_p^a$. With this and our abstract commutativity conditions, we may freely derive concrete conditions for $\mathtt{cpush}$, $\mathtt{cpop}$, $\mathtt{cpeek}$, and $\mathtt{csize}$. One interesting case is the $cpop/csize$ condition, which demands a nonempty stack with size 0. This cannot occur with a correctly initialized stack and counter, meaning in practice $\mathtt{cpop}$ and $\mathtt{csize}$ never commute.

8   Related Work

To the best of our knowledge there are no existing works on commutativity reasoning specifically geared toward heap-based programs, aside from preliminary work by Pincus [30] which is limited to deterministic programs. We discuss some works in Sect. 1; and here discuss those and others in more detail.

Commutativity without the Heap. Some prior works focused on verifying and even inferring commutativity properties, though without a heap memory model. These include aforementioned work such as Kim and Rinard [24] who verified commutativity properties in two steps: verifying an implementation satisfies its ADT specs, and then verifying commutativity of the ADT spec. Further in that direction, Bansal *et al.* [2, 3] showed that commutativity conditions of ADT specs could be synthesized, and introduced the tool SERVOIS, which was later improved as SERVOIS2 [7]. Chen *et al.* [6] adapted these approaches to perform commutativity analysis on (non-heap) imperative programs for parallelization. Koskinen and Bansal [25] also verified commutativity, but not of heap-based programs.

Commutativity with the Heap. Pîrlea *et al.* [32] describe the COSPLIT tool, which performs a commutativity analysis on a smart contract language. Their analysis determines commutativity of heap operations, but only when commutativity can be determined based on heap ownership. Thus, their approach could not handle the examples in this paper including the non-negative counter, which involves commutative updates to the same heap cell. Eilers *et al.* [13] leverage commutativity modulo user-specified abstractions to prove information flow security in the space of relational concurrent separation logic.

Exploiting Commutativity for Verification. Commutativity is a widely used abstraction in many verification tools, which use commutativity specifications as user-provided inputs. In the context of concurrent programs, QED [14] and later CIVL [26] both use commutativity (more specifically, left/right movers) to build atomic sections. The ANCHOR [20] verifier also uses commutativity for a more automated approach of verifying concurrent programs. Commutativity is also used for proofs of parameterized programs [18], operational-style proofs of concurrent objects [15] and termination of concurrent programs [27]. Abstract commutativity relations have also been used in reductions for verification [17]. A summary of the use of commutativity in verification was given by [16].

Abstract Domains for the Heap. Other works have focused on the intersection of abstract interpretation and heap logics, although they do not specifically target commutativity, and generally seek to abstract the semantics of separation logic rather than reasoning about particular allocated structures. Sims [37] constructs a detailed abstract domain for representing separation logic heap predicates. Calcagno *et al.* [4] introduce separation algebras, which generalize the semantics of separation logic.

9   Conclusion

We have demonstrated how commutativity analysis on concrete heap programs can be reduced to much simpler reasoning about mathematical objects in an abstract space. We have designed our abstract domain to account for allocation nondeterminism, concrete observational equivalence, and improperly allocated heap structures. We formalized this domain, laid out the abstract semantics and program soundness relation, established a sound commutativity theorem, and introduced composition of abstractions for multi-structure program settings. We then worked through several examples illustrating the convenience of analyzing abstract program commutativity and deriving concrete results. Our work has been implemented in Coq and is available in the supplement.

In future work we plan to pursue automation by implementing an abstract interpreter, and exploring how it can be combined with existing commutativity synthesis techniques in the abstract domain [2,7]. Furthermore, we will explore how our model of data structure abstraction could be applicable to various kinds of heap-style reasoning other than commutativity. We are also interested in augmenting our abstract domain to feature an abstract value subset lattice; this would admit nondeterministic abstract programs, and enable reasoning about bounded divergence commutativity.

Acknowledgments. We thank Marco Gaboardi, David Naumann, VFC, and the anonymous reviewers for their feedback on earlier versions of this draft. Both authors were partially supported by NSF award #2008633. Koskinen was partially supported by NSF award #2315363. Pincus was partially supported by NSF award #1801564.

Disclosure of Interests. The authors have no competing interests to declare.

References

1. Antonopoulos, T., Koskinen, E., Le, T.C., Nagasamudram, R., Naumann, D.A., Ngo, M.: An algebra of alignment for relational verification. Proc. ACM Program. Lang. **7**(POPL), 573–603 (2023). https://doi.org/10.1145/3571213
2. Bansal, K., Koskinen, E., Tripp, O.: Automatic generation of precise and useful commutativity conditions (extended version). CoRR (2018). http://arxiv.org/abs/1802.08748
3. Bansal, K., Koskinen, E., Tripp, O.: Synthesizing precise and useful commutativity conditions. J. Autom. Reason. **64**(7), 1333–1359 (2020). https://doi.org/10.1007/S10817-020-09573-W
4. Calcagno, C., O'Hearn, P.W., Yang, H.: Local action and abstract separation logic. In: 22nd Annual IEEE Symposium on Logic in Computer Science (LICS 2007), pp. 366–378 (2007). https://doi.org/10.1109/LICS.2007.30
5. Charguéraud, A.: Separation Logic Foundations, Software Foundations, vol. 6. Electronic textbook (2023). http://softwarefoundations.cis.upenn.edu, version 2.0
6. Chen, A., Fathololumi, P., Koskinen, E., Pincus, J.: Veracity: Declarative multi-core programming with commutativity. Proc. ACM Program. Lang. **6**(OOPSLA2) (2022). https://doi.org/10.1145/3563349

7. Chen, A., Fathololumi, P., Nicola, M., Pincus, J., Brennan, T., Koskinen, E.: Better predicates and heuristics for improved commutativity synthesis. In: André, É., Sun, J. (eds.) Automated Technology for Verification and Analysis - 21st International Symposium, ATVA 2023, Singapore, 24-27 October 2023, Proceedings, Part II. LNCS, vol. 14216, pp. 93–113. Springer (2023). https://doi.org/10.1007/978-3-031-45332-8_5

8. Clements, A.T., Kaashoek, M.F., Zeldovich, N., Morris, R.T., Kohler, E.: The scalable commutativity rule: designing scalable software for multicore processors. In: Proceedings of the Twenty-Fourth ACM Symposium on Operating Systems Principles, SOSP 2013, pp. 1–17. Association for Computing Machinery, New York (2013). https://doi.org/10.1145/2517349.2522712

9. Cortesi, A., Costantini, G., Ferrara, P.: A survey on product operators in abstract interpretation. Electr. Proc. Theoret. Comput. Sci. **129**, 325–336 (sep 2013). https://doi.org/10.4204/eptcs.129.19

10. Cousot, P., Cousot, R.: Abstract interpretation: A unified lattice model for static analysis of programs by construction or approximation of fixpoints. In: Proceedings of the 4th ACM SIGACT-SIGPLAN Symposium on Principles of Programming Languages. pp. 238–252. POPL '77, Association for Computing Machinery, New York, NY, USA (1977). https://doi.org/10.1145/512950.512973

11. Cousot, P., Cousot, R.: Abstract interpretation and application to logic programs. J. Logic Program. **13**(2), 103–179 (1992). https://doi.org/10.1016/0743-1066(92)90030-7

12. Dickerson, T., Gazzillo, P., Herlihy, M., Koskinen, E.: Adding concurrency to smart contracts. In: Proceedings of the ACM Symposium on Principles of Distributed Computing, PODC 2017, pp. 303–312. ACM, New York (2017). https://doi.org/10.1145/3087801.3087835

13. Eilers, M., Dardinier, T., Müller, P.: Commcsl: proving information flow security for concurrent programs using abstract commutativity. Proc. ACM Program. Lang. **7**(PLDI), 1682–1707 (2023). https://doi.org/10.1145/3591289

14. Elmas, T., Qadeer, S., Tasiran, S.: A calculus of atomic actions. ACM SIGPLAN Notices **44**(1), 2–15 (2009)

15. Enea, C., Koskinen, E.: Scenario-based proofs for concurrent objects. Proc. ACM Program. Lang. (to appear) (OOPSLA2) (2024)

16. Farzan, A.: Commutativity in automated verification. In: LICS, pp. 1–7 (2023). https://doi.org/10.1109/LICS56636.2023.10175734

17. Farzan, A., Klumpp, D., Podelski, A.: Stratified commutativity in verification algorithms for concurrent programs. Proc. ACM Program. Lang. **7**(POPL), 1426–1453 (2023). https://doi.org/10.1145/3571242

18. Farzan, A., Klumpp, D., Podelski, A.: Commutativity simplifies proofs of parameterized programs. Proc. ACM Program. Lang. (POPL) (2024)

19. Farzan, A., Mathur, U.: Coarser equivalences for causal concurrency. Proc. ACM Program. Lang. **8**(POPL), 911–941 (2024). https://doi.org/10.1145/3632873

20. Flanagan, C., Freund, S.N.: The anchor verifier for blocking and non-blocking concurrent software. Proc. ACM Program. Lang (OOPSLA) **4**, 1–29 (2020)

21. Giacobazzi, R., Ranzato, F.: The reduced relative power operation on abstract domains. Theoret. Comput. Sci. **216**(1), 159–211 (1999). https://doi.org/10.1016/S0304-3975(98)00194-7

22. Giacobazzi, R., Ranzato, F., Scozzari, F.: Making abstract domains condensing. ACM Trans. Comput. Logic **6**(1), 33–60 (2005). https://doi.org/10.1145/1042038.1042040

23. Giacobazzi, R., Scozzari, F.: A logical model for relational abstract domains. ACM Trans. Program. Lang. Syst. **20**(5), 1067–1109 (1998). https://doi.org/10.1145/293677.293680

24. Kim, D., Rinard, M.C.: Verification of semantic commutativity conditions and inverse operations on linked data structures. In: Proceedings of the 32nd ACM SIGPLAN Conference on Programming Language Design and Implementation, PLDI 2011, pp. 528–541. ,Association for Computing Machinery, New York (2011). https://doi.org/10.1145/1993498.1993561

25. Koskinen, E., Bansal, K.: Decomposing data structure commutativity proofs with *mn*-differencing. In: Henglein, F., Shoham, S., Vizel, Y. (eds.) VMCAI 2021. LNCS, vol. 12597, pp. 81–103. Springer, Cham (2021). https://doi.org/10.1007/978-3-030-67067-2_5

26. Kragl, B., Qadeer, S.: The CIVL verifier. In: 2021 Formal Methods in Computer Aided Design (FMCAD), pp. 143–152. IEEE (2021)

27. Lette, D., Farzan, A.: Commutativity for concurrent program termination proofs. In: Enea, C., Lal, A. (eds.) Computer Aided Verification - 35th International Conference, CAV 2023, Paris, France, 17-22 July 2023, Proceedings, Part I. LNCS, vol. 13964, pp. 109–131. Springer (2023). https://doi.org/10.1007/978-3-031-37706-8_6

28. Nagasamudram, R., Naumann, D.A.: Alignment completeness for relational hoare logics. In: 36th Annual ACM/IEEE Symposium on Logic in Computer Science, LICS 2021, Rome, Italy, 29 June - 2 July, 2021. pp. 1–13. IEEE (2021). https://doi.org/10.1109/LICS52264.2021.9470690

29. Parkinson, M., Bierman, G.: Separation logic and abstraction. In: Proceedings of the 32nd ACM SIGPLAN-SIGACT Symposium on Principles of Programming Languages, pp. 247–258 (2005)

30. Pincus, J.: Commutativity Reasoning for the Heap. Master's thesis, Stevens Institute of Technology (2022). https://www.proquest.com/docview/2681771819

31. Pincus, J., Koskinen, E.: An abstract domain for heap commutativity (extended version) (2024). https://doi.org/10.48550/arXiv.2411.12857

32. Pîrlea, G., Kumar, A., Sergey, I.: Practical Smart Contract Sharding with Ownership and Commutativity Analysis, pp. 1327–1341. Association for Computing Machinery, New York (2021). https://doi.org/10.1145/3453483.3454112

33. Prabhu, P., Ghosh, S., Zhang, Y., Johnson, N.P., August, D.I.: Commutative set: a language extension for implicit parallel programming. In: Proceedings of the 32nd ACM SIGPLAN Conference on Programming Language Design and Implementation, pp. 1–11 (2011). https://doi.org/10.1145/1993316.1993500

34. Reynolds, J.: Separation logic: a logic for shared mutable data structures. In: Proceedings 17th Annual IEEE Symposium on Logic in Computer Science, pp. 55–74 (2002). https://doi.org/10.1109/LICS.2002.1029817

35. Rinard, M.C., Diniz, P.C.: Semantic foundations of commutativity analysis. In: Bougé, L., Fraigniaud, P., Mignotte, A., Robert, Y. (eds.) Euro-Par 1996. LNCS, vol. 1123, pp. 414–423. Springer, Heidelberg (1996). https://doi.org/10.1007/3-540-61626-8_55

36. Shapiro, M., Preguiça, N., Baquero, C., Zawirski, M.: A comprehensive study of convergent and commutative replicated data types. Ph.D. thesis, Inria–Centre Paris-Rocquencourt; INRIA (2011)

37. Sims, E.J.: An abstract domain for separation logic formulae. In: Proceedings of the 1st International Workshop on Emerging Applications of Abstract Interpretation (EAAI 2006), pp. 133–148. ENTCS, Vienna, Austria (2006)

38. Weihl, W.E.: Data-dependent concurrency control and recovery (extended abstract). In: Proceedings of the second annual ACM symposium on Principles of distributed computing (PODC 1983), pp. 63–75. ACM Press, New York (1983). https://doi.org/10.1145/800221.806710

39. Yang, H.: Relational separation logic. Theor. Comput. Sci. **375**(1–3), 308–334 (2007). https://doi.org/10.1016/J.TCS.2006.12.036

A Static Analysis of Entanglement

Nicola Assolini$^{(\boxtimes)}$ [ID], Alessandra Di Pierro [ID], and Isabella Mastroeni [ID]

Dipartimento di Informatica, Università di Verona, Verona, Italy
{nicola.assolini,alessandra.dipierro,
isabella.mastroeni}@univr.it

Abstract. Managing quantum variables in quantum programs presents specific challenges due to the possible occurrence of *entanglement*, the quantum mechanical phenomenon for which two variables can reach a state where they cannot be separated into two distinct individual states. Such a phenomenon may lead to critical issues due to unintended measurements, which may alter the outcome of computations involving entangled variables. To address this problem, we propose a static analysis based on the abstract interpretation framework to soundly and automatically detect entanglement occurring in quantum programs. By constructing an abstract domain for the entanglement property, our analysis identifies cases where side effects from quantum operations may produce unwanted entanglement, thus reducing the possibility of unintended computational side effects.

1 Introduction

Quantum programming languages play an important role in quantum software development and are essential for effectively using a quantum computer for problem-solving [13,17]. Quantum programming language design and implementation provide crucial support to the rapid technological progress driven by the efforts of a number of companies striving to build large-scale quantum computers. The nature of quantum computation introduces specific challenges when working with these languages [3]. One such challenge comes from the fact that, unlike classical variables, quantum variables represent information encoded in quantum states, i.e., vectors within a Hilbert space; this inevitably implies the need for specific approaches, different from the classical case, for their abstraction.

Quantum computation is characterised by two fundamental principles: superposition and entanglement. Unlike classical systems, where a state exists in a single, definite configuration, a quantum state can simultaneously exist in a superposition of multiple classical states. Entanglement occurs when some particles become so strongly correlated through a computation that they form an 'inseparable' state, i.e., a state where they are no longer identifiable in their individual states. For program variables, this means that every time we act on a variable q, we also alter the state of the other variables entangled with q. Entanglement is crucial for many applications such as quantum communication

K. Shankaranarayanan et al. (Eds.): VMCAI 2025, LNCS 15530, pp. 50–71, 2025.
https://doi.org/10.1007/978-3-031-82703-7_3

protocols (e.g. quantum teleportation [20, Chapter 5]), but can be problematic in quantum programming, particularly when combined with the principle of implicit measurement [22, Section 4.4]. Therefore, analysing and tracking entanglement is essential for effectively reasoning about quantum programs. A particularly important task is identifying sets of entangled variables, namely sets of variables, such that whenever we operate on one variable, we may alter other variables in the same set. Crucial for this task is to identify which variables are in a quantum state and which are not.

In this work, we introduce a new abstract domain for entanglement, which refines the abstract domain introduced in [2] by incorporating additional labels that abstract the quantum states of variables. These labels allow us to define a static analysis, which is able to identify entangled variables and approximate the variable's state. Our static analysis not only determines the sets of entangled variables but also distinguishes a particular relation, which we call *direct inseparability*, between entangled variables capturing the case of entangled states of the form $\alpha |00 \dots 0\rangle + \beta |11 \dots 1\rangle$. These states called GHZ from the names of their inventors[1], are particularly interesting because they can be 'disentangled' in a very easy way. Our work improves the accuracy of existing methods that use a similar but less refined domain, such as the approach in [27], while avoiding the computational challenges associated with more precise abstract domains that grow exponentially with the number of qubits [18]. Additionally, our analysis is flexible enough to be applied to all imperative quantum languages (e.g. Quipper [15], QWire [24], Qiskit [1], qrisp [31] and Guppy [28]) rather than being limited to specific quantum circuits as in [29,30].

In the following sections, we will first provide an overview of the essential background of quantum computation (Sect. 2) and introduce the programming language used to define our analysis (Sect. 3). We then refine the properties introduced in our previous work [2] (Sect. 4) and define a new abstract domain (Sect. 5). In Sect. 6, we present the abstract semantics, and in Sect. 7, we show how to compute the analysis on a control flow graph. Finally, Sect. 8 discusses related work and Sect. 9 concludes the paper with a summary of our findings and potential directions for future research.

2   Quantum Computation

This section briefly recalls the main aspects of quantum computation related to the entanglement phenomenon. In doing so, we will refer to the circuit model of computation. In a quantum circuit, wires represent quantum bits, or qubits, rather than bits. Thus, a qubit replaces the classical unit of information (the bit) in the quantum computation model, generalising the two only possible values 0 and 1 of a bit to any vector in a complex Hilbert space (the quantum system), with 0 and 1 as basis vector. The typical notation of such vectors (or states

---

[1] Danny Greenberg, Mike Horne, and Anton Zeilinger experimentally created this three-particle entanglement showing that quantum mechanics is not compatible with Einstein's theory of 'hidden variables'.

of a qubit) is the Dirac *ket* notation, according to which $|0\rangle = (1,0)^T$ and $|1\rangle = (0,1)^T$ indicate the basis states 0 and 1 and, in general, $|\psi\rangle = \alpha\,|0\rangle + \beta\,|1\rangle$ denotes a linear combination or *superposition* state. The numbers α and β are complex numbers called probability amplitudes since, from them, we can infer the probability of the state resulting in 0 or 1 after measuring the system. Such probabilities are obtained as $|\alpha|^2$ and $|\beta|^2$, which explains why quantum states must be unitary, i.e., $|\alpha|^2 + |\beta|^2 = 1$ must hold.

Implementing significant and powerful quantum algorithms requires performing quantum computation on circuits that are more complex than a single qubit operation and involve n qubit states with $n > 1$. A n qubit state corresponds to a unitary vector in the 2^n-dimensional Hilbert space ($\mathcal{H}^{2^n}$), obtained by composing by tensor product ($\otimes$) the vector space of the single qubits, each living in a 2-dimensional complex Hilbert space ($\mathcal{H}^2$) [22, Chapter 2]. For instance, the space of two qubits is $\mathcal{H}^4 = \mathcal{H}^2 \otimes \mathcal{H}^2$ and a generic state $|\psi\rangle$ in $\mathcal{H}^4$ can be written as $|\psi\rangle = \alpha_0\,|00\rangle + \alpha_1\,|01\rangle + \alpha_2\,|10\rangle + \alpha_3\,|11\rangle$ where all α_i are complex numbers.

2.1   Measurement

Quantum measurement is an operation that allows us to extract a classical result from a quantum superposition $|\psi\rangle$. This operation transforms the quantum state into a classical one by breaking the quantum coherence (and so the quantum nature) of the state. Therefore, measurement is typically applied as the last operation in a quantum circuit to get the final (classical and probabilistic) result of the coherent (i.e., in superposition) evolution of the quantum system represented by the circuit. Formally, quantum measurement on the state space of the quantum system is represented using measurement operators $\{M_m\}$, where m corresponds to the possible outcomes of the measurement. If the system is in the quantum state $|\psi\rangle$ before the measurement, the probability of obtaining outcome m is given by $p(m) = \|M_m\,|\psi\rangle\,\|^2$, where $\|\cdot\|$ is the vector norm[2], and the system state after the measurement is $\frac{M_m|\psi\rangle}{\sqrt{p(m)}}$. For instance, given one qubit, the measurement operators are $M_0 = |0\rangle\langle 0|$ and $M_1 = |1\rangle\langle 1|$, corresponding to the outcomes 0 and 1. If the state of the qubit before the measurement is $|\psi\rangle = \alpha\,|0\rangle + \beta\,|1\rangle$, the probability of measuring 0 is: $p(0) = \|M_0\psi\|^2 = |\alpha|^2$ and the probability of measuring 1 is $p(1) = \|M_1\psi\|^2 = |\beta|^2$. After the measurement, if outcome 0 is observed, the state collapses to $\frac{M_0|\psi\rangle}{|\alpha|} = |0\rangle$ while if outcome 1 is observed, the state collapses to $\frac{M_1|\psi\rangle}{|\beta|} = |1\rangle$.

---

[2] We refer here to the Hilbert space vector norm defined as $\|\,|\psi\rangle\,\| = \sqrt{\langle\psi\rangle}$, where $\langle\psi|$ is the conjugate transpose of $|\psi\rangle$ and $\langle x\rangle y$ is the inner product between vector $|x\rangle$ and vector $|y\rangle$.

2.2   Entanglement

The behaviour of quantum circuits is determined by the laws of quantum mechanics and undergoes the effect of an important quantum phenomenon with no classical counterparts, namely *entanglement*. This can be intuitively described as an application of the superposition principle to a system composed of two or more subsystems. It occurs when statistically correlated measurement outcomes are observed as the effect of two subsequent quantum measurements, one on each subsystem. More concretely, the term entanglement describes a situation in which two particles, designated as x and y, which form a composite system, become strongly correlated. This occurs as a result of a computational process that generates a superposition of product states for both particles. This superposition implies that the state of the composite system cannot be described without considering the other particle's state. Consequently, if measurements are made on an entangled state $ab + cd$, where a and c are two possible states of x and b and d are two possible states of y, then if x is found in state a, y must be in state b; similarly if x is found in state c, y must be in state d.

As an example, the state $1/\sqrt{2}(|00\rangle + |11\rangle)$ in the Hilbert space $\mathcal{H}^2 \otimes \mathcal{H}^2$ is entangled because it cannot be expressed as a tensor product of the individual states of the two-component qubits. In this state, if one qubit is measured and found to be in the state $|0\rangle$, the other qubit will instantaneously collapse to the state $|0\rangle$ as well, and similarly for the state $|1\rangle$. In some cases, measuring a qubit of an entangled pair alters the other, keeping it in a quantum state. For instance, consider the entangled state $1/2(|00\rangle + |01\rangle + |10\rangle - |11\rangle)$. If the first qubit is measured and found in the state $|0\rangle$, the other qubit will instantaneously collapse to the state $1/\sqrt{2}(|0\rangle + |1\rangle)$ and, similarly, if the first qubit after the measurement collapses to $|1\rangle$, then the other one will be in state $1/\sqrt{2}(|0\rangle - |1\rangle)$.

2.3   Quantum Variables

A quantum variable is the high-level abstraction of the state of a quantum register. Therefore, its type is the dimension of the Hilbert space to which those states belong, namely 2^n for a n-qubit quantum register. Thus, the abstraction of a qubit is a quantum variable q of type 2, whose states are vectors in a two-dimensional complex Hilbert space $\mathcal{H}_q$. Following [37], we construct the space of values for a set $Q = \{q_i\}$ of quantum variables q_i as the Hilbert space $\mathcal{H}_Q = \bigotimes_i \mathcal{H}_{q_i}$ obtained by composing via tensor product the space of each variable. Given a quantum program characterised by a set Q of quantum variables, we say that the Hilbert space $\mathcal{H}_Q$ is the program space, and the semantics of the program can be described using vectors and operators in the space $\mathcal{H}_Q$ [37].

We write $|\psi\rangle_{q_i}$ to indicate that q_i represents the state $|\psi\rangle$ in $\mathcal{H}_{q_i}$. For entangled states, such for example $1/\sqrt{2}(|01\rangle + |10\rangle)$, we write $(1/\sqrt{2}(|01\rangle + |10\rangle))_{p,q}$ to indicate that p and q represent, respectively, the first and the second variable of the entangled pair. In this case, the state is an inseparable vector in the space $\mathcal{H}_p \otimes \mathcal{H}_q$. We use the same notation to represent linear operators on the program Hilbert space. Given an operator U in the Hilbert space $\mathcal{H}_Q$ and a set of variables

$v \subseteq Q$, we write U_v to indicate the operator acts as U on the variables in v and acts as the identity on the other variables of Q. For instance, H_q is the unitary operator in $\mathcal{H}_Q$ that acts as the Hadamard gate(H) on q and as the identity on the other variables. In the same way, $CX_{p,q}$ is the operator corresponding to the control-not operators on p and q and the identity on the other variables.

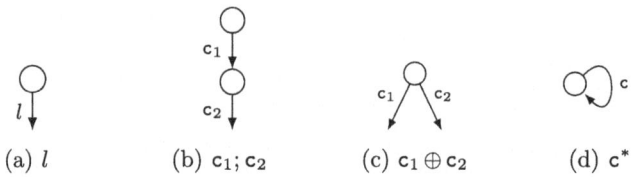

(a) l (b) $c_1; c_2$ (c) $c_1 \oplus c_2$ (d) c^*

Fig. 1. Graphical representation of the language c constructs. The CFG (a), where $l \in$ label, represents a single statement, the CFG (b) corresponds to the sequential composition, (c) represents the branching c $\oplus$ c and (d) corresponds to the iteration c^*.

3   Control Flow Graph Language

Static analysis is usually performed using the control flow graph (CFG) representation of programs [32]. We follow [3,6,23] and consider the CFG language defined by the syntax:

$$c ::= \text{label} \mid c; c \mid c \oplus c \mid c^*, \tag{1}$$

where the term c; c represents sequential composition; the term c $\oplus$ c represents branching; the term c^* is the Kleene closure of the n time composition $c^n = c; \ldots c;$, where $n \in \mathbb{N}$, and the term label represents the labels that correspond to the statements of the analysed language. The CFG associated with a program is a graph with a start node corresponding to the program entry point, an end node corresponding to the exit point, and all other nodes corresponding to intermediate points in the execution of the program; each edge of the graph has a label that represents the change produced by the execution of an instruction of the language that is analysed. Hence, label defines the language of the programs we are analysing, while the other elements in the syntactic category c describe how we compose the edges of the program CFG depending on their labels as displayed in Fig. 1. If N is the set of program points, the CFG will be defined as sets of edges, with a label $l \in$ label, between nodes in N, i.e., as subsets of $N \times$ label $\times N$. In this way, we can define a new analysis simply by providing abstract semantics for the instructions defining label.

We will define our analysis by referring to a minimal quantum language characterised by quantum statements on quantum variables with a control flow based on measurement. This means that the branching in the CFG is guided by the probabilistic result of a quantum measurement. Given a finite set of quantum variables Q and $q, p \in Q$ we define the language label as follows:

$$\text{label} ::= \texttt{NonZero}(q) \big| \texttt{Zero}(q) \mid \texttt{skip} \mid \texttt{h}(q) \big| \texttt{t}(q) \big| \texttt{cx}(p, q), \tag{2}$$

where $\texttt{Zero}(q)$ ($\texttt{NonZero}(q)$) indicates that the measurement of q returns 0 (1); the statements, $\texttt{h}$, $\texttt{t}$, and $\texttt{cx}$ indicate, respectively, the Hadamard gate, the T gate and the control-not gate [22, Chapter 4]. Considering only these three quantum operations is not a limitation because they allow us to cover all possible quantum computations [5]. Moreover, without loss of generality, we can assume that every variable $q_i \in Q$ corresponds to a 1-qubit register initialised to the state $|0\rangle$. In our language, we can write both $\texttt{while}$ and $\texttt{if then else}$ statements [35, Chapter 14, Exercise 14.4]:

$$\texttt{if } (q) \texttt{ then } c_1 \texttt{ else } c_2 \equiv (\texttt{NonZero}(q); c_1) \oplus (\texttt{Zero}(q); c_2)$$
$$\texttt{while } (q) \texttt{ do } c \equiv (\texttt{NonZero}(q); c)^*; \texttt{Zero}(q)$$

Thus, our language is equivalent to the quantum while language that is used in [25,36,37], and to the one that is used to define the other two entanglement analyses based on abstract semantics [18,27].

3.1 Collecting Semantics

Let $Q = \{q_i\}_n$ be the set of variables, we call $\mathcal{H}_Q = \bigotimes_i^n \mathcal{H}_{q_i}$ the n-qubit Hilbert space, i.e., a space of dimension 2^n. Let V_Q be the set of all vectors $|\psi\rangle \in \mathcal{H}_Q$, we define the collecting semantics as a function $[\![\cdot]\!] : \wp(V_Q) \to \wp(V_Q)$. First, given a set $v \in \wp(V_Q)$, we define the collecting semantics for each instruction of the language label as follows:

$$[\![\texttt{skip}]\!]v = v$$
$$[\![\texttt{h}(q)]\!]v = \{ H_q |\psi\rangle \mid |\psi\rangle \in v \}$$
$$[\![\texttt{t}(q)]\!]v = \{ T_q |\psi\rangle \mid |\psi\rangle \in v \}$$
$$[\![\texttt{cx}(p, q)]\!]v = \{ CX_{p,q} |\psi\rangle \mid |\psi\rangle \in v \}$$
$$[\![\texttt{NonZero}(q)]\!]v = \left\{ \frac{M_{1_q} |\psi\rangle}{\|M_{1_q} |\psi\rangle\|} \mid \|M_{1_q} |\psi\rangle\|^2 > 0, |\psi\rangle \in v \right\}$$
$$[\![\texttt{Zero}(q)]\!]v = \left\{ \frac{M_{0_q} |\psi\rangle}{\|M_{0_q} |\psi\rangle\|} \mid \|M_{0_q} |\psi\rangle\|^2 > 0, |\psi\rangle \in v \right\}$$

where H_q and T_q are the unitary operators in $\mathcal{H}_Q$ that correspond to the gate Hadamard and T applied to q, $CX_{p,q}$ is the unitary that corresponds to control-not on p and q (where p is the controller and q the target) and the identity on the other variables while $\texttt{NonZero}(q)$ and $\texttt{Zero}(q)$ corresponds to measurement 1 and 0 on q. For instance $[\![\texttt{NonZero}(q)]\!]\{|1\rangle_q\} = \{|1\rangle_q\}$ while $[\![\texttt{Zero}(q)]\!]\{|1\rangle_q\} = \varnothing$.

Finally, we can define the collecting semantics for the whole language:

$$\begin{aligned}
[\![\, c_1 ; c_2 \,]\!] v &= [\![\, c_2 \,]\!] ([\![\, c_1 \,]\!] v) \\
[\![\, c_1 \oplus c_2 \,]\!] v &= [\![\, c_1 \,]\!] v \cup [\![\, c_2 \,]\!] v \\
[\![\, c^* \,]\!] v &= \bigcup_n [\![\, c^n \,]\!] v.
\end{aligned} \tag{3}$$

With this collecting semantics, we only want to represent the set of all states that a program can return as a result. For this reason, when we consider the measurement, we follow a conservative approach by collecting all possible results, ignoring the probability with which these results occur.

4   Characterising Entanglement

In this section, we recall the properties of *separability* and of *direct insepara-bility*[3] [2], and the abstract domain proposed to represent them. Here, the idea is to refine this domain to make it suitable for performing a static analysis for soundly detecting entanglement. To define an abstract domain which is able to capture the entanglement property of quantum variables, we introduce a characterisation of this property by means of an equivalence relation. For bipartite systems (e.g. two qubits), entanglement and separability are dual concepts. In fact, a composite quantum state $|\psi\rangle_{q_1,q_2} \in \mathcal{H}_{q_1} \otimes \mathcal{H}_{q_2}$ is separable if and only if it can be written as a tensor product $|\psi\rangle_{q_1,q_2} = |\phi_{q_1}\rangle \otimes |\phi_{q_2}\rangle$ for some states $|\phi_{q_1}\rangle \in \mathcal{H}_{q_1}$ and $|\phi_{q_2}\rangle \in \mathcal{H}_{q_2}$. A state $|\psi\rangle_{q_1,q_2}$ is entangled if and only if it is *not* separable. However, the scenario becomes more complex when considering systems consisting of three or more subsystems. Entanglement in such systems corresponds to inseparability across the entire system, but various degrees of entanglement can occur within subsystems.

Some metrics have been introduced to analyse the entanglement of these systems, such as *entanglement monotones* and *entanglement measures* [33], which quantify entanglement between subsystems. In [10], the entanglement of two subsystems S_1, S_2 is measured in terms of an entanglement monotone function $E_{|\psi\rangle}(S_1, S_2)$, such that $E_{|\psi\rangle}(S_1, S_2) = 0$ if and only if S_1 and S_1 taken in isolation are not entangled in the global system state $|\psi\rangle$. For instance, the 3-qubits state $|\psi\rangle_{q_1,q_2,q_3} = {}^1\!/\!{}_2(|000\rangle + |001\rangle + |011\rangle + |111\rangle)_{q_1,q_2,q_3}$ is *fully inseparable* since it cannot be decomposed via the tensor product ($|\psi\rangle = |\phi_1\rangle \otimes |\phi_2\rangle$ for any $|\phi_1\rangle$ and $|\phi_2\rangle$). In fact, the entanglement monotone $E_{|\psi\rangle}(q_i, \{q_j, q_k\})$ always differs from zero for all $i, j, k \in \{1, 2, 3\}$. However, if we measure the entanglement between pairs of qubits, we have $E_{|\psi\rangle}(q_1, q_2) > 0$, $E_{|\psi\rangle}(q_2, q_3) > 0$ but $E_{|\psi\rangle}(q_1, q_3) = 0$, that is the qubits q_1 and q_3 taken in isolation are a separable subsystem. This example shows that the entanglement is not transitive, i.e., the fact that q_1 is entangled with q_2 and q_2 with q_3 does not imply that q_1 is entangled with q_3.

In our analysis, we are interested in understanding when a set of variables is fully inseparable or whether the variables are separable in some way. When two

---

[3] In [2] we call the direct inseparability property as being at the same level.

variables are separable (and thus not entangled), we know we can measure one without altering the other. Thus, in our analysis, we speak about *separability* and we consider its dual notion *inseparability* instead of entanglement. Let us formally define the separability of two variables in a multi-variable state.

Definition 1 (Separability). *Let Q be the set of variables in a state $|\psi\rangle_Q$. Two variables $q_1, q_2 \in Q$ are separable if the state $|\psi\rangle_Q$ can be written as $|\psi\rangle_Q = |\phi_1\rangle_{Q_1} \otimes |\phi_2\rangle_{Q_2}$, where $Q_1, Q_2 \subset Q$, $q_1 \in Q_1$ and $q_2 \in Q_2$. Otherwise, we say that q_1, q_2 are inseparable. Given a set $v \in \wp(V_Q)$, two variables q_1, q_2 are inseparable in v if they are inseparable in at least one state $|\psi\rangle_Q \in v$.*

There exists a particular relation between inseparable variables. For instance, let us consider the state $|\psi\rangle_{a,b,c} = (|000\rangle + |110\rangle + |001\rangle - |111\rangle)_{a,b,c}$, where a, b, c are inseparable. On an intuitive level, it can be seen that the three variables are not related in the same way. In fact, a and b are more closely related to each other than either a with c or b with c: if we measure a we obtain one of the two states: $(|00\rangle + |01\rangle)_{b,c}$ or $(|10\rangle - |11\rangle)_{b,c}$, where b has collapsed to a base state $(0$ or $1)$ in both states while c is still in superposition. Instead, if we measure c we obtain: $(|00\rangle + |11\rangle)_{a,b}$ or $(|00\rangle - |11\rangle)_{a,b}$ where a and b are in a entangled and superpose state. When two variables are related as a and b in this example, we say that they are *directly inseparable*.

Definition 2 (Direct Inseparability).
Given a and $b \in Q$ in a state $|\psi\rangle_Q$, we say that a and b are directly inseparable (d-inseparable) if, by measuring one of them, the other also collapses to a base state. Two variables $q_1, q_2 \in Q$ are d-inseparable in $v \in \wp(V_Q)$ if they are d-inseparable in all states $|\psi\rangle_Q \in v$.

Being d-inseparable is a useful property when reasoning about entanglement. In fact, if we apply a controlled not (CX) between two d-inseparable variables, we will always 'disentangle' the target variable. For instance, if we apply $CX_{a,b}$, where a is the controller and b is the target, the state $|\psi\rangle_{a,b,c}$ defined above, we obtain:

$$CX_{a,b}(|\psi\rangle_{a,b,c}) = (|000\rangle + |100\rangle + (|001\rangle - |101\rangle))_{a,b,c} =$$
$$= ((|0\rangle + |1\rangle)|0\rangle + (|0\rangle - |1\rangle)|1\rangle)_{a,c} \otimes |0\rangle_b. \qquad (4)$$

In other words, we have separated b from the other variables. Instead if we apply $CX_{a,c}$ we obtain:

$$CX_{a,c}(|\psi\rangle_{a,b,c}) = (|000\rangle + |111\rangle + |001\rangle - |110\rangle)_{a,b,c}, \qquad (5)$$

and we do not 'disentangle' a since c and a are not d-inseparable.

As shown in [2], inseparability and d-inseparability are equivalence relations.

5   An Abstract Domain for Entanglement

The generation of entanglement depends on the values of the variables, which, therefore, must be taken into account when defining the elements of our abstract domain. Thus, we define an abstract state as consisting of two parts: the first is based on sets of variables representing inseparability and d-inseparability. In contrast, the second consists of a function that associates each variable with a specific label indicating the variable's state.

The Inseparability and D-Inseparability Domain. Inseparability and d-inseparability are both equivalence relations; thus, given a set of variables Q, we can represent both properties on Q by partitions of Q. For instance, consider the state $|\psi\rangle_{a,b,c,d} = ((|00\rangle + |11\rangle)|0\rangle + (|00\rangle - |11\rangle)|1\rangle)_{a,b,c} \otimes |1\rangle_d$. We can build the partition $(\{a, b, c\}\{d\})$ that represents the inseparable variables in $|\psi\rangle_{a,b,c,d}$ and another partition $(\{a, b\}, \{c\}, \{d\})$ that identifies the d-inseparable variables. Since being d-inseparable implies being inseparable, the d-inseparable partition is always included in the inseparable one. We encode this information in our abstract states by representing them as a list of numbered sets (e.g., $[(\{a, b\}, 0), (\{c\}, 0), (\{d\}, 1)]$), where the smaller partition $((\{a, b\}, \{c\}, \{d\}))$ represents which variables are d-inseparable, while by merging sets with the same number, we obtain the partition representing inseparability $((\{a, b, c\}\{d\}))$.

Definition 3. *Given a set of quantum variables Q, we define the abstract state $\mathcal{E}^Q$ as the set of tuples*

$$\mathcal{E}^Q = \left\{ (e, k) \,\middle|\, e \in \wp(Q) \text{ and } k \in \mathbb{N} \right\},$$

where $\forall (e, k), (e', k') \in \mathcal{E}^Q, e \cap e' = \varnothing$ and $\bigcup_{(e,k) \in \mathcal{E}^Q} e = Q$. In other words, the sets e form a partition of Q.

We call $\mathbb{E}^Q \subset \wp(\wp(Q) \times \mathbb{N})$ the abstract domain of all possible $\mathcal{E}^Q$.

   To better refer to the abstract state, we introduce the following notation. Given an abstract state $\mathcal{E}^Q$, we write E_k, using a capital letter, to refer to the set $E_k = \bigcup_{(e,k) \in \mathcal{E}^Q} e$, i.e., the union of all e with the same index k. For instance, if $\mathcal{E}^{\{a,b,c,d,e\}} = [(\{a, b\}, 0), (\{c\}, 0), (\{d, e\}, 1)]$, $E_0 = \{a, b, c\}$ while $E_1 = \{d, e\}$. Using this notation, we introduce a partial order in $\mathbb{E}^Q$.

Definition 4 ($\mathbb{E}^Q, \leqslant_\mathbb{E}$). *Given $\mathcal{E}_1^Q, \mathcal{E}_2^Q \in \mathbb{E}^Q$. $\mathcal{E}_1^Q \leqslant_\mathbb{E} \mathcal{E}_2^Q$ iff $\forall (e_2, k) \in \mathcal{E}_2^Q, \exists (e_1, k') \in \mathcal{E}_1^Q$ such that $e_2 \subseteq e_1$ and $\forall E_k \in \mathcal{E}_1^Q, \exists E_h \in \mathcal{E}_2^Q$ such that $E_k \subseteq E_h$.*

We write $\vee_\mathbb{E}$ and $\wedge_\mathbb{E}$ to refer to the least upper bound (lub) and the greatest lower bound (glb) induced by $\leqslant_\mathbb{E}$, and the resulting domain is a complete lattice. For instance, $[(\{a, b, c\}, 0), (\{d\}, 1)] \leqslant_\mathbb{E} [(\{a, b\}, 0), (\{c\}, 0), (\{d\}, 1)] \leqslant_\mathbb{E} [(\{a\}, 0), (\{b\}, 0), (\{c\}, 0), (\{d\}, 1)] \leqslant_\mathbb{E} [(\{a\}, 0), (\{b\}, 0), (\{c\}, 0), (\{d\}, 0)]$, and $[(\{a, b\}, 0), (\{c\}, 1)] \vee_\mathbb{E} [(\{a\}, 0), (\{b, c\}, 1)] = [(\{a\}, 0), (\{b\}, 0), (\{c\}, 0)]$.

   To ensure soundness, we overestimate the non-separabilities and determine which variables are potentially inseparable. On the other hand, since being d-inseparable implies special effects in relation to control-not and measurement, we underestimate the d-inseparability property to make sure we do not introduce errors in the abstract semantics.

We have defined an abstract domain that allows us to represent inseparability and d-inseparability. However, we need a final ingredient to define the abstract semantics: some elements abstracting the variables' state.

Labels: We introduce some labels that represent some specific states that are relevant to entanglement abstraction. The CX gate does not introduce entanglement if the controller is in a classical state ($|0\rangle$ or $|1\rangle$) or the target is in a uniform superposition ($1/\sqrt{2}|0\rangle \pm |1\rangle$). To track these two states, we introduce two labels that represent two sets of states: $Z = \{\phi|b\rangle\}$ (the set of classical values), and $X = \{\phi(1/\sqrt{2}|b\rangle \pm 1/\sqrt{2}|\bar{b}\rangle)\}$ (the set of values in uniform superposition), where ϕ represents a global phase, b is a binary string and $\bar{b}$ is the negation of b (e.g. if $b = 0$ then $\bar{b} = 1$ and if $b = 010$ then $\bar{b} = 101$). Moreover, we introduce three other labels:

$$P = \{\phi(1/\sqrt{2}|b\rangle \pm^{i+1}/\sqrt{2}|\bar{b}\rangle)\} \qquad Y = \{\phi(1/\sqrt{2}|b\rangle \pm^{i}/\sqrt{2}|\bar{b}\rangle)\}$$
$$R = \{\phi(1/\sqrt{2}|b\rangle \pm^{i-1}/\sqrt{2}|\bar{b}\rangle)\}.$$

We need these labels to represent the semantics of the gate t. In particular, $tX = P$, $tP = Y$, $tY = R$ and $tR = X$ while $tZ = Z$. Then, we add a final label to represent states that are not classical, i.e., those that are definitely in superposition: $S = \{\phi(\alpha|b\rangle \pm \beta|\bar{b}\rangle) \mid |\alpha|^2 + |\beta|^2 = 1$ and $\alpha \neq 0 \wedge \beta \neq 0\}$. Finally, we define $\bot_{\mathcal{L}}$ as the empty set and $\top_{\mathcal{L}}$ as the set of all possible vectors. We can order these labels by inclusion, constructing a lattice $(\mathcal{L}, \leqslant_{\mathcal{L}})$, represented in Fig. 2.

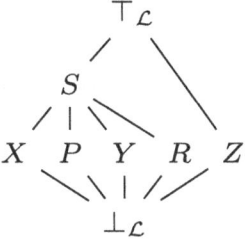

Fig. 2. Lattice $(\mathcal{L}, \leqslant_{\mathcal{L}})$

5.1 Put All Together: The Abstract Domain

Now, we include the labels in the definition of the abstract domain. When variables are inseparable, their state cannot be described as a combination of individual states. Recall that, in an abstract state $\mathcal{E}^Q$, a set of inseparable variables is the union of all sets e with the same index k. We introduce a labelling function to associate each set of inseparable variables with a label. Formally, we define the labelling function as $\lambda : \mathbb{N} \to \mathcal{L}$, and we call $\Lambda = \mathbb{N} \times \mathcal{L}$ the domain of the labelling function. Consequently, we reformulate the definition of the abstract state, including the labelling function, as follows.

Definition 5. *Given a set of quantum variables Q, let be $\mathcal{E}^Q \in \mathbb{E}^Q$ and $\lambda \in \Lambda$, we define an abstract state as the pair $(\mathcal{E}^Q, \lambda)$.*

We define $\mathbb{A}^Q = \mathbb{E}^Q \times \Lambda$ as the abstract domains. For instance, let us consider a set of variables $Q = \{a, b, c, d\}$ in the state

$$|\psi\rangle_Q = (1/2(|0\rangle + |1\rangle) |0\rangle + i/2(|0\rangle - |1\rangle) |1\rangle)_{a,b} \otimes 1/\sqrt{2}(|10\rangle + |01\rangle)_{c,d}.$$

First, we construct the partition corresponding to the sets of inseparable variables, i.e., the sets $\{a, b\}$ and $\{c, d\}$. Once these two sets have been identified, we construct the abstract state by identifying which variables are d-inseparable, obtaining the abstract state: $[(\{a\}, 0)(\{b\}, 0)(\{d, c\}, 1)]$. The last step is to label the sets of inseparable variables. In particular, given a set of variables in a state $|\phi\rangle$, we choose the smallest label L such that $|\phi\rangle \in L$. We see that the state of the variables a, b is contained only in $\top_\mathcal{L}$ while the state of c, d is contained in $X, S, \top_\mathcal{L}$, consequently the final state will be equal to: $(\mathcal{E}^Q, \lambda) = ([[(\{a\}, 0)(\{b\}, 0)(\{c, d\}, 1)], \{0 : \top_\mathcal{L}, 1 : X\}.$

At this point, if we want to compute the set of concrete states represented by an abstract state, starting from the obtain $(\mathcal{E}^Q, \lambda)$, we consider $\mathcal{E}^Q$. In this case, we know that there are two sets $\{a, b\}, \{c, d\}$ of inseparable variables, so the abstract state corresponds to a set of concrete states in the form $|\phi_1\rangle_{a,b} \otimes |\phi_2\rangle_{c,d}$. Then, we can get more precise information about $|\phi_1\rangle$ and $|\phi_2\rangle$ by checking which variables are d-inseparable, that is, c and d. In particular, we know that the concrete states can be expressed by the set $\{|\Psi\rangle \mid |\Psi\rangle = (|\phi_1\rangle_{a,b} \otimes (|b\rangle + |\bar{b}\rangle)_{c,d}\}$. Now, we complete the concretisation by checking the information contained in the labels. By the labels we know that a, b can be in any quantum state while c, d are in the form of $1/\sqrt{2}(|b\rangle + |\bar{b}\rangle)_{c,d}$. Finally, by intersecting this information with the previous one, we obtain the set: $\{|\psi\rangle \mid |\psi\rangle = (|\phi_1\rangle_{a,b} \otimes 1/\sqrt{2}(\alpha |b\rangle + \beta |\bar{b}\rangle)_{c,d}\}$, which represents the concretisation $\gamma((\mathcal{E}^Q, \lambda))$ of $(\mathcal{E}^Q, \lambda)$ and, of course, $|\psi\rangle \in \gamma((\mathcal{E}^Q, \lambda))$. Note that labels represent the state of a single variable or a state of n d-inseparable variables. So, if a set of inseparable variables is not labelled as $\top_\mathcal{L}$, Z or $\bot_\mathcal{L}$, it means that it corresponds either to a single variable or to all d-inseparable variables.

Based on the partial order $\leqslant_\mathbb{E}$ defined in Definition 4, we can define an ordering in $\mathbb{A}^Q$.

Definition 6 ($\mathbb{A}^Q, \sqsubseteq$). *Given $(\mathcal{E}_1^Q, \lambda_1), (\mathcal{E}_2^Q, \lambda_2) \in \mathbb{A}^Q$, $(\mathcal{E}_1^Q, \lambda_1) \sqsubseteq (\mathcal{E}_2^Q, \lambda_2)$ if and only if*

$$\mathcal{E}_1^Q \leqslant_\mathbb{E} \mathcal{E}_2^Q \ \wedge \ \forall E_h \in \mathcal{E}_2, \begin{cases} \lambda(k) \leqslant_\mathcal{L} \lambda(h) & \text{if } \exists E_k \in \mathcal{E}_1 \text{ such that } E_h = E_k \\ \lambda(h) = \top_\mathcal{L} & \text{otherwise} \end{cases}$$

We call $\sqcup$ and $\sqcap$ the lub and glb induced by the order. Since the lattices $(\mathbb{E}^Q, \leqslant_{\mathbb{E}})$ and $(\mathcal{L}, \leqslant_{\mathcal{L}})$ are finite and defined by inclusion operators, they are complete lattices. Thereby, also the lattice $((\mathcal{E}^Q, \lambda), \sqsubseteq)$ is complete.

For instance, consider two abstract states $(\mathcal{E}_1^Q, \lambda_1) = ([(\{p, q\}, 0)], 0 : X)$ and $(\mathcal{E}_2^Q, \lambda_2) = ([(\{p, q\}, 0)], 0 : Y)$; the lub between them is $(\mathcal{E}_3^Q, \lambda_3) = ([(\{p, q\}, 0)], 0 : S)$. In this case, since p, q are d-inseparable in both states, they are also in the lub, and we label the partition by the lub between the labels (we are in the 'if' case of the Definition 6). In fact, $(\mathcal{E}_1^Q, \lambda_1)$ represent the set of states $\{|\psi\rangle \mid \phi(1/\sqrt{2}|b\rangle \pm 1/\sqrt{2}|\bar{b}\rangle)\}$ and $(\mathcal{E}_2^Q, \lambda_2)$ represent the set of states $\{|\psi\rangle \mid \phi(1/\sqrt{2}|b\rangle \pm i/\sqrt{2}|\bar{b}\rangle)\}$, where $b \in \{00, 01, 10, 11\}$, and the abstraction of the union of these two sets is $(\mathcal{E}_3^Q, \lambda_3)$.

On the other hand, if we consider the states $(\mathcal{E}_4^Q, \lambda_4) = ([(\{p, q\}, 0), (\{t\}, 1)], \{0 : X, 1 : X\})$ and $(\mathcal{E}_5^Q, \lambda_5) = ([(\{p\}, 0), (\{q, t\}, 1)], \{0 : X, 1 : X\}$, the lub between them is $(\mathcal{E}_6^Q, \lambda_6) = ([(\{p\}, 0), (\{q\}, 0), (\{t\}, 0)], \{0 : \top_{\mathcal{L}}\}$. In fact, $(\mathcal{E}_4^Q, \lambda_4)$ represent the set of states $\{|\psi\rangle \mid \phi(1/\sqrt{2}|b\rangle \pm 1/\sqrt{2}|\bar{b}\rangle)_{p,q} \otimes 1/\sqrt{2}|0\rangle \pm 1/\sqrt{2}|1\rangle)_t\}$ and $(\mathcal{E}_5^Q, \lambda_5)$ represent the set of states $\{|\psi\rangle \mid \phi(1/\sqrt{2}|0\rangle \pm 1/\sqrt{2}|1\rangle)_p \otimes 1/\sqrt{2}|b\rangle \pm 1/\sqrt{2}|\bar{b}\rangle)_{q,t}\}$. If we join these sets, we obtain a set of states where p, q, t are inseparable. However, p and q are only d-inseparable in some states, while q and t are d-inseparable in others. So, in general, we can only say that p, q, t are inseparable and not d-inseparable, and the only possible label for these states is $\top_{\mathcal{L}}$.

Concretisation Function. Now, we can introduce some formalism to define the concretisation function $\gamma : \mathbb{A} \to \wp(V_Q)$.

Definition 7. *Let Q be a set of variables, and $\pi \subset \wp(Q)$ a partition of Q. Given a state $|\psi\rangle_Q$, we say that the variables in Q are π-separable if $|\psi\rangle_Q = \bigotimes_{p \in \pi} |\phi\rangle_p$, i.e., their joint state can be decomposed into a product of states across a partition π of the variables.*

For instance, given a state $|\psi\rangle_{q_1, q_2, q_3}$ and a partition $\pi = \{\{q_1, q_2\}\{q_3\}\}$, q_1, q_2, q_3 are π-separable if and only if we can write $|\psi\rangle_{q_1, q_2, q_3}$ as $|\phi_1\rangle_{q_1, q_2} \otimes |\phi_2\rangle_{q_3}$.

Given an abstract state $(\mathcal{E}^Q, \lambda)$, we write $\{E_k\}$ to indicate the sets that are obtained by $\mathcal{E}^Q = \{(e, k)\}$ joining the sets e with the same k. Recall that $\{E_k\}$ is a partition that represents the sets of inseparable variables.

Definition 8. *Given a set of variables Q, we say that $|\psi\rangle_Q \triangleright (\mathcal{E}^Q, \lambda)$ (that is, $|\psi\rangle_Q$ is abstracted by $(\mathcal{E}^Q, \lambda)$) if and only if*

- $|\psi\rangle_Q$ *is $\{E_k\}$-separable;*
- $\forall (e, k) \in \mathcal{E}^Q$, $q_i, q_j \in e \Rightarrow q_j$ *and q_i are d-inseparable in $|\psi\rangle_Q$;*
- *given $|\psi\rangle_Q = \bigotimes_k |\psi\rangle_{E_k}$, for all k, $|\psi\rangle_{E_k} \in \lambda(k)$.*

In other words, we say that an abstract state $\mathcal{E}^Q$ abstracts a concrete state if and only if the abstract state over-approximates the set of inseparable variables, under-approximates the d-inseparability, and all separable sub-states that compose the state are represented by labels. Now we have all the elements to define the concretisation function $\gamma : \mathbb{A} \to \wp(V_Q)$:

$$\gamma(\mathcal{E}^Q, \lambda) = \{|\psi\rangle_Q \mid |\psi\rangle_Q \triangleright (\mathcal{E}^Q, \lambda)\}.$$

Theorem 1. *Given a set of abstract states* $\{(\mathcal{E}_j^Q, \lambda_j)\}_j$, *let* $I = \bigcap_j \gamma(\mathcal{E}_j^Q, \lambda_j)$, $\exists$ *an abstract state* $(\mathcal{E}_I^Q, \lambda_I)$ *such that* $\gamma((\mathcal{E}_I^Q, \lambda_I)) = I$.

Proof (Sketch). $\gamma(\mathcal{E}_j^Q, \lambda_j)$ is the set of state in $\mathcal{H}_Q$ that are abstracted by $(\mathcal{E}_j^Q, \lambda_j)$, so I is the set of states that are abstracted by all $(\mathcal{E}_j^Q, \lambda_j)$. Consequently, we if $(\mathcal{E}_I^Q, \lambda_I) = \prod_j (\mathcal{E}_j^Q, \lambda_j)$ then $\gamma((\mathcal{E}_I^Q, \lambda_I)) = I$.

Theorem 1 proves that $\mathbb{A}$ is isomorphic to a Moore family of $\wp(V_Q)$. This means that exist a function $\alpha_l : \wp(V_Q) \to \mathbb{A}$ such that $\langle \mathbb{A}, \alpha_l, \gamma_l, \wp(V_Q) \rangle$ forms a Galois Insertion [7,8,34].

6   An Abstract Semantics

In this section, we define the abstract semantics for our analysis. We first need to introduce some additional notation to better represent the operations on abstract states. Let us consider a generic abstract state $(\mathcal{E}^Q, \lambda)$, where $\mathcal{E}^Q = \{(e_i, k_i)\}_i$. We write $\mathcal{E}^Q(q)$ to refer to the pair $(e, k) \in \mathcal{E}^Q$ such that $q \in e$, while we have $\mathcal{E}^Q\{q\}$ to refer to set $E_k \in \mathcal{E}^Q$ that contains q. For instance, if $\mathcal{E}_1 = [(\{q\}, 0), (\{t\}, 0), (\{r\}, 1)]$ then $\mathcal{E}_1(q) = (\{q\}, 0)$ and $\mathcal{E}_1\{q\} = \{q, t\}$. We write $\mathcal{E}^Q[q_1 + q_2]$ to mark q_1 and q_2 inseparable in $\mathcal{E}^Q$ (e.g., $\mathcal{E}_1[r + q]$ is equal to $[(\{q\}, 1), (\{t\}, 1), (\{r\}, 1)]$). Note that if we mark q and r as inseparable, this also affects t due to the transitivity of the inseparability. $\mathcal{E}^Q[q_1 \uplus q_2]$ means that we marked q_1 and q_2 as d-inseparable, so given $\mathcal{E}_1$ from above, $\mathcal{E}_1[q \uplus t] = [(\{q, t\}, 0), (\{r\}, 1)]$. Note that $\mathcal{E}[q_1 \uplus q_2]$ implies applying also $\mathcal{E}[q_1 + q_2]$. $\mathcal{E}^Q[\sim q]$ denotes the removal of q from the set of variables d-inseparable from itself, and $\mathcal{E}^Q[\neg q]$ denotes that we make q separable from the rest of the variables. For instance, given $\mathcal{E}_2 = [(\{q, r\}, 0), (\{t\}, 0)]$, $\mathcal{E}_2[\sim q]$ is equal to $[(\{r\}, 0), (\{q\}, 0), (\{t\}, 0)]$ and $\mathcal{E}_2[\neg t]$ correspond to $[(\{q, r\}, 0), (\{t\}, 1)]$. Of course, $\mathcal{E}_2[\neg q_1]$ implies $\mathcal{E}_2[\sim q_1]$. Finally, to ease the use of the labelling function, given a variable q, let $\mathcal{E}^Q(q) = (e, k)$, we write $\lambda(q)$ to refer to $\lambda(k)$. Additionally, we write $\lambda[q \leftarrow L]$ to state that we change the label referred to the index k associated with q, setting it equal to L. For instance, given $(\mathcal{E}^{q,p,t}, \lambda) = ([(\{q\}, 0)(\{p\}, 0), (\{t\}, 1)], \{0 : \top_\mathcal{L}, 1 : X\})$, $\lambda(q) = \lambda(p)$ correspond to $\lambda(0) = \top_\mathcal{L}$, $\lambda(t) = X$, since it corresponds to $\lambda(1)$, and $(\mathcal{E}^{q,p,t}, \lambda[q \leftarrow Y])$ is equal to $([(\{q\}, 0)(\{p\}, 0), (\{t\}, 1)], \{0 : Y, 1 : X\})$. In general, when we speak about a label associated with a variable q, we implicitly refer to the label associated with the index associated with q.

Before defining the abstract semantics of the language, we define four abstract operations $H_q^\sharp, T_q^\sharp, CX_{p,q}^\sharp, M_q^\sharp : \mathbb{A}^Q \to \mathbb{A}^Q$, that correspond to the abstraction of the unitary operation H, T, CX and the measurement respectively.

For all the abstract operations it holds that if $(\mathcal{E}^Q, \lambda) = \bot$, then $G_v^\sharp(\mathcal{E}^Q, \lambda) = (\mathcal{E}^Q, \lambda)$. If fact $\gamma(\bot) = \varnothing$ and $G_v \varnothing = \varnothing$.

T gate ($T_q^\sharp$) This gate does not make a difference if we apply it to a single or a group of d-inseparable variables. Formally,

$$T_q^\sharp(\mathcal{E}^Q, \lambda) = (\mathcal{E}^Q, \lambda[q \leftarrow V]),$$

where V can be derived from the red arrows in Fig. 3.

Hadamard Gate ($H_q^\sharp$.) The Hadamard gate distinguishes whether q is entangled with other variables. Formally, the abstract semantics is defined as follows:

$$H_q^\sharp(\mathcal{E}^Q, \lambda) = \begin{cases} (\mathcal{E}^Q, \lambda[q \leftarrow V]) & |\mathcal{E}^Q\{q\}| = 1, \\ (\mathcal{E}^Q(\sim q), \lambda[q \leftarrow \top_{\mathcal{L}}]) & \text{otherwise.} \end{cases}$$

In particular, if q is separable from the other variables (i.e., $|\mathcal{E}^Q\{q\}| = 1$), V can be derived from the blue arrows in Fig. 3. If q is inseparable (i.e., $|\mathcal{E}^Q\{q\}| > 1$), applying Hadamard to it produces a state that we can only label with $\top_{\mathcal{L}}$ and the variable q is no longer d-inseparable. For instance, given $|\psi\rangle_{p,q,t} = (1/\sqrt{2}(|000\rangle + |111\rangle))_{p,q,t}$, p, q, t are d-inseparable and their state can be labelled as X. Then, applying $H_q |\psi\rangle_{p,q,t} = (|000\rangle + |010\rangle + |101\rangle + |111\rangle)_{p,q,t}$, q is not is no longer d-inseparable from p, t, and the state is only labelable by $\top_{\mathcal{L}}$.

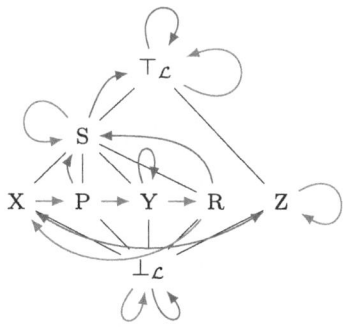

Fig. 3. The red arrow indicates the semantics of the abstract operation $T_q^\sharp$ while the blue one indicates the semantics of the abstract operation $H_q^\sharp$.

Controlled-Not gate ($CX_{c,t}^\sharp$.) The CX gate can introduce or nullify entanglement, so we need to consider different cases according to the state of c and t. Given an abstract state $(\mathcal{E}^Q, \lambda)$, we can define the abstract semantics as follows:

– if $\lambda(c) = Z$, i.e., the controller is in a base value, the CX corresponds to a classical controlled not, so it does not introduce or nullify entanglement and it does not change labels, thus:

$$CX_{c,t}^{\sharp}(\mathcal{E}^Q, \lambda) = (\mathcal{E}^Q, \lambda);$$

– if t is separable from other variables (i.e. $|\mathcal{E}^Q\{t\}| = 1$) we check the state of c and t:

$$CX_{c,t}^{\sharp}(\mathcal{E}^Q, \lambda) = \begin{cases} (\mathcal{E}^Q, \lambda) & \text{if } \lambda(t) = X, \\ (\mathcal{E}[c \uplus t], \lambda) & \text{if } \lambda(c) \neq \top_{\mathcal{L}} \wedge \lambda(t) = Z, \\ ((\mathcal{E}[\sim t])[c + t], \lambda[t \leftarrow \top_{\mathcal{L}}]) & \text{otherwise.} \end{cases}$$

In particular, if the target is in uniform superposition, the CX gate corresponds to the identity.

If the controller is surely in superposition (i.e. $\lambda(c) \neq \top_{\mathcal{L}}$ and $\lambda(c) \neq Z$) and the target is a classical value, the CX gate makes t d-inseparable from c. Moreover, when we set t d-inseparable from c, t automatically gets the label of c. For instance, if we have three variables, q, c, t in the state $|\psi_1\rangle = 1/\sqrt{2}(|00\rangle + |11\rangle)_{q,c} \otimes |1\rangle_t$, applying $CX_{c,t}|\psi_1\rangle$ we obtain $1/\sqrt{2}(|001\rangle + |110\rangle)_{q,c,t}$ in which q, c, t are d-inseparable. Then, given $Q = \{q, c, t\}$ and $(\mathcal{E}_1^Q, \lambda_1) = \{(\{q, c\}, 0), (\{t\}, 1)\}, \{0 : X, 1 : Z\})$ such that $|\psi_1\rangle \in \gamma(\mathcal{E}_1^Q, \lambda_1)$, $CX_{c,t}^{\sharp}(\mathcal{E}_1^Q, \lambda_1) = (\{\{q, c, t\}, 0)\}, \{0 : X\})$ and $CX_{c,t}|\psi_1\rangle \in \gamma(CX_{c,t}^{\sharp}(\mathcal{E}_1^Q, \lambda_1))$. Finally, if none of the above conditions is fulfilled, we need to approximate the relation between c and t. To maintain the soundness, we mark c and t as inseparable without setting c and t d-inseparable. Moreover, we mark c and t equal to $\top_{\mathcal{L}}$ (recall that, since t and c are inseparable, they are related to the same k, so writing $\lambda[t \leftarrow \top_{\mathcal{L}}]$ or $\lambda[c \leftarrow \top_{\mathcal{L}}]$ produce the same effects).

– if t is inseparable from other variables, we need to check if the CX gate nullifies some entanglements:

$$CX_{c,t}^{\sharp}(\mathcal{E}^Q, \lambda) = \begin{cases} (\mathcal{E}[\neg t], \lambda[t \leftarrow Z]) & \mathcal{E}^Q(c) = \mathcal{E}^Q(t) \\ (\mathcal{E}[\sim t], \lambda[t \leftarrow \top_{\mathcal{L}}]) & \text{otherwise} \end{cases}.$$

In particular, if c and t are d-inseparable ($\mathcal{E}^Q(c) = \mathcal{E}^Q(t)$), we 'disentangle' t, as we see in Eq. 4, and, we set the disentangled variable separable from the rest, labelling it as Z. Otherwise, we do not know the exact effect of the gate since, as we have shown in Eq. 5, the CX alters the entangled state. Thus, to maintain soundness, we mark t as not d-inseparable from the other variables and label it as $\top_{\mathcal{L}}$. For instance, given $(\mathcal{E}_1^Q, \lambda_1) = ([(\{a, b\}, 0), (\{c\}, 0)], \{0 : \top_{\mathcal{L}}\})$, $CX_{a,b}^{\sharp}(\mathcal{E}_1^Q, \lambda_1) = ([(\{a\}, 0), (\{c\}, 0), (\{b\}, 1)], \{0 : \top_{\mathcal{L}}, 1 : Z\})$ while $CX_{c,a}^{\sharp}(\mathcal{E}_1^Q, \lambda_1) = ([(\{a\}, 0), (\{c\}, 0), (\{b\}, 0)], \{0 : \top_{\mathcal{L}}\})$. Given $|\psi\rangle_{a,b,c}$, $CX_{a,b}|\psi\rangle_{a,b,c}$ and $CX_{c,a}|\psi\rangle_{a,b,c}$ from Eq. 4 and Eq. 5, note that $|\psi\rangle_{a,b,c} \in \gamma((\mathcal{E}_1^Q, \lambda_1))$, $CX_{a,b}|\psi\rangle_{a,b,c} \in \gamma(CX_{c,a}^{\sharp}(\mathcal{E}_1^Q, \lambda_1))$ and $CX_{c,a}|\psi\rangle_{a,b,c} \in \gamma(CX_{a,b}^{\sharp}(\mathcal{E}_1^Q, \lambda_1))$.

Measurement $(M_q^\sharp)$. In this case, we need to approximate which variables may be affected by the measurement. The semantics of measurement is formally defined as:

$$M_q^\sharp(\mathcal{E}^Q, \lambda) = (\mathcal{E}[\neg Q], L[Q \leftarrow Z, T \leftarrow \top_{\mathcal{L}}]),$$

where $Q = \mathcal{E}^Q(q)$ and $T = (\mathcal{E}^Q\{q\} \setminus Q)$. In particular, when a variable q is separable, $Q = \{q\}$ and $T = \varnothing$ and the measurement simply makes q collapse to a base state. Instead, if q is inseparable from other variables, all variables d-inseparable from q collapse to Z (making them separable), while all other variables inseparable and not d-inseparable from q are altered in a way that cannot be modelled given an abstract state. For this reason, we can only label these variables with $\top_{\mathcal{L}}$.

Language Abstract Semantics. Now we have all the ingredients to define the abstract semantics of the language, which we define as a function $[\![\cdot]\!]^\sharp : \mathbb{A}^Q \to \mathbb{A}^Q$, for each instruction of our language:

$$
\begin{aligned}
[\![skip]\!]^\sharp(\mathcal{E}^Q, \lambda) &= (\mathcal{E}^Q, \lambda) \\
[\![\mathsf{h}(q)]\!]^\sharp(\mathcal{E}^Q, \lambda) &= H_q^\sharp(\mathcal{E}^Q, \lambda) \\
[\![\mathsf{t}(q)]\!]^\sharp(\mathcal{E}^Q, \lambda) &= T_q^\sharp(\mathcal{E}^Q, \lambda) \\
[\![\mathsf{cx}(p,q)]\!]^\sharp(\mathcal{E}^Q, \lambda) &= CX_{p,q}^\sharp(\mathcal{E}^Q, \lambda) \\
[\![\mathsf{NonZero}(b)]\!]^\sharp(\mathcal{E}^Q, \lambda) &= M_q^\sharp(\mathcal{E}^Q, \lambda) \\
[\![\mathsf{Zero}(b)]\!]^\sharp(\mathcal{E}^Q, \lambda) &= M_q^\sharp(\mathcal{E}^Q, \lambda) \\
[\![\mathsf{c}_1; \mathsf{c}_1]\!]^\sharp(\mathcal{E}^Q, \lambda) &= [\![\mathsf{c}_2]\!]^\sharp([\![\mathsf{c}_1]\!]^\sharp(\mathcal{E}^Q, \lambda)) \\
[\![\mathsf{c}_1 \oplus \mathsf{c}_2]\!]^\sharp(\mathcal{E}^Q, \lambda) &= [\![\mathsf{c}_1]\!]^\sharp \sqcup [\![\mathsf{c}_2]\!]^\sharp \\
[\![\mathsf{c}^*]\!]^\sharp(\mathcal{E}^Q, \lambda) &= \bigsqcup_n [\![\mathsf{c}^n]\!]^\sharp(\mathcal{E}^Q, \lambda).
\end{aligned}
\tag{6}
$$

where $H_q^\sharp, T_q^\sharp, CX_{p,q}^\sharp, M_q^\sharp : \mathbb{A}^Q \to \mathbb{A}^Q$ represent the abstract semantics of gates and measurement.

Finally, we formulate the soundness of our abstraction in terms of the concretisation function γ [8].

Proposition 1 (Soundness). *Let Q be a set of variables, $\forall l \in$ label, $\forall(\mathcal{E}^Q, \lambda) \in \mathbb{A}$, $[\![l]\!] \circ \gamma(\mathcal{E}^Q, \lambda) \subseteq \gamma \circ [\![l]\!]^\sharp(\mathcal{E}^Q, \lambda)$.*

Evidence in support of this proposition is shown by the examples in Sect. 7.1. Since every label is sound, by induction on c we can prove that $[\![\mathsf{c}]\!] \circ \gamma(\mathcal{E}^Q, \lambda) \subseteq \gamma \circ [\![\mathsf{c}]\!]^\sharp(\mathcal{E}^Q, \lambda)$.

7   Computing the Analysis

To compute the analysis on the CFG, we need to compute the abstract state $(\mathcal{E}^Q, \lambda)$ for each node of the CFG, namely at each program point of the analysed program [32]. The analysis we propose is forward; namely, the state $(\mathcal{E}^Q, \lambda)$ at node u, denoted $(\mathcal{E}^Q, \lambda)[u]$, depends on the pairs $\{(\mathcal{E}_i^Q, \lambda_i)\}$ of its predecessors and the label semantics of the edges entering in u. Given a CFG G, for all node u in G (program points of the represented program), we define the following system of equations:

$$(\mathcal{E}^Q, \lambda)[u] = \begin{cases} (\mathcal{E}_0^Q, \lambda_0) & \text{if } u = \textbf{start} \\ \bigsqcup \left\{ [\![l]\!]^\sharp (\mathcal{E}^Q, \lambda)[u] \,\middle|\, (u, l, v) \in G \right\} & \text{otherwise} \end{cases} \tag{7}$$

where $(\mathcal{E}_0^Q, \lambda_0)$ is the initial state. Since we assume that all variables are initialised to $|0\rangle$, the initial state is the state where all variables are separable and labelled as Z. So if $Q = \{a, b, c\}$ then $(\mathcal{E}_0^Q, \lambda_0) = ([(\{a\}, 0), (\{b\}, 1), (\{c\}, 2)], \{0, 1, 2, : Z\})$.

This system can be solved by the least fixed point obtaining the best solution for each program point [21].

Proposition 2. *For all statement $l \in$ label, the abstract semantics $[\![l]\!]^\sharp : \mathbb{A} \to \mathbb{A}$ is monotonic w.r.t. $\sqsubseteq$.*

Since the semantics is monotonic, it is granted that we reach the fix-point.

We provide a prototype of our procedure[4] that analyses the quantum language used to present the analysis. Together with the prototype, we provide examples showing how our analysis works in various scenarios. In particular, we analyse the examples contained in [30](superdense coding [4], Deutsch algorithmn [9] and the Creation and disentanglement of the GHZ state) and in [27] (teleportation circuit and GHZ), obtaining the same results as [30] and improving [27]. We also provide some examples showing how we lose precision in the presence of control flow, showing when our analysis is sound but incomplete.

7.1   Showing the Analysis

Consider the example in Fig. 4. In Fig. 4, we show the CFG to the program $dGHZ ::= \text{h}(a); \text{cx}(a, b); \text{cx}(a, c); \text{cx}(c, b); \text{t}(b); \text{cx}(c, a)$ displaying for each program point the concrete state in blue and the abstract state in black. This program creates the GHZ states up to node **3**, then 'disentangles' b with the first cx, then changes the relatives phase with the t gate and then disentangles c with the last cx. In the abstract domain up to node **3**, we construct the state in which $\{a, b, c\}$ are d-inseparable. Then we are able to keep track in the abstract state the entanglement cancellation made by the cx gate in edges $(\textbf{3}, \text{cx}(a, b), \textbf{4})$ and $(\textbf{5}, \text{cx}(c, a), \textbf{6})$ and the phase change made by the t gate in edge $(\textbf{4}, \text{t}(b), \textbf{5})$. We show how our analysis works with control flow in Fig. 5. We consider the program $prog ::= \text{h}(a); \text{cx}(a, b); \text{h}(c); \text{if } (b) \text{ then } \{\text{cx}(a, c)\} \text{ else } \{\text{cx}(c, b)\}$, writing $|\phi\rangle$ to indicate the state $1/\sqrt{2}(|0\rangle + |1\rangle)$ and $|\varphi\rangle$ to indicate $1/\sqrt{2}(|0\rangle - |1\rangle)$.

---

[4] The following GitHub repository NicolaAssolini98/EntaglementAnalysis contains our prototype implemented in Python.

In this example, we start in node **1** with a state where a and b form a Bell state and are therefore abstracted as being entangled d-inseparable. In nodes **2** and **3**, due to the measurement of b, both a and b collapse to a basis state. Since a and b are d-inseparable, they are both labelled with Z. In node **4**, the concrete state is obtained by merging the semantics of the two paths, while the abstract state is the lub between $CX^{\sharp}_{a,c}([(\{a\},0),(\{b\},1),(\{c\},2)],\{(0,1):Z,2:X\})$ and $CX^{\sharp}_{c,b}([(\{a\},0),(\{b\},1),(\{c\},2)],\{(0,1):Z,2:X\})$, which correspond to: $([(\{a\},0),(\{b\},1),(\{c\},2)],\{(0,1):Z,2:X\}) \sqcup ([(\{a\},0),(\{b,c\},1)],\{(0):Z,1:X\})$ $= [(\{a\},0),(\{c\},1),(\{b\},1)],\{0:Z,1:\top_{\mathcal{L}}\}$. In both examples, using the labels, we can approximate the variable's state during the execution of the program.

8 Related Works

Yu et al. [38] propose an abstraction of quantum domains using the abstract interpretation formalism to achieve an abstract simulation of quantum circuits. Another work [11] explores the relation between quantum Hoare logic, quantum incorrectness logic and abstract interpretation in the context of quantum programs.

	Concrete state	Abstract state
Start	$\{\|000\rangle_{a,b,c}\}$	$[(\{a\},0),(\{b\},1),(\{c\},2)],\{(0,1,2):Z\})$
1	$\{^1/\sqrt{2}(\|0\rangle + \|1\rangle)_a \|00\rangle_{b,c}\}$	$[(\{a\},0),(\{b\},1),(\{c\},2)],\{0:X,(1,2):Z\})$
2	$\{^1/\sqrt{2}(\|00\rangle + \|11\rangle)_{a,b} \|0\rangle_c\}$	$[(\{a,b\},0),(\{c\},2)],\{0:X,2:Z\})$
3	$\{^1/\sqrt{2}(\|000\rangle + \|111\rangle)_{a,b,c}\}$	$[(\{a,b,c\},0)],\{0:X\})$
4	$\{^1/\sqrt{2}(\|00\rangle + \|11\rangle)_{a,c} \|0\rangle_b\}$	$[(\{a,c\},0),(\{b\},1)],\{0:X,1:Z\})$
5	$\{^{i+1}/\sqrt{2}(\|00\rangle + \|11\rangle)_{a,c} \|0\rangle_b\}$	$[(\{a,c\},0),(\{b\},1)],\{0:P,1:Z\})$
6	$\{^{i+1}/\sqrt{2}(\|0\rangle + \|1\rangle)_c \|0\rangle_b \|0\rangle_c\}$	$[(\{c\},0),(\{b\},1),(\{a\},2)],\{0:P,(1,2):Z\})$

(a)   (b)

Fig. 4. The $dGHZ$ CFG (a), and a table representing concrete and abstract states for each node (b).

An entanglement analysis was introduced in [26,27] for a simple while language that uses abstract semantics based on partitions. In this work, Perdrix uses partitions to represent entangled variables and two labels (X and Z) to support the analysis. Moreover, in this approach, the labels are related to single variables, not to partitions, and every entangled variable is labelled as $\top$. In this way, this approach fails to detect when a gate or a measurement removes entanglement, to track the state of entangled variables, and to approximate the T gate. For instance, in the examples in Fig. 4 from the previous section, the

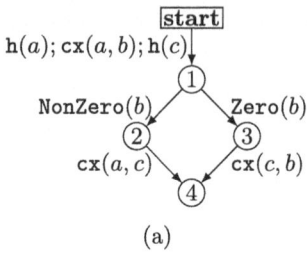

(a)

	Concrete state	Abstract state
Start	$\{|000\rangle_{a,b,c}\}$	$[(\{a\},0),(\{b\},1),(\{c\},2)],\{(0,1,2):Z\})$
1	$\{1/\sqrt{2}(|00\rangle+|11\rangle)_{a,b}|\phi\rangle_c\}$	$[(\{a,b\},0),(\{c\},1)],\{0:X,1:Z\})$
2	$\{|11\phi\rangle_{a,b,c}\}$	$[(\{a\},0),(\{b\},1),(\{c\},2)],\{(0,1):Z,2:X\})$
3	$\{|00\phi\rangle_{a,b,c}\}$	$[(\{a\},0),(\{b\},1),(\{c\},2)],\{(0,1):Z,2:X\})$
4	$\{|11\varphi\rangle_{a,b,c},(|0\rangle_a\,1/\sqrt{2}|00\rangle+|11\rangle)_{a,b,c}\}$	$[(\{a\},0),(\{c\},1),(\{b\},1)],\{0:Z,1:\top_{\mathcal{L}}\})$

(b)

Fig. 5. The *prog* CFG (a), and a table representing concrete and abstract states for each node (b), where $|\phi\rangle = 1/\sqrt{2}(|0\rangle + |1\rangle)$ and $|\varphi\rangle = 1/\sqrt{2}(|0\rangle - |1\rangle)$

Perdrix analysis computes in node **3** the abstract state $(\{a,b,c\},\{a,b,c \to \top_{\mathcal{L}}\})$ that as no information. Consequently, after node 4, the three variables will always be considered entangled regardless of the gates applied.

Other systems have been developed to detect entanglement. Honda's approach [18] is based on an abstract domain that uses abstract density matrices. In this way, it is possible to abstract more information about entanglement, but the space of the abstract states explodes with the program's size. In [29,30], Rand introduces a type system based on Gottesman's [14] representation of Clifford gates (H,S,CX). This approach proposes an entanglement analysis, although working at the circuit level (i.e., no control flow) and limited to Clifford gates and measurement. This approach cannot be applied to the program in Fig. 5 due to the presence of the if statement and to the program in Fig. 4 due to the presence of the T gate. Since our approach is based on abstract interpretation instead of type systems, it can be easily integrated with other analyses to improve precision.

The design of the language Twist [39] includes the verification of the separability of states, which is based on a type system with annotation (that must be inserted manually) and dynamic checking (that uses classical simulation). Also, the Scaffold compiler [19] includes an analysis of entanglement, which works at the circuit level, considering only the circuit composed by controlled-NOT gates.

9   Conclusion

In this paper, we have introduced a static analysis for quantum programming languages, building on the abstract domain for entanglement analysis developed in [2]. By extending the abstract domain with additional labels, we provide a more precise and practical method for analysing quantum entanglement. Our static analysis framework not only identifies entangled variables but also provides an abstract description of the program state during a computation, which can potentially be used as a basis for other analyses. Our approach improves other analyses proposed in the literature, such as [27], by enhancing precision while avoiding the exponential growth in computational complexity [18]. Additionally, the adaptability of our method within a while-language framework makes it a versatile tool for various quantum programming scenarios. In summary, our work contributes to the effective reasoning about quantum programs by providing a robust framework for entanglement analysis and quantum state approximation.

As a future development, we aim to implement our analysis so that it can be used in real quantum programming languages such as Qiskit [1], Qrisp [31], Guppy [28] or Isq [16]. To this end, we plan to integrate our approach into existing analysis tools, such as LiSA [12]. Future research will also explore further refinements to our analysis framework, e.g. by considering abstractions of the entanglement property along the lines of [10]. Moreover, we plan to define a probabilistic version of our analysis to better deal with the probabilistic control flow.

Acknowledgment. Alessandra Di Pierro was supported by INdAM - GNCS Project CUP_E53C22001930001, Isabella Mastroeni by the project SERICS (PE00000014) under the MUR National Recovery and Resilience Plan funded by the European Union - NextGenerationEU and by PRIN2022PNRR "RAP-ARA" (PE6) - MUR: P2022HXNSC.

References

1. Aleksandrowicz, G., et al.: Qiskit: an Open-source Framework for Quantum Computing (Jan 2019). https://doi.org/10.5281/zenodo.2562111
2. Assolini, N., Di Pierro, A., Mastroeni, I.: Abstracting entanglement. In: Proceedings of the 10th ACM SIGPLAN International Workshop on Numerical and Symbolic Abstract Domains (NSAD 2024) (2024). https://doi.org/10.1145/3689609.3689998
3. Assolini, N., Di Pierro, A., Mastroeni, I.: Static analysis of quantum programs. In: Static Analysis: 31th International Symposium, SAS 2024, Pasadena, USA, 20-22 October 2024. Proceedings. Springer (2024)
4. Bennett, C.H., Wiesner, S.J.: Communication via one- and two-particle operators on einstein-podolsky-rosen states. Phys. Rev. Lett. **69**, 2881–2884 (1992). https://doi.org/10.1103/PhysRevLett.69.2881, https://link.aps.org/doi/10.1103/PhysRevLett.69.2881
5. Boykin, P., Mor, T., Pulver, M., Roychowdhury, V., Vatan, F.: A new universal and fault-tolerant quantum basis. Inform. Process. Lett. **75**(3), 101–107 (2000). https://doi.org/10.1016/S0020-0190(00)00084-3, https://www.sciencedirect.com/science/article/pii/S0020019000000843
6. Bruni, R., Giacobazzi, R., Gori, R., Ranzato, F.: A correctness and incorrectness program logic. J. ACM **70**(2) (2023). https://doi.org/10.1145/3582267

7. Cousot, P., Cousot, R.: Abstract interpretation: a unified lattice model for static analysis of programs by construction or approximation of fixpoints. In: Proceedings of the 4th ACM SIGACT-SIGPLAN Symposium on Principles of Programming Languages, POPL 1977, pp. 238-252. Association for Computing Machinery, New York (1977). https://doi.org/10.1145/512950.512973, https://doi.org/10.1145/512950.512973

8. Cousot, P., Cousot, R.: Systematic design of program analysis frameworks. In: Proceedings of the 6th ACM SIGACT-SIGPLAN Symposium on Principles of Programming Languages. p. 269-282. POPL '79, Association for Computing Machinery, New York, NY, USA (1979). https://doi.org/10.1145/567752.567778, https://doi.org/10.1145/567752.567778

9. Deutsch, D., Penrose, R.: Quantum theory, the church-turing principle and the universal quantum computer. Proc. Royal Soc. London. A. Mathe. Phys. Sci. **400**(1818), 97 117 (1985). https://doi.org/10.1098/rspa.1985.0070, https://royalsocietypublishing.org/doi/abs/10.1098/rspa.1985.0070

10. Di Pierro, A., Mancini, S., Memarzadeh, L., Mengoni, R.: Homological analysis of multi-qubit entanglement. Europhy. Lett. **123**(3), 30006 (2018). https://doi.org/10.1209/0295-5075/123/30006

11. Feng, Y., Li, S.: Abstract interpretation, Hoare logic, and incorrectness logic for quantum programs. Inform. Comput. **294**, 105077 (2023). https://doi.org/10.1016/j.ic.2023.105077, https://www.sciencedirect.com/science/article/pii/S0890540123000809

12. Ferrara, P., Negrini, L., Arceri, V., Cortesi, A.: Static analysis for dummies: experiencing LiSA. In: Proceedings of the 10th ACM SIGPLAN International Workshop on the State Of the Art in Program Analysis, pp. 1-6. SOAP 2021. Association for Computing Machinery, New York (2021). https://doi.org/10.1145/3460946.3464316

13. Fürntratt, H., Schnabl, P., Krebs, F., Unterberger, R., Zeiner, H.: Towards higher abstraction levels in quantum computing. In: International Conference on Service-Oriented Computing, pp. 162–173. Springer (2023). https://doi.org/10.1007/978-981-97-0989-2_13

14. Gottesman, D.: The heisenberg representation of quantum computers. arXiv preprint quant-ph/9807006 (1998)

15. Green, A.S., Lumsdaine, P.L.F., Ross, N.J., Selinger, P., Valiron, B.: An introduction to quantum programming in quipper. In: Dueck, G.W., Miller, D.M. (eds.) RC 2013. LNCS, vol. 7948, pp. 110–124. Springer, Heidelberg (2013). https://doi.org/10.1007/978-3-642-38986-3_10

16. Guo, J., et al.: isq: an integrated software stack for quantum programming. IEEE Trans. Quantum Eng. **4**, 1–16 (2023). https://doi.org/10.1109/TQE.2023.3275868

17. Heim, B., et al.: Quantum programming languages. Nat. Rev. Phys. **2**(12), 709–722 (2020). https://doi.org/10.1038/s42254-020-00245-7

18. Honda, K.: Analysis of quantum entanglement in quantum programs using stabilizer formalism. arXiv preprint arXiv:1511.01572 (2015)

19. JavadiAbhari, A., et al.: Scaffcc: a framework for compilation and analysis of quantum computing programs. In: Proceedings of the 11th ACM Conference on Computing Frontiers, CF 2014, Association for Computing Machinery, New York(2014). https://doi.org/10.1145/2597917.2597939

20. Kaye, P., Laflamme, R., Mosca, M.: An introduction to quantum computing. OUP Oxford (2006)

21. Khedker, U.P., Sanyal, A., Sathe, B.: Data Flow Analysis: Theory and Practice. CRC Press, 1st edn. (2009). https://doi.org/10.1201/9780849332517

22. Nielsen, M.A., Chuang, I.L.: Quantum Computation and Quantum Information: 10th Anniversary Edition. Cambridge University Press (2010)
23. O'Hearn, P.W.: Incorrectness logic. Proc. ACM Program. Lang. 4(POPL) (2019). https://doi.org/10.1145/3371078
24. Paykin, J., Rand, R., Zdancewic, S.: Qwire: a core language for quantum circuits. SIGPLAN Not. 52(1), 846–858 (2017). https://doi.org/10.1145/3093333.3009894
25. Peng, Y., Ying, M., Wu, X.: Algebraic reasoning of quantum programs via non-idempotent kleene algebra. In: Proceedings of the 43rd ACM SIGPLAN International Conference on Programming Language Design and Implementation, PLDI 2022, pp. 657-670. Association for Computing Machinery, New York (2022). https://doi.org/10.1145/3519939.3523713, https://doi.org/10.1145/3519939.3523713
26. Perdrix, S.: Quantum patterns and types for entanglement and separability. Electr. Notes Theoret. Comput. Sci. 170, 125–138 (2007). https://doi.org/10.1016/j.entcs.2006.12.015, https://www.sciencedirect.com/science/article/pii/S157106610700059X, proceedings of the 3rd International Workshop on Quantum Programming Languages (QPL 2005)
27. Perdrix, S.: Quantum entanglement analysis based on abstract interpretation. In: Alpuente, M., Vidal, G. (eds.) SAS 2008. LNCS, vol. 5079, pp. 270–282. Springer, Heidelberg (2008). https://doi.org/10.1007/978-3-540-69166-2_18
28. Quantinuum: guppylang (2024). https://github.com/CQCL/guppylang
29. Rand, R., Sundaram, A., Singhal, K., Lackey, B.: Extending gottesman types beyond the clifford group. In: The Second International Workshop on Programming Languages for Quantum Computing (PLanQC 2021) (2021)
30. Rand, R., Sundaram, A., Singhal, K., Lackey, B.: Gottesman types for quantum programs. arXiv preprint arXiv:2109.02197 (2021)
31. Seidel, R., et al.: Qrisp: A framework for compilable high-level programming of gate-based quantum computers (2024). https://arxiv.org/abs/2406.14792
32. Seidl, H., Wilhelm, R., Hack, S.: Compiler Design: Analysis and Transformation. Springer (2012)
33. Vidal, G.: Entanglement monotones. J. Mod. Optics 47(2-3), 355–376 (2000). https://doi.org/10.1080/09500340008244048, https://www.tandfonline.com/doi/abs/10.1080/09500340008244048
34. Ward, M.: The closure operators of a lattice. Annals Math. 43(2), 191–196 (1942). https://doi.org/10.2307/1968865, http://www.jstor.org/stable/1968865
35. Winskel, G.: The formal semantics of programming languages: an introduction. MIT press (1993)
36. Ying, M.: Floyd–hoare logic for quantum programs. ACM Trans. Program. Lang. Syst. 33(6) (2012). https://doi.org/10.1145/2049706.2049708
37. Ying, M.: Foundations of quantum programming. Morgan Kaufmann (2016)
38. Yu, N., Palsberg, J.: Quantum abstract interpretation. In: Proceedings of the 42nd ACM SIGPLAN International Conference on Programming Language Design and Implementation, PLDI 2021, pp. 542-558. Association for Computing Machinery, New York (2021). https://doi.org/10.1145/3453483.3454061, https://doi.org/10.1145/3453483.3454061
39. Yuan, C., McNally, C., Carbin, M.: Twist: sound reasoning for purity and entanglement in quantum programs. Proc. ACM Program. Lang. 6(POPL) (2022). https://doi.org/10.1145/3498691

Synthesis

Synthesis of Parametric Locally Symmetric Protocols from Abstract Temporal Specifications

Ruoxi Zhang[1]($\boxtimes$), Richard Trefler[1], and Kedar S. Namjoshi[2]

[1] University of Waterloo, Waterloo, ON, Canada
{r378zhan,trefler}@uwaterloo.ca
[2] Nokia Bell Labs, Murray Hill, NJ, USA
kedar.namjoshi@nokia-bell-labs.com

Abstract. Scalable distributed systems are typically parametric in design. The key parameter is the number of isomorphic components, K. A second important parameter is the number of neighbors, k, of each component process. In this work, we describe a methodology that uses an automated synthesis procedure to construct parametric system instances where both K and k can vary arbitrarily, extending prior work on synthesis for a fixed k. The methodology relies crucially on locality, symmetry, and abstraction. The first step is to eliminate K by refining a general, system-wide specification to a local temporal specification for a generic process in its parameterized neighborhood. Next, the local process specification is abstracted to remove its dependence on k. These steps are done by hand. The given synthesis procedure then automatically constructs an abstract process from the abstract local specification with a worst-case cost exponential in the length of the abstract local specification. We show that, for any k, the concretized abstract process meets the local specification. We then show that instantiating the abstract process with different k and K forms system instances that satisfy the system-level specification. The worst-case cost of instantiation is linear in K. We use this method to synthesize an atomic snapshots protocol on fully connected networks and a dining philosophers protocol on hypercubes.

1 Introduction

Scalable distributed systems, such as network protocols, web services, and multi-core processors, are typically parameterized. In this work, we focus on parametric systems composed of K isomorphic processes communicating with neighboring

R. Trefler and R. Zhang were supported, in part, by an Individual Discovery Grant from the Natural Sciences and Engineering Research Council of Canada. Kedar Namjoshi was supported in part by DARPA under contract HR001120C0159. The views, opinions, and/or findings expressed are those of the author(s) and should not be interpreted as representing the official views or policies of the Department of Defense or the U.S. Government.

K. Shankaranarayanan et al. (Eds.): VMCAI 2025, LNCS 15530, pp. 75–96, 2025.
https://doi.org/10.1007/978-3-031-82703-7_4

processes according to an underlying network topology. We further assume that each component process in a parametric system has k neighbors. We suppose that K and k can both take on arbitrarily large values.

Fully automated verification and synthesis of parametric systems are undecidable [3,26]. We present a semi-automated method for parametric synthesis:

1. Reduce a global specification to a local one that describes the requirements of a single, generic process P. See [2,9,21] for examples of such reductions.
2. Abstract the unbounded neighborhoods of P and rewrite the local specification under this abstraction.
3. Automatically synthesize from the abstract local specification a model for P that satisfies this specification under interference transitions from isomorphic neighbors. Automatically extract an abstract P from the model.

Our proposed method guarantees that the synthesized P meets the global specification for arbitrarily large k and K. The abstract P forms a template that can be replicated on nodes of arbitrary networks.

Prior work [29] shows how to synthesize parametric systems with fixed neighborhoods (e.g., rings, where each node has two neighbors). This work extends [29] to parametric systems with neighborhoods of unbounded, varying sizes, i.e., it adds another layer of parameterization. A key difficulty is that, in general, abstraction does not preserve realizability. An abstract specification may have a model while the concrete specification does not. Hence, the abstraction process must carry side-information that makes concretization possible. We detail the method in this paper.

An example abstraction is that of counting the number of neighbors for which certain variables are true. We assume that the k neighbors of P can be divided into a fixed number of non-empty partitions according to how P interacts with its neighbors. We define a bounded number of Boolean predicates to approximate the *counts* for each partition and manually rewrite the parametric local specification using these predicates to remove the dependence on k.

For example, consider a parametric property of the generic process P that describes 'for each neighbor, if unread, then all next steps of P read the port shared with the neighbor.' In addition, assume that a port once read, stays read. We rewrite the parametric property as an abstract property stating that 'if not all neighbors are read, then in all next steps of P, all neighbors of P are read.'

In this example, the abstract property is equivalent to the original. We also handle the case where it is impossible to rewrite a parametric property into an equivalent abstract property using a given set of abstract variables. Hence, we allow the abstract property to over-approximate the parametric property. The precondition in the abstract property describes a set of concrete states. We define *context* as auxiliary information to describe a subset of the over-approximated state set so that the update in the parametric property only applies to this subset. Although the process indices of neighbors are omitted in the abstract property, the context connects the current and next states that P shares with each neighbor during concretization.

We define a restricted parametric specification format, e.g., parameterized eventualities are not allowed. We then present a synthesizer that takes as input an abstract local specification and produces a model that can be concretized for any given k and K (c.f. [13,14,29]). Our synthesizer has time complexity exponential in the size of the abstract local specification. However, once P is synthesized, instances can be constructed in time linear in K by replicating P.

The synthesizer has been implemented and used to automatically synthesize parametric protocols. We detail two protocols of interest: an atomic snapshots protocol [1] on fully connected networks and a dining philosophers protocol [8] on hypercubes. Automatically constructing abstract representative processes for both protocols takes only a few minutes.

2   Preliminaries

Networks and Processes. A *network* (N, E) is a directed graph, where N is a set of nodes, and E is a set of edges. The size of the network, denoted K, is the number of nodes in N. Each *node* $n \in N$ is connected to a set of *edges* $E_n \subseteq E$. The connection is towards n, away from n, or both. Two nodes with a common connected edge are *neighbors*. The set of neighbors of n is denoted by $nbr(n)$, and $k := |nbr(n)|$ denotes the size of the neighborhood.

Processes are deployed on nodes, e.g., P_n denotes the process deployed on n. Edges are assigned *external* variables, where incoming and outgoing connections of a node represent the *read* and *write* permissions on the corresponding variables, respectively. An external variable assigned to a common connected edge is called a *shared* variable between the pair of neighboring processes.

Parametric Tiles. This work focuses on *uniform* networks, where all nodes are associated with the same parametric tile process $P_n := \{P_n(k)\}$. Note that P_n is a family of concrete tile processes, and in any system instance, a node with k neighbors is assigned a replica of $P_n(k)$. The *parametric tile* describes the variables shared between n and each $m \in nbr(n)$, where n refers to the representative of all nodes in all networks induced by the parametric tile.

For example, Fig. 1a shows the parametric tile for the dining philosophers protocol that works for scalable k and K. Variable f_{nm} (resp. f_{mn}) represents that the fork between n and m is owned by P_n (resp. P_m) . Note that f_{nm}, from the perspective of n, and f_{mn}, from the perspective of m, are isomorphic, and $f_{nm} \equiv \neg f_{mn}$ because the fork is either owned by P_n or P_m. Similarly, possession of the request token for the fork shared by n and m is denoted by t_{nm} and t_{mn}. Variables d_{nm} and d_{mn}, with equal values, represent whether the shared fork is dirty or clean. As shown by the bidirectional connections between n and the edges, P_n has read/write access to all these variables.

Protocol instances can be constructed by deploying the tile process on underlying networks. For example, a concrete network with $K = 8$, $k = 3$ is formed by replicating the tile in Fig. 1a based on the 3-dimensional cube in Fig. 1b.

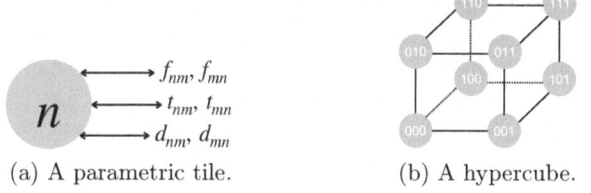

(a) A parametric tile.   (b) A hypercube.

Fig. 1. Obtaining instances by copying a tile process according to underlying networks.

Concrete Processes and Instances. Semantically, an *instance* $P_0||...||P_{K-1}$ is a concurrent program consisting of K sequential processes running in a non-deterministic interleaving manner. The behavior of an instance is defined as a *global state transition system*, denoted G. Each component process $P_{i\in[0,K)}$ in an instance interacts with k neighbors through shared memory.

If the neighborhood pattern defined by the tile is fixed, then a single representative process P_n^c suffices for all instances where K varies. We use the superscript c (for 'concrete') to indicate the case where the neighborhood is fixed.

The state machine for P_n^c is a tuple $(S_n, S_n^0, T_n, \lambda_n)$, where S_n is the set of local states, S_n^0 is the set of initial states, $T_n \subseteq S_n \times S_n$ is the transition relation, and $\lambda_n : S_n \to 2^{\Sigma_n^c}$ is the labeling function that labels each state $s \in S_n$ with variables true in s. All transitions in T_n are labeled with n and are therefore called *n-transitions*. Here, Σ_n^c denotes the set of atomic propositions (i.e., internal and external variables) of n. The *internal states* (resp. *shared states*) are the projection of state labels to internal (resp. shared) valuations.

Interference Transitions. For each $m \in nbr(n)$, a *joint state* (s, t) is a pair of local states $s \in S_n$ and $t \in S_m$ such that s and t have the same value on each variable shared between n and m. A *joint transition* is a transition from (s, t) to the joint state (s', t'), labeled with n if $(s, s') \in T_n$ or m if $(t, t') \in T_m$.

A joint transition caused by a neighbor interferes with the local states of a process by changing the common shared state. The transition from local state s to s', caused by P_m^c through a joint transition labeled m, is called an *interference transition* in the state space of the interfered process, P_n^c. The internal state of the interfered process remains unchanged by the interference transition.

The *local state transition system* of n, denoted H_n^θ, describes the local state space of P_n^c with interference transitions. Here, $\theta = \forall i \in [0, K) : \theta_i^c$ denotes a compositional invariant of any instance, where each local invariance assertion θ_i^c defines a set of local states for P_i^c. A transition (s, s') in H_n^θ is either an n-transition by P_n^c where $\theta_n^c(s)$ holds or an interference transition caused by P_m^c via a joint transition from (s, t) to (s', t') where $\theta_n^c(s)$ and $\theta_m^c(t)$ hold.

Global-Local Reduction. Theorem 1 shows the relationship between H_n^θ and any instance G built from copies of P_n^c extracted from H_n^θ. See [24] for proof. Here, G_n is obtained by replacing the transition labels in G except n with τ.

Theorem 1. *([24]) Let the scheduling of processes be unconditionally fair. With outward-facing interaction, H_n^θ and G_n are stuttering-bisimular.*

Informally, P_n^c is outward-facing if its interference with each neighbor P_m^c depends solely on the shared state between n and m. Formally, as shown in [24], two local states of P_n^c are related by a relation B_{mn} on H_n^θ if every common shared variable has the same value. If for each $m \in nbr(n)$, B_{mn} is a stuttering bisimulation, then P_n^c is *outward-facing* in the interactions with its neighbors.

We follow the standard definition of stuttering bisimulation [6]. A *stuttering simulation* from $M_1 := (S_1, S_1^0, T_1, \lambda_1)$ to $M_2 := (S_2, S_2^0, T_2, \lambda_2)$ is a binary relation B from S_1 to S_2 satisfying the following conditions. (B is a *stuttering bisimulation* if B^{-1} is also a stuttering simulation.)

1. $(s_1^0, s_2^0) \in B$ for each pair of initial states $s_1^0 \in S_1^0$ and $s_2^0 \in S_2^0$;
2. If $(s_1, s_2) \in B$ for states $s_1 \in S_1$ and $s_2 \in S_2$, then the common variables in labels $\lambda_1(s_1)$ and $\lambda_2(s_2)$ have the same value;
3. If $(s_1, s_2) \in B$, then for each state t_1 reachable from s_1 through a finite path π_1 with labels $\tau^*; a$, there exists t_2 reachable from s_2 through π_2 with labels $\tau^*; a$ such that $(t_1, t_2) \in B$ and $(u_1, v_2) \in B$ for every pair of intermediate states u_1 on π_1 and v_2 on π_2. Here, a is a transition label, and $\tau \neq a$.

The work in [29] introduced a decision procedure for concrete local specifications in Fair CTL. The input specification, denoted φ_n^c, describes the behavior of P_n^c over a fixed neighborhood. The synthesizer constructs a model for φ_n^c, denoted H_n^c, as shown in Theorem 2. The representative P_n^c is derived from H_n^c by eliminating interference transitions in H_n^c.

Theorem 2. *([29]) A labeled system H_n^c synthesized from a concrete local specification φ_n^c using the decision procedure in [29] satisfies φ_n^c, is closed under neighboring interference, and unravels into an outward-facing P_n^c.*

Partition the Neighbors. The neighbors of P_n are partitioned according to how P_n interacts with them. Let P_m and P_r be two neighbors of P_n. We define the partitioning of neighbors using a bijection on the local state space of P_n for any k. The bijection, denoted δ, swaps the shared state between P_n and P_m with the shared state between P_n and P_r. We say (P_m, δ, P_r) holds if, for each joint transition $((s_n, s_m), (t_n, t_m)) \in T_n$, there exists another joint transition in T_n from $\delta(s_n, s_m)$ to $\delta(t_n, t_m)$. A *partition*, denoted $nbr(n)_i$, is defined as the maximal set of neighbors such that (P_m, δ, P_r) and (P_r, δ^{-1}, P_m) for each pair of neighbors $P_m, P_r \in nbr(n)_i$.

Unless otherwise stated, we assume that all neighbors are in a single, monolithic partition, which is the case for both example protocols in this paper. We can also handle the case of multiple neighbor partitions, i.e., $\cup_i nbr(n)_i = nbr(n)$, and these partitions are mutually disjoint and non-empty. The token-ring protocol is an example of having two partitions, where tokens only move from left to right. We extend the definition of parametric tiles accordingly. The parametric tiles we are interested in have a fixed number of neighbor partitions.

Fair CTL. We write specifications as *Fair CTL* [14] formulas. Fair CTL requires that any path quantifier, A (for all paths) or E (there exists a path), be immediately followed by a linear-time temporal operator, X_n (indexed strong next-time), Y_n (indexed weak next-time), G (always), F (sometime), U (until), or W (weak until) to form a Fair CTL *modality*.

The grammar of Fair CTL is as follows. If $p \in \Sigma_n$, then p is a Fair CTL formula. If f and g are Fair CTL formulas, then so are $\neg f$, $f \lor g$, $AY_n f$, $EX_n f$, $AfUg$, and $EfUg$. The syntax is extended with abbreviations $f \land g \equiv \neg(\neg f \lor \neg g)$, $f \Rightarrow g \equiv \neg f \lor g$, $A(fWg) \equiv \neg E(\neg fU\neg g)$, $E(fWg) \equiv \neg A(\neg fU\neg g)$, $AFf \equiv A(\top Uf)$, $EFf \equiv E(\top Uf)$, $AGf \equiv A(\bot Wf)$, and $EGf \equiv E(\bot Wf)$. A formula is called an *eventuality* if its modality is one of AU, EU, AF, EF, or EG.

The local specification φ_n is interpreted over a system $M := (S, S^0, T, \lambda)$. Let $M, s \models f$ denote that f holds at state s in M under a fairness condition. A formula f is *satisfiable* iff there exists an M such that $M, s \models f$ for some state s in M. The semantics of Fair CTL are defined in the usual way (see the extended version).

We assume that the network scheduling of processes is *unconditionally fair*, i.e., each process is selected to run infinitely often. Locally, fairness is denoted by $\Phi := \bigwedge_{m \in nbr(n)} GFex_m \land GFex_n$, where $M, \pi \models GFg$ iff for every $i \geq 0$, there exists $j \geq i$, such that $M, \pi^j \models g$. A *path* π is a sequence of states $(s_0, s_1, ...)$ such that $(s_i, s_{i+1}) \in T$ for all i, and its *suffix* $\pi^j = (\pi_j, \pi_{j+1}, ...)$. A fair *full path* is an infinite path that satisfies Φ.

3   Our Approach

We summarize the semi-automated approach into four steps and illustrate these steps using a property taken from the dining philosophers protocol.

Step 1 (cf. [29]***).*** Given a *global correctness specification*, $\varphi = \bigwedge_{i \in [0,K)} \varphi_i$, describing the behavior of any instance $\|_{i \in [0,K)} P_i$, the purpose of step 1 is to obtain a specification independent of K. To achieve this, we manually reduce φ to a *local correctness specification* φ_n describing the behavior of a single representative process P_n in its parametric neighborhood. Note that P_n can be viewed as a function that outputs a concrete representative process P_n^c for each k in the range described by the protocol. An example of a local property is:

$$eg_1 := \bigwedge_{m \in nbr(n)} AG((\underbrace{f_{nm} \land d_{nm} \land t_{nm}}_{\alpha_{nm}}) \Rightarrow AY_n((\underbrace{\neg f_{nm} \land \neg d_{nm} \land t_{nm}}_{\beta_{nm}}))).$$

Property eg_1 represents that 'for all neighbors P_m of P_n, if P_n holds the dirty fork and the request token shared with P_m, then in any next step of n, P_n cleans and sends the fork to P_m.' In this example, α_{nm} and β_{nm} denote the parametric

An extended version of this paper is available at https://github.com/rzhang378/LocalSynth.git.

precondition and postcondition of eg_1, respectively. We also assume that for any m, if α_{nm} is false, each variable not explicitly updated remains unchanged in all the next steps of P_n described by eg_1.

We use eg_1 as a running example. Given any abstract precondition α_n^a describing that α_{nm} holds for some m, we rewrite eg_1 as an abstract property $\mathsf{AG}(\alpha_n^a \Rightarrow (\mathsf{AY}_n\beta_n^a \wedge \mathsf{EX}_n\gamma_n^a))$, where β_n^a and γ_n^a are abstract postconditions.

Step 2, in Section 4. We consider φ_n as a Fair CTL formula of the form $\wedge_j f_j$, which can be split into a finite list of subformulas using $\wedge$ as the separator. A subformula f_{nm} is *parameterized by* k if $\wedge_{m \in nbr(n)} f_{nm}$ is in φ_n, i.e., there are k replicas of f_{nm} in φ_n, where each replica corresponds to a different neighbor. We call φ_n the *parametric local specification* because φ_n contains subformulas parameterized by k. We manually rewrite φ_n into an *abstract local specification* independent of k, denoted φ_n^a.

In this work, we accomplish the rewriting through a *counting abstraction*. For each set of related variables $\cup_{m \in nbr(n)}\{p_{nm}\} \subseteq \Sigma_n$ are replaced by abstract variables that approximate the number of neighbors such that p_{nm} holds. Example abstract variables include p_n^{A} for '$\forall m : p_{nm}$,' and p_n^{E} for '$\exists m : p_{nm}$.'

The abstract specification φ_n^a is created using variables in the abstract alphabet Σ_n^a while preserving the behavior of P_n described in φ_n. The specification transformation only applies to φ_n subject to certain format restrictions, i.e., (roughly) the formulas in φ_n describe the current or next-time behavior of P_n as invariants, and we only allow restricted formats of eventualities. For example, eventualities parameterized by k, such as $\wedge_{m \in nbr(n)}\mathsf{AF}p_m$, are not allowed.

In eg_1, β_{nm} indicates a reduction by one in the number of forks owned by P_n and the number of dirty forks adjacent to P_n. This count update applies to every $m \in nbr(n)$ that meets α_{nm}. Hence, for any abstract precondition composed of at least one neighbor in α_{nm}, the abstract postconditions must reflect the results of applying the corresponding updates.

Given, for example, $\Sigma_n^a = \{f_n^{\mathsf{E}}, f_n^{\mathsf{A}}, d_n^{\mathsf{E}}, d_n^{\mathsf{A}}, t_n^{\mathsf{E}}, t_n^{\mathsf{A}}\}$, the count update in eg_1 affects all abstract preconditions containing $f_n^{\mathsf{E}} \wedge t_n^{\mathsf{E}} \wedge d_n^{\mathsf{E}}$. Let $\alpha_n^a = f_n^{\mathsf{E}} \wedge f_n^{\mathsf{A}} \wedge d_n^{\mathsf{E}} \wedge \neg d_n^{\mathsf{A}} \wedge t_n^{\mathsf{E}} \wedge \neg t_n^{\mathsf{A}}$, representing '$P_n$ owns all forks and some but not all request tokens, and some but not all forks adjacent to P_n are dirty,' be one of the affected abstract preconditions. The abstract property for α_n^a is formulated as:

$$eg_1^a := \mathsf{AG}(\alpha_n^a \Rightarrow (\mathsf{AY}_n(\alpha_n^a \vee \beta_1^a \vee \beta_2^a) \wedge \mathsf{EX}_n\alpha_n^a \wedge \mathsf{EX}_n\beta_1^a \wedge \mathsf{EX}_n\beta_2^a))$$

where $\beta_1^a = f_n^{\mathsf{E}} \wedge \neg f_n^{\mathsf{A}} \wedge d_n^{\mathsf{E}} \wedge \neg d_n^{\mathsf{A}} \wedge t_n^{\mathsf{E}} \wedge \neg t_n^{\mathsf{A}}$ and $\beta_2^a = f_n^{\mathsf{E}} \wedge \neg f_n^{\mathsf{A}} \wedge \neg d_n^{\mathsf{E}} \wedge \neg d_n^{\mathsf{A}} \wedge t_n^{\mathsf{E}} \wedge \neg t_n^{\mathsf{A}}$. For simplicity, the subscript n is omitted from β_1^a and β_2^a.

Formula eg_1^a covers all transitions from concrete states satisfying α_n^a. When α_{nm} is false for all $m \in nbr(n)$, transitions of n leave α_n^a unchanged. Otherwise, when α_{nm} is triggered for some m, α_n^a changes to β_1^a or β_2^a through an n-transition depending on how α_n^a is satisfied. Each set of related shared valuations satisfying α_n^a is expressed as a combination of parametric preconditions. For combinations containing α_{nm} and $(f_{nm} \wedge d_{nm} \wedge \neg t_{nm})$, α_n^a becomes β_1^a because 'some but not all dirty forks owned by P_n are requested by the neighbors' in α_n^a.

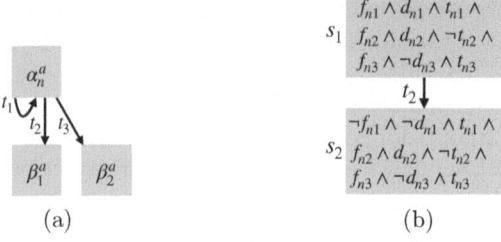

Fig. 2. An example of abstract structure and concrete structure.

For combinations containing α_{nm} but not $(f_{nm} \wedge d_{nm} \wedge \neg t_{nm})$, α_n^a becomes β_2^a because 'all dirty forks owned by P_n are requested by the neighbors' in α_n^a.

For any abstract precondition α_n^a, we save, as *context*, the combinations that satisfy α_n^a and the count updates for each parametric precondition in the combinations. The context is used to track changes in shared variables between n and individual neighbors. The context is not included in φ_n^a but will be used as auxiliary information to concretize the abstract model constructed from φ_n^a.

Step 3, in Section 5. Sect. 5 describes the steps and complexity for automatically constructing an abstract model H_n^a from a satisfiable φ_n^a.

Figure 2a shows an abstract structure constructed from eg_1^a rooted in a state labeled α_n^a. This state has three outgoing n-transition: t_1, t_2, and t_3. Interference transitions are inferred from these existing n-transitions. In this example, a neighbor m, performing an isomorphic transition, affects the states of P_n whose label contains $\neg f_n^A \wedge d_n^E \wedge \neg t_n^A$ by increasing the number of forks owned by P_n and the number of clean forks adjacent to P_n.

Step 4, in Section 6. We explain the concretization of the synthesized model and give the overall correctness theorem in Sect. 6. We show that the abstract model, H_n^a, can be concretized into a concrete model, H_n^c of φ_n^c, for any k. Based on the global-local reduction in Theorem 1, the abstract process P_n^a extracted from H_n^a can be instantiated as programs on nodes of underlying networks to form instances that satisfy the global specification.

An example of concretization is as follows. When $k = 3$, Fig. 2b shows the transition of a given concrete state s_1 derived from the abstract model in Fig. 2a. State s_1 is in the concrete interpretation of α_n^a, and by examining the context of α_n^a, t_2 in Fig. 2a is the only transition whose starting combination matches that of s_1. Hence, s_2 is derived from β_1^a by applying the count update to m_1.

4   Local Specification Transformation

In this section, we form the abstract alphabet Σ_n^a, formulate restrictions on the parametric local specification φ_n, explain the connection between φ_n and the abstract local specification φ_n^a resulting from its rewriting, give an example of rewriting φ_n into φ_n^a, and describe the manual rewriting steps in detail.

4.1   Defining Abstract Variables

Parametric Variables, Terms, and Properties. In this paper, we stick to the following naming convention for variables. In the parametric alphabet Σ_n, the internal variables of n are indexed n, and the parametric external variables shared between n and $m \in nbr(n)$ are indexed nm, mn, or simply m.

A *parametric term* is of the form $\vee_w t_w$ or $\wedge_w t_w$, where w ranges over $nbr(n)$, and t_w comprises (optional) logical connectives and shared variables between n and w. Examples of parametric terms are $\vee_w p_w$ and $\wedge_w (p_w \vee q_w)$.

We define *parametric local properties* as $(\wedge_{m \in nbr(n)} f_{nm}) \mid g_n$, where f_{nm} is a Fair CTL formula containing parametric terms, internal variables, and shared variables between n and m. The definitions of g_n and f_{nm} are almost identical, but g_n does not contain any shared variables other than those that contained in the parametric terms in g_n.

Abstract Variables, Terms, and Properties. An *abstract shared variable* represents the quantification over a set of parametric shared variables. An *abstract local property* is a Fair CTL formula whose alphabet Σ_n^a is derived by replacing all parametric variables in Σ_n with abstract shared variables. An *abstract term* comprises logical connectives and variables in Σ_n^a.

We describe how to form Σ_n^a. To begin with, for each set of related variables $\cup_{m \in nbr(n)}\{p_m\}$ in Σ_n, we define two abstract variables, p_n^E and p_n^A, to approximate the number of neighbors for which p_m is true. The abstract variables are subscripted n because these variables count the number of neighbors centered on n.

We relax the definition of parametric terms by allowing their indices to range over subsets of $nbr(n)$. For example, in the atomic snapshots protocol, we specify that 'exactly one unread and unselected port is selected to be read next' as:

$$\wedge_{m \in nbr(n)} \mathsf{AG}((\neg r_m \wedge \neg s_m) \Rightarrow \mathsf{EX}_n(\neg r_m \wedge s_m \wedge (\wedge_{w \in nbr(n) \setminus \{m\}} \neg s_w)))$$

where r_m (resp. s_m) indicates that the port shared by n and m is read (resp. selected) by n, and $\wedge_{w \neq m} \neg s_w$ ranges over all neighbors of n except m.

Accordingly, more abstract variables can be added to Σ_n^a. For example, in addition to s_n^E and s_n^A, we can express the number of selected neighbors more precisely by adding an abstract variable s_n^{One} that means 's_m is true for exactly one neighbor m.' However, we are only interested in abstract alphabets whose size is independent of k.

When the neighbors of n are divided into multiple partitions, and the behavior of P_n interacting with neighbors in each partition $nbr(n)_i$ is specified separately, we create abstract shared variables separately for each $nbr(n)_i$. For example, suppose half of the neighbors are in $nbr(n)_1$, and the other half are in $nbr(n)_2$. The variables in $\cup_{m \in nbr(n)}\{s_m\}$ are divided into $\cup_{m_1 \in nbr(n)_1}\{s_{m_1}\}$ and $\cup_{m_2 \in nbr(n)_2}\{s_{m_2}\}$. Assume that in $nbr(n)_1$, one unread port is selected at a time, but in $nbr(n)_2$, all unread ports are selected simultaneously. Instead of defining s_n^E and s_n^A for $nbr(n)$, we create separate abstract variables for each partition, e.g., $s_n^{E_i}$, which means 'there exists a neighbor $m_i \in nbr(n)_i$ such that s_{m_i} is true,' and $s_n^{A_i}$, which means 's_{m_i} is true for all neighbors in $nbr(n)_i$.'

Concrete vs. Abstract Interpretation. For any k, let $M^c := (S^c, S^{c,0}, T^c, \lambda^c)$ be a concrete system defined over the language of φ_n^c. Let t be a parametric term that defines a set of concrete states S_t^c in M^c, called the *concrete interpretation* of t. Given any abstract system $M^a := (S^a, S^{a,0}, T^a, \lambda^a)$, let h be an abstract term whose *abstract interpretation* is a set of abstract states S_h^a in M^a. Here, h also defines a set of concrete states S_h^c. If for all k, $S_t^c \subseteq S_h^c$, we say h *includes* t.

For example, for all multi-process models, p_n^E includes $\vee_{w \in nbr(n)} p_w$, and $p_n^A \vee q_n^A \vee (p_n^E \wedge q_n^E)$ includes $\wedge_{w \in nbr(n)} (p_w \vee q_w)$. If for every state $s \in S^c$, $M^c, s \models t$ iff $M^c, s \models h$, then we say $t \equiv h$, i.e., t and h are *equivalent*.

4.2   Format Restrictions

In this subsection, we describe the format restrictions on φ_n. We assume that a more complex specification is refined (manually, in most cases) into this restricted format, e.g., [9] contains several examples of such refinements.

The parametric local specifications of interest are in the form of *init-spec* ∧ *other-spec*, where *init-spec* specifies a single initial state. If there are multiple initial states, we create a copy of φ_n for each initial state such that each copy has a different *init-spec* but the same *other-spec*.

We further require that every parametric property of the form $\wedge_{m \in nbr(n)} f_{nm}$ in *other-spec* be either $\wedge_m AG(\alpha_{nm} \Rightarrow AY_n \beta_{nm})$ or $\wedge_m AG(\alpha_{nm} \Rightarrow EX_n \gamma_{nm})$. Among the remaining properties in *other-spec*, for any property g_n containing parametric terms, either g_n is restricted to $AG(\alpha_n \Rightarrow AY_n \beta_n)$ or $AG(\alpha_n \Rightarrow EX_n \gamma_n)$, or we require that for every parametric term t in g_n, there is an equivalent abstract term h based on Σ_n^a. Here, the preconditions and postconditions are terms consisting of logical connectives and variables.

If every parametric term t in g_n is exactly interpreted as an abstract term h, we rewrite g_n into an equivalent property by replacing each t with h. For example, $AG((\vee_{w \in nbr(n)} p_w) \Rightarrow AF(\wedge_{w \in nbr(n)} p_w))$ is rewritten as $AG(p_n^E \Rightarrow AFp_n^A)$. An eventuality is allowed only if it is in a formula g_n that can be exactly rewritten.

Finding an exact interpretation for every parametric term in g_n may not be possible. This is the case for $AG((\wedge_w (p_w \vee q_w)) \Rightarrow AF((\wedge_w p_w) \wedge (\wedge_w q_w)))$, where $w \in nbr(n)$, because $\wedge_w (p_w \vee q_w)$ cannot be exactly described by any abstract term given $\Sigma_n^a = \{p_n^E, p_n^A, q_n^E, q_n^A\}$. We can add more variables to Σ_n^a to ensure that the requirement for exact interpretation is met, such as $z_n^A := \wedge_w z_w$, which represents the count of a new parametric variable $z_w := p_w \vee q_w$. The eventuality is rewritten into $AG(z_n^A \Rightarrow AF(p_n^A \wedge q_n^A))$. Except for those properties that can be directly rewritten as equivalent abstract properties, the remaining properties require a more complex rewriting approach, as described in Sect. 4.5.

4.3   Parametric vs. Abstract Local Specifications

In this subsection, we elaborate on the goals of the specification rewriting transformation. Subsequently, we describe in Sect. 4.5 how to perform concrete-to-abstract specification rewriting to achieve the goals. We explain in Sect. 6 how to concretize an abstract model into concrete models.

For each precondition α_{nm} in parametric properties of the form $\wedge_{m \in nbr(n)} f_{nm}$ in φ_n, there is a universal postcondition β_{nm} and a set of existential postconditions $\{\gamma_{nm}\}$. For any k, $S^c_{\alpha_{nm}}$ denotes the concrete interpretation of α_{nm}, i.e., $S^c_{\alpha_{nm}}$ consists of the concrete states that satisfy α_{nm} for at least one m. Given an abstract alphabet Σ^a_n, let α^a_n be an abstract term whose interpretation, denoted $S^c_{\alpha^a_n}$, is a superset of $S^c_{\alpha_{nm}}$. The rewriting guarantees that the abstract postconditions of α^a_n preserve the following property:

Lemma 1. *For any k, for each state $s^c \subset S^c_{\alpha^a_n}$ and neighbor $m \in nbr(n)$, if $M^c, s^c \models \alpha_{nm}$, then $M^c, s^c \models \mathsf{AY}_n \beta_{nm}$, and for each γ_{nm}, $M^c, s^c \models \mathsf{EX}_n \gamma_{nm}$.*

Proof. When $S^c_{\alpha^a_n} \subseteq S^c_{\alpha_{nm}}$, the rewriting ensures Lemma 1 by establishing the following on an abstract system M^a. For each abstract state $s^a \in S^a_{\alpha^a_n}$, $M^a, s^a \models \mathsf{AY}_n \beta^a_n$, and $M^a, s^a \models \mathsf{EX}_n \gamma^a_n$ for each γ^a_n, where $S^a_{\alpha^a_n}$ is the abstract interpretation of α^a_n. Here, β^a_n is obtained by applying the update from α_{nm} to β_{nm} to each applicable m, and $\{\gamma^a_n\}$ is obtained by applying each update from α_{nm} to γ_{nm} to one of the applicable neighbors.

When $S^c_{\alpha^a_n} \setminus S^c_{\alpha_{nm}} \neq \emptyset$, there are multiple ways to satisfy α^a_n. Each way of satisfying α^a_n is represented as a combination of parametric preconditions, denoted $\hat{A}_i$, where $S^c_{\alpha^a_n} \cap S^c_{\alpha_{nm}}$ corresponds to the combinations containing α_{nm}. The rewriting establishes that, for each abstract state $s^a \in S^a_{\alpha^a_n}$, $M^a, s^a \models \mathsf{AY}_n(\vee_i \beta^a_{n,i})$, and $M^a, s^a \models \mathsf{EX}_n(\beta^a_{n,i} \wedge \gamma^a_{n,i})$ for each $\gamma^a_{n,i}$, where i ranges over the combinations in α^a_n.

The concrete successor states of each $s^c \in S^c_{\alpha^a_n}$ are obtained by concretizing a subset of the abstract successor states of u^a_n. These abstract successor states meet the parametric postconditions, as do the concrete successor states.

Note that α^a_n, in its disjunctive normal form, can be split into multiple abstract preconditions $\cup_j \{\alpha^a_j\}$, where each α^a_j is a conjunctive clause in α^a_n. The postconditions of each α^a_j must guarantee the aforementioned conditions for each concrete state $s^c \in S^c_{\alpha^a_j}$, where $S^c_{\alpha^a_j}$ is the concrete interpretation of α^a_j.

Similarly, for each precondition α_n in parametric properties g_n of the form $\mathsf{AG}(\alpha_n \Rightarrow \mathsf{AY}_n \beta_n)$ or $\mathsf{AG}(\alpha_n \Rightarrow \mathsf{EX}_n \gamma_n)$, we define an abstract term α^a_n for α_n such that $S^c_{\alpha_n} \subseteq S^c_{\alpha^a_n}$, where $S^c_{\alpha_n}$ is the concrete interpretation of α_n. The rewriting guarantees the following property: (See the extended version for proof.)

Lemma 2. *For any k, for each state $s^c \in S^c_{\alpha^a_n}$, if $M^c, s^c \models \alpha_n$, then $M^c, s^c \models \mathsf{AY}_n \beta_n$, and for each γ_n, $M^c, s^c \models \mathsf{EX}_n \gamma_n$.*

4.4 Specification Rewriting: Example

We take part of the atomic snapshots protocol as an example to illustrate the rewriting procedure: (We specify that any variable that is not explicitly updated remains unchanged in the next state.) The informal, full description of the protocol is given in Sect. 7.

c1 [Initial Condition] An odd-numbered pass begins, and all shared ports are unread and unselected by P_n: $o_n \land (\land_m \neg r_m) \land (\land_m \neg s_m)$

c2 P_n stays on the current pass until all ports are read:
$\mathsf{AG}((o_n \land (\lor_m \neg r_m)) \Rightarrow \mathsf{AY}_n o_n)$

c3 When all ports are read, P_n starts the next pass (an even-numbered pass):
$\mathsf{AG}((o_n \land (\land_m r_m)) \Rightarrow \mathsf{AY}_n(\neg o_n \land (\land_m \neg r_m) \land (\land_m \neg s_m)))$

c4 P_n selects one unread port at a time:
$\land_m \mathsf{AG}((\neg r_m \land \neg s_m) \Rightarrow \mathsf{EX}_n(\neg r_m \land s_m \land (\land_{w \neq m} \neg s_w)))$

c5 P_n reads and deselects the selected port:
$\land_m \mathsf{AG}((\neg r_m \land s_m \land (\land_{w \neq m} \neg s_w)) \Rightarrow \mathsf{AY}_n(r_m \land \neg s_m))$

In this example, indices $m \in nbr(n)$ and $w \in nbr(n) \setminus \{m\}$. Variables r_m (i.e., read port m) and s_m (i.e., select port m) are shared variables between n and m. Variable o_n (i.e., the current pass number is odd) is internal to n.

Properties c1-c3 are directly rewritten into equivalent abstract properties a1-a3, respectively. Properties $\mathsf{AG}(\neg s_n^{\mathsf{E}} \Rightarrow \neg s_n^{\mathsf{A}})$ and $\mathsf{AG}(s_n^{\mathsf{A}} \Rightarrow s_n^{\mathsf{E}})$ for s, are part of the abstract specification, as are similar properties for r. Next, we rewrite c4-c5.

a1 [Initial Condition] $o_n \land \neg r_n^{\mathsf{E}} \land \neg s_n^{\mathsf{E}}$
a2 $\mathsf{AG}((o_n \land \neg r_n^{\mathsf{A}}) \Rightarrow \mathsf{AY}_n o_n)$
a3 $\mathsf{AG}((o_n \land r_n^{\mathsf{A}}) \Rightarrow \mathsf{AY}_n(\neg o_n \land \neg r_n^{\mathsf{E}} \land \neg s_n^{\mathsf{E}}))$

Let $\alpha_1 := \neg r_m \land \neg s_m$ and $\alpha_2 := \neg r_m \land s_m \land (\land_{w \neq m} \neg s_w)$ be preconditions in c4 and c5, respectively. For each neighbor, either α_1 is true, α_2 is true, or neither (further specified in the full protocol). The consistent combinations of α_1 and α_2 are $\{\alpha_1\}$, $\{\alpha_2\}$, and $\{\alpha_1, \alpha_2\}$, where $\{\alpha_1\}$ stands for 'α_1 holds for all m,' $\{\alpha_2\}$ stands for 'α_2 holds for one m, requiring $k = 1$,' and $\{\alpha_1, \alpha_2\}$ stands for 'α_2 holds for one m, and α_1 holds for every $w \neq m$, requiring $k \geq 2$.'

These combinations are rewritten into abstract terms: $\{\alpha_1\}$ is rewritten into $\neg r_n^{\mathsf{E}} \land \neg s_n^{\mathsf{E}}$, $\{\alpha_2\}$ is rewritten into $\neg r_n^{\mathsf{E}} \land s_n^{\mathsf{E}} \land s_n^{\mathsf{A}} \land s_n^{\mathsf{One}}$, and $\{\alpha_1, \alpha_2\}$ is rewritten into $\neg r_n^{\mathsf{E}} \land s_n^{\mathsf{E}} \land \neg s_n^{\mathsf{A}} \land s_n^{\mathsf{One}}$. For each abstract term, we create an abstract property that is premised on that term:

a4 $\mathsf{AG}((\neg r_n^{\mathsf{E}} \land \neg s_n^{\mathsf{E}}) \Rightarrow (\mathsf{EX}_n(\neg r_n^{\mathsf{E}} \land s_n^{\mathsf{E}} \land \neg s_n^{\mathsf{A}} \land s_n^{\mathsf{One}}) \land \mathsf{EX}_n(\neg r_n^{\mathsf{E}} \land s_n^{\mathsf{E}} \land s_n^{\mathsf{A}} \land s_n^{\mathsf{One}})))$
a5 $\mathsf{AG}((\neg r_n^{\mathsf{E}} \land s_n^{\mathsf{E}} \land s_n^{\mathsf{A}} \land s_n^{\mathsf{One}}) \Rightarrow \mathsf{AY}_n(r_n^{\mathsf{A}} \land \neg s_n^{\mathsf{E}}))$
a6 $\mathsf{AG}((\neg r_n^{\mathsf{E}} \land s_n^{\mathsf{E}} \land \neg s_n^{\mathsf{A}} \land s_n^{\mathsf{One}}) \Rightarrow (\mathsf{AY}_n(r_n^{\mathsf{E}} \land \neg s_n^{\mathsf{A}}) \land \mathsf{EX}_n(\neg r_n^{\mathsf{A}} \land s_n^{\mathsf{E}} \land s_n^{\mathsf{One}})))$

For each abstract precondition α^a, the postconditions are derived as follows. First, we record the change to α_1 in c4 and the change to α_2 in c5 as count updates. The existential update in c4, denoted s_+^{e}, represents the change from $\neg s_m$ to s_m applied to a single m that meets α_1. We also append to the update any abstract terms in the postcondition, e.g. $\land_{w \neq m} \neg s_w$. Whereas the update in c5, denoted $(r_+, s_-)^{\mathsf{u}}$, is universal and applies to all neighbors that meet α_2.

Next, for each combination in α^a, we apply the count updates to the parametric preconditions in the combination and rewrite the resulting combinations as abstract postconditions of α^a. We take a6 as an example. The abstract precondition in a6 has a single combination, $\{\alpha_1, \alpha_2\}$. The result of applying s_+^{e} is that $\neg r_m \land s_m \land (\land_{w \neq m} \neg s_w)$ holds for exactly one m. The result is included in the existential abstract postcondition in a6. The result of applying $(r_+, s_-)^{\mathrm{u}}$ is that $r_m \wedge \neg s_m$ holds for the m whose precondition is α_2. The result is included in the universal abstract postcondition in a6.

4.5   Specification Rewriting: Full Procedure

The input to the manual specification rewriting procedure, i.e., the parametric local specification φ_n can be viewed as a list of properties, denoted $\cup_{i \in [0,l)} \{\psi_i\}$, where ψ_i is either a non-parametric property or a parametric property of the form $\wedge_{m \in nbr(n)} f_{nm}$ or g_n. We focus on cases where l and k are independent. In this subsection, we describe the rewriting procedure in three stages.

Step 1. Grouping Properties. We group the properties in φ_n according to preconditions. Let α_{ψ_i} be the precondition in property ψ_i and $\mathcal{P}(\cup_{i \in [0,l)} \{\alpha_{\psi_i}\})$ be the power set of the preconditions for all properties in φ_n. We construct the set of parametric preconditions, denoted A, as follows. Starting with $A = \emptyset$, for each $\alpha \in \mathcal{P}(\cup_{i \in [0,l)} \{\alpha_{\psi_i}\})$, α is appended to A if α is propositionally consistent, and there is no superset of α in A. Expressions are inconsistent if they contradict each other, such as p_m and $\wedge_{w \in nbr(n)} \neg p_w$. Parametric variables are inconsistent if they cannot both hold for the same neighbor, such as p_m and $\neg p_m$.

Step 2. Obtaining Abstract Preconditions. We perform case splitting to obtain the abstract preconditions. Starting with $\mathcal{G} = \mathcal{P}(A)$, we remove from $\mathcal{G}$ any combination of parametric preconditions $\hat{A} \in \mathcal{G}$ if the concrete interpretation of $\hat{A}$ contains concrete states that violate the following conditions: (1) for each $m \in nbr(n)$, one of the parametric preconditions in $\hat{A}$ holds, (2) each parametric precondition $\alpha \in \hat{A}$ holds for at least one $m \in nbr(n)$, and (3) internal values and parametric terms hold for n. The resulting $\mathcal{G}$ is the set of combinations.

Each $\hat{A} \in \mathcal{G}$ is then transformed into an abstract term h that includes $\hat{A}$. If h is in disjunctive normal form, h is further split into multiple conjunctive terms. Each term is an abstract precondition. Depending on the abstract alphabet Σ_n^a, an abstract precondition may correspond to multiple combinations, where each combination represents a different way of satisfying the abstract precondition.

Step 3. Deriving Abstract Postconditions. We conduct a case analysis to derive the abstract postconditions for each abstract precondition.

First, we extract the updates to a parametric precondition α from φ_n. If $AY_n\beta$ (resp. $EX_n\gamma$) is a next-time property of α, then the update from α to β (resp. γ) is *universal* (resp. *existential*). There is no update for variables that remain unchanged. Updates to the variables in $\cup_{m \in nbr(n)} \{p_m\}$ are considered changes in the number of neighbors for which p_m is true. Therefore, updates are either incremental, denoted p_+, decremental, denoted p_-, or propositionally inconsistent (e.g., the postcondition contains p_m and $\neg p_m$ for the same neighbor). The updates of α is the union of the updates of each $\alpha_{\psi_{i \in [0,L)}} \in \alpha$. The updates of α are divided into a universal subset, B, and an existential subset, Γ.

Following that, let α^a be an abstract precondition. For each combination $\hat{A}$ included in α^a, we do the following. For each $\alpha \in \hat{A}$, we apply the universal updates in B (to all neighbors satisfying α_{nm}) and express the result as an abstract term β^a that includes the updated valuation. The term β^a is the universal postcondition of $\hat{A}$. Similarly, for each $\alpha \in \hat{A}$ and each existential update $\gamma \in \Gamma$, we express the result of applying γ (to one neighbor satisfying α_{nm}) as abstract terms $\{\gamma^a\}$, where each term γ^a is an existential postcondition of $\hat{A}$.

If α^a includes a single combination (e.g., a4-a6 in Sect. 4.4) or multiple combinations but with the same universal postcondition β^a, then we create an $\mathsf{AY}_n\beta^a$ subformula for β^a and an $\mathsf{EX}_n\gamma_j^a$ subformula for each existential postcondition γ_j^a of α^a. The resulting subformulas are filled in to create an abstract property $\mathsf{AG}(\alpha^a \Rightarrow (\mathsf{AY}_n\beta^a \wedge (\wedge_j\mathsf{EX}_n\gamma_j^a)))$ for α^a.

However, if the universal postconditions of the combinations in α^a are different from each other (as in the example in Sect. 3), we create an abstract property $\mathsf{AG}(\alpha^a \Rightarrow (\mathsf{AY}_n(\vee_i\beta_i^a) \wedge (\wedge_{ij}\mathsf{EX}_n(\beta_i^a \wedge \gamma_{ij}^a))))$ for α^a. Here, i ranges over the combinations in α^a, and j ranges over the existential postconditions of each combination $\hat{A}_i$.

We can take a shortcut to generate φ_n^a by directly rewriting the properties that satisfy the exact interpretation requirements (e.g., c2-c3 in Sect. 4.4) into equivalent abstract properties and then rewriting the remaining properties (e.g., c4-c5 in Sect. 4.4) using the detailed procedure.

Generate Contexts. We introduce context as auxiliary information to maintain the connection between parametric and abstract properties. The *context* of an abstract precondition α^a is a list of tuples, where each tuple consists of a combination $\hat{A}$ in α^a, and the count updates for each parametric precondition $\alpha \in \hat{A}$. For each update, we indicate whether it is universal or existential. For example, the context of property a4 in Sect. 4.4 is a tuple $(\{\alpha_1\}, \alpha_1 : (s_+, \wedge_{w \neq m}\neg s_w)^{\mathsf{e}})$, where $\{\alpha_1\}$ is the only combination in the abstract precondition of a4, and $(s_+, \wedge_{w \neq m}\neg s_w)^{\mathsf{e}}$ is the only update of α_1.

5   Synthesis

The work in [29] describes a tableau-based, iterative decision procedure for concrete local specifications φ_n^c written in Fair CTL. We modify the algorithm to take an abstract local specification φ_n^a as input and automatically build an abstract model H_n^a of φ_n^a. We describe the modified algorithm and its complexity.

A *tableau* of n is denoted by $\mathcal{T}_n := (V, R, L)$, where V is a set of nodes, R is a left-total transition relation, and $L : V \to 2^{cl(\varphi_n^a)}$ is a labeling function using $cl(\varphi_n^a)$ to represent the Fischer-Ladner closure of φ_n^a [4,17], i.e., the negation, subset, and fixpoint closure of Fair CTL modalities in φ_n^a. The formulas are in *negation normal form*, i.e., negation is driven to the proposition-level.

In $\mathcal{T}_n$, $V := C \cup D$ and $R \subseteq (C \times D) \cup (D \times C)$, where C is a set of AND-nodes (i.e., the potential states of P_n^a), and D is a set of OR-nodes. AND-OR transitions are labeled with n or the letter m, whereas OR-AND transitions are

Table 1. The expansion rules (cf. [13,14]).

$a = f \wedge g$	$a_1 = f$	$a_2 = g$
$a = \mathsf{A}(f\mathsf{W}g)$	$a_1 = g$	$a_2 = f \vee \mathsf{AYA}(f\mathsf{W}g)$
$a = \mathsf{E}(f\mathsf{W}g)$	$a_1 = g$	$a_2 = f \vee \mathsf{EXE}(f\mathsf{W}g)$
$a = \mathsf{AG}g$	$a_1 = g$	$a_2 = \mathsf{AYAG}g$
$a = \mathsf{EG}g$	$a_1 = g$	$a_2 = \mathsf{EXEG}g$
$a = \mathsf{AY}g$	$a_1 = \mathsf{AY}_n g$	$a_2 = \mathsf{AY}_m g$
$b = f \vee g$	$b_1 = f$	$b_2 = g$
$b = \mathsf{A}(f\mathsf{U}g)$	$b_1 = g$	$b_2 = f \wedge \mathsf{AYA}(f\mathsf{U}g)$
$b = \mathsf{E}(f\mathsf{U}g)$	$b_1 = g$	$b_2 = f \wedge \mathsf{EXE}(f\mathsf{U}g)$
$b = \mathsf{AF}g$	$b_1 = g$	$b_2 = \mathsf{AYAF}g$
$b = \mathsf{EF}g$	$b_1 = g$	$b_2 = \mathsf{EXEF}g$
$b = \mathsf{EX}g$	$b_1 = \mathsf{EX}_n g$	$b_2 = \mathsf{EX}_m g$

unlabeled. Since the process indices of the neighbors are hidden, when applying the abstraction, m denotes a generic neighbor of n without an explicit index.

Starting from an initial tableau $\mathcal{T}_n^0$, the algorithm iteratively constructs $\mathcal{T}_n^{i+1}$ from $\mathcal{T}_n^i$ until a fixpoint tableau is reached. For simplicity, we assume that the input specification specifies a single initial condition. The root of $\mathcal{T}_n^0$ is an OR-node labeled $\{\varphi_n^a\}$, and $\mathcal{T}_n^0$ is constructed by adding successors to leaf nodes.

Constructing the Initial Tableau (cf. [13,14]***).*** The successors of an OR-node d describe various ways of satisfying the formulas in $L(d)$. The algorithm computes the successors of d by expanding $L(d)$ according to the rules in Table 1. In Table 1, the conjunctive rules expand formula a into $a_1 \wedge a_2$, and the disjunctive rules expand formula b into $b_1 \vee b_2$. Each successor of d is an AND-node c, corresponding to an expansion of $L(d)$.

We show how to derive the expansions of $L(d)$, i.e., the label of each successor c of d. Starting from $L(c) = L(d)$, if $a \in L(c)$, both a_1 and a_2 are added to $L(c)$. If $b \in L(c)$, one of b_1 and b_2 is added to $L(c)$. Each formula is expanded at most once. The expansion continues until all unexpanded formulas in $L(c)$ are *elementary* formulas, i.e., each unexpanded formula is either a variable p, its negation $\neg p$, or a formula whose modality is indexed AY or EX.

The successors of an AND-node c are created to satisfy the next-time formulas indexed by n in $L(c)$. For each $\mathsf{EX}_n g \in L(c)$, an OR-node d is created as a successor of c such that $g \in L(d)$. For each $\mathsf{AY}_n f \in L(c)$, $f \in L(d)$ for each d. If there are no $\mathsf{EX}_n g$ in $L(c)$, a single successor d is created for c such that for every $\mathsf{AY}_n f$ in $L(c)$, f is in $L(d)$. If there is no next-step formula in $L(c)$, the successor d copies the label of c and points back to c to form a self-loop of c.

Table 2. The deletion rules for tableau pruning (cf. [29]).

deleteP	Delete any node whose label is propositionally inconsistent.
deleteOR	Delete any OR-node all of whose successors have been deleted.
deleteAND	Delete any AND-node one of whose successors has been deleted.
deleteEU	Delete any node v if $\mathsf{E}(f\mathsf{U}g) \in L(v)$, and there is no AND-node c' reachable from v via a finite path π such that $g \in L(c')$ and $f \in L(c)$ for each AND-node c on π except c'.
deleteAU	Delete any node v if $\mathsf{A}(f\mathsf{U}g) \in L(v)$, and there is no subdag $\mathcal{U}$ rooted at v such that $g \in L(c')$ for each leaf AND-node c' in $\mathcal{U}$, and $f \in L(c)$ for each internal AND-node c of $\mathcal{U}$.
deleteEG	Delete any node v if $\mathsf{EG}g \in L(v)$, and there is no fair full path π starting from v such that $g \in L(c)$ for each AND-node c on π.
deleteJoint	Delete any AND-node c_n if every AND-node e_n forming the joint state (e_n, c_m) has been deleted, where c_m is isomorphic to c_n.
deleteUpdate	Delete any AND-node c_n if one of its updates is inconsistent.

The transition from c to each d is labeled n. The modified decision procedure extends the labels of AND-OR transitions to include the corresponding context. Let α_n^a be an abstract precondition in $L(c)$, and β_n^a (resp. γ_n^a) be an abstract postcondition satisfied in $L(d)$. The algorithm appends the combinations and updates corresponding to β_n^a (resp. γ_n^a) in the context of α_n^a to the label of (c, d).

Duplicate nodes of the same type and label are merged. When there are no more leaf nodes, an initial tableau consisting of nodes reachable from the root via n-transitions is constructed.

Constructing the Fixpoint Tableau (cf. [29]***).*** The next tableau, $\mathcal{T}_n^{i+1}$, is constructed based on $\mathcal{T}_n^i$. The algorithm captures the updates in the labels of the n-transitions in $\mathcal{T}_n^i$ and creates isomorphic updates (caused by m) as interference transitions for the AND-nodes affected by these updates.

Let (c_n, c_n') be an n-transition from AND-node c_n to c_n' in $\mathcal{T}_n^i$, where c_n and c_n' have different shared states by updating the parametric precondition a_{nm} that holds in c_n. When a neighbor of n executes an isomorphic transition (c_m, c_m'), an AND-node e_n in $\mathcal{T}_n^i$ is affected by the m-transition if a parametric precondition isomorphic to α_{nm} is in one of the combinations in e_n. Note that an interference update is existential for n. The update only changes the values of variables shared with the neighbor that initiated the transition.

Let y be a shared state, and $\cup_i\{y_i'\}$ be a set of m-successor shared states of y under the interference of a generic neighbor m. The algorithm computes the successors of each affected AND-node e_n as follows. For each y_i', an OR-node whose label contains y_i' is created as an m-successor of e_n. For each $\mathsf{EX}_m g \in L(e_n)$, an OR-node whose label contains g and $\vee_i y_i'$ is also added to e_n as an m-successor. For each $\mathsf{AY}_m f \in L(e_n)$, $f \in L(d)$ for every m-successor d of c.

After adding interference transitions to $\mathcal{T}_n^i$, the decision procedure adds successors to each leaf node via n-transitions. The resulting tableau is $\mathcal{T}_n^{i+1}$. When $\mathcal{T}_n^{i+1} = \mathcal{T}_n^i$, the fixpoint tableau, denoted $\mathcal{T}_n^*$, is reached. As with the algorithm in [29], an error message is raised if it is detected that $\mathcal{T}_n^*$ may produce a non-outward-facing model. If no error message is issued, $\mathcal{T}_n^*$ is pruned by repeatedly applying the deletion rules in Table 2. If the root of $\mathcal{T}_n^*$ is deleted, φ_n^a is unsatisfiable. Otherwise, the pruned $\mathcal{T}_n^*$ is unraveled into a model H_n^a, from which the representative process P_n^a is extracted by removing interference transitions.

Checking Fulfillment of Eventualities. We elaborate on the differences between the pruning steps for abstract and concrete fixpoint tableaux.

Note that an abstract precondition α_n^a can include multiple combinations of parametric preconditions. We compute successor combinations for α_n^a by applying updates to each combination in α_n^a. These successor combinations, written as abstract terms, are considered preconditions for subsequent transitions. Thus, we say that a path $(\alpha_0^a, \alpha_1^a, \alpha_2^a)$ of abstract shared states *includes concrete paths* if the combinations produced by transition (α_0^a, α_1^a) intersect with the combinations required by transition (α_1^a, α_2^a).

We require that the fulfilling path π in deletion rules *deleteEU* and *deleteEG* (in Table 2) be a path that includes concrete paths. Since the fulfillment of abstract eventualities must be independent of k, we further require that π be a valid path for all k. For example, suppose π contains a transition whose precondition shows that a shared variable p_{nm} holds for multiple neighbors. Since π does not apply to the case of only one neighbor, either π is not a path to verify the fulfillment of the eventuality or k cannot take the value one.

The construction of subdags that verify the fulfillment of AU is also modified accordingly. In *deleteAU*, *subdag* $\mathcal{U}$ is a directed graph extracted from tableau $\mathcal{T}$. The paths in $\mathcal{U}$ are abstract paths that include concrete paths. Beyond that, $\mathcal{U}$ contains no fair full path, i.e., all cycles in $\mathcal{U}$ must be finite. For example, because k is unbounded but not infinite, a cycle is finite if it contains p_+ as the only update for $\cup_{m \in nbr(n)} \{p_m\}$ but does not contain any node labeled p_n^A.

Rule *deleteUpdate* removes nodes with inconsistent updates. The rule ensures that no propositional inconsistencies are in the values of concrete variables shared with any neighbor, even if the values of abstract variables are consistent.

Correctness and Complexity. We show that the modified decision procedure satisfies Lemma 3. See the extended version for proof.

Lemma 3. *The constructed system H_n^a satisfies the abstract local specification φ_n^a, is closed under neighboring interference, and unravels into an outward-facing abstract representative process P_n^a.*

The modified decision procedure constructs a fixpoint tableau $\mathcal{T}_n^*$ whose size is bounded by $2^{|\varphi_n^a|}$. The cost of constructing H_n^a from φ_n^a is in time polynomial in the size of $\mathcal{T}_n^*$. Therefore, the synthesis of P_n^a can be done in time $O(2^{|\varphi_n^a|})$, independent of k and K. The synthesized P_n^a is instantiated on the network nodes. The run time of instantiating P_n^a to form a protocol instance is $O(K)$.

6   Concretization and Soundness

The synthesizer constructs an abstract model H_n^a from an abstract local specification φ_n^a. We show that, for any k, a concrete system H_n^c is derived from H_n^a by *concretization*, and H_n^c is a model of the concrete local specification φ_n^c.

We show how concretization works. First, we create the root of H_n^c and label it with the concrete *init-spec* in φ_n^c. The label of the concrete root is included in the label of the abstract root in H_n^a. Next, we add successor states to each concrete leaf state. Upon termination, the resulting system is H_n^c.

Let s^c be a concrete leaf state and s^a in H_n^a be the abstract state corresponding to s^c. The steps to attach successor states $\cup_i \{t_i^c\}$ to s^c are as follows. If s^c is in the concrete interpretation of a combination $\hat{A}$ in the context of s^a, then for each successor combination $\hat{A}'$ of $\hat{A}$, there is an abstract state t^a such that transition $(s^a, t^a) \in T_n^a$ and $\hat{A}'$ is included in t^a. Based on transition $(\hat{A}, \hat{A}')$, a concrete state t_i^c is created and attached to s^c as a successor. The label of t_i^c is obtained by replacing the abstract shared state in $\lambda(t^a)$ with the concrete shared state derived by applying the updates of $(\hat{A}, \hat{A}')$ to $\lambda(s^c)$.

Note that, by symmetry, there are multiple concrete ways to apply an existential update. For example, when $\neg r_m \wedge \neg s_m$ is the parametric precondition for existential update s_+^e, different choices of selecting one from all candidate neighbors that meet the precondition yield different concrete successor states.

Theorem 3. *If an abstract model H_n^a is synthesized from φ_n^a, then for any k, a concrete system H_n^c derived from H_n^a by concretization satisfies φ_n^c.*

By Theorem 3 (see the extended version for proof), the abstract process P_n^a, extracted from H_n^a, is instantiated at each node of a given network graph to form a protocol instance. By Theorem 1, all instances satisfy the global specification.

7   Experiment

We implemented the modified decision procedure in Python. We constructed models for two example parametric protocols (informally described as follows) using the synthesizer on a machine with a 2.5 GHz CPU and 16 GB of memory. The model of the atomic snapshots protocol has 19 states and takes about 50 s to construct. The dining philosophers model has a total of 44 states, and its construction takes about 290 s. See the extended version for details.

Atomic Snapshots. We describe an example atomic snapshots protocol [1]. An atomic snapshot object is a group of K processes with an operation called 'snapshot' that returns the simultaneous state of all processes. In a fully connected network, an atomic process shares a port with each neighbor, so that the process writes to the port (values indicating the current state of the process and sequence numbers that distinguish the order of write operations), and the neighbor reads from the port. That is, a process reads k ports, however, a process can read only one port at a time. The objective of a snapshots algorithm is to implement an atomic snapshot object using these read/write variables.

To implement the objective, a process reads all ports once and then reads them all again. If the two read passes yield the same outcome, then the process takes as a snapshot the list of value-sequence number pairs read from the ports. That is, for each neighbor, the process picks a point between the end of the first pass and the start of the second pass and records the data in the port as the state of the neighbor. Otherwise, the process restarts the snapshot operation.

When a process is not running the snapshot operation, it can simultaneously write to all shared ports. The algorithm is wait-free, but the snapshot operation may never complete because ports read by the process are prone to changes made by its neighbors, thus interfering with the creation of a valid snapshot.

Dining Philosophers. We apply the classic dining philosophers protocol to hypercubes. For each increase of one in the number of neighbors, k goes to $k+1$. The hypercube size grows from $K = 2^k$ to $K = 2^{k+1}$.

A total of K philosophers are seated at the nodes of a k-dimensional cube, with a fork shared between each pair of neighbors. In local states, philosophers are either 'thinking,' 'hungry,' or 'eating.' A thinking philosopher becomes hungry in finite time. A hungry philosopher enters into eating by holding all adjacent forks. A pair of philosophers cannot hold a particular fork at the same time, which prevents any neighbors from simultaneously accessing the eating state.

To pursue fair access to the eating state so that no philosopher is permanently hungry, we use the hygienic solution introduced in [8], where the conflict between hungry neighbors over who eats is resolved using a precedence graph. The evolution of the precedence graph is implemented using forks, dirty/clean status of forks, and request tokens for forks. Each fork (resp. token) is held by one of its two adjacent philosophers. Initially, all philosophers are thinking, all forks are dirty, and tokens and forks are possessed in such a way that the precedence graph is acyclic (see the extended version for details). A hungry philosopher sends its tokens to request unowned forks. Upon receiving a token, if the requested fork is dirty, the philosopher cleans the fork and gives it to the neighbor. A hungry philosopher turns into eating when it owns all adjacent forks, and each fork is either clean or not requested. An eating philosopher dirties all its forks at once. The described local specification enforces the absence of starvation.

8   Related Work and Conclusion

To solve the parameterized synthesis problem, Emerson and Srinivasan [12] use counting arguments for fully symmetric systems. Emerson and Attie [5] construct a pair-system, which is the product of two connected processes obtained from each other by swapping the indices. Jacobs and Bloem [18] apply cutoff results and attack the distributed synthesis problem on token-ring networks with a semi-decision procedure for bounded synthesis. Ehlers and Finkbeiner [11] take an automata-theoretic approach on rotation-symmetric architectures, where the symmetry property is partially encoded in the specification and partially ensured by post-processing an obtained computation tree. Klinkhamer and Ebnenasir [19] synthesize the code for symmetric processes in self-stabilizing parameterized unidirectional rings, where a set of legitimate states of the template process are provided as inputs. Bollig et al. [7] focus on round-bounded parameterized systems and turn the synthesis problem into multi-pushdown games, where a winning strategy is computed in non-elementary time.

This work extends an approach that reduces the parameterized synthesis problem to the problem of synthesizing representatives of equivalence classes of balanced, locally symmetric processes [29]. Our approach applies to network families constructed from abstract tiling patterns, including but not limited to rings, tori, rectangular meshes, fully connected networks, and hypercubes. Assuming a uniform system and unconditionally fair scheduling, the approach finds an outward-facing representative P_n in worst-case time exponential in the length of the abstract local specification. The approach is incomplete because the formulation of local specifications for P_n is done manually and has formatting restrictions. Parametric synthesis of reactive, distributed systems is generally undecidable [26]. Furthermore, the approach is not fully automated as global information and proofs are required. The choice of abstract alphabets is up to the designer, taking into account parametric local specifications and formatting guidelines.

Emerson and Clarke [13], Manna and Wolper [22] addressed the *satisfaction* problem and proposed a decision procedure that determines if there is a set of processes such that their joint behavior satisfies a temporal logic formula. The implementation is reactive but not *open* because the environment is assumed to be cooperative instead of adversarial. Pnueli and Rosner [25,27] introduced the *realization* problem that finds out whether there exists a system process implementation against all possible inputs obtained from the environment. As shown in [26], the realization problem for propositional temporal specifications in general architectures is undecidable (recursively enumerable complete) even for finite-state machines. Finkbeiner and Schewe [15] characterized undecidable architectures as cases when processes have incomparable information from the environment. Even for decidable cases, the size of the global state space, i.e., the automata-theoretic product of K sequential processes, is exponential in K.

The distributed synthesis problem has been addressed in many different ways, including the automata-theoretic approach [20], game-theoretic approach [23], bounded synthesis [16,28], and symbolic synthesis [10]. Our method is based on the tableau construction in [13]. The main difference is that we repeatedly add neighboring interference until we get a fixpoint tableau. Our approach is not open, but it is also not closed because the way that neighbors interfere with the local states of P_n is not specified as a property of P_n.

Future Work. Parametric systems can contain several connectivity templates and process types. Our approach can be modified to apply to systems with multiple equivalence classes, such as red/black protocols, where each red (resp. black) process has only black (resp. red) neighbors (cf. [24]), and we are applying the work to other protocols (cf. [9]). We are investigating automation in deriving abstract local specifications, applying the work using general abstract domains, and extending the work to fault-tolerant systems (cf. [4]).

References

1. Afek, Y., Dolev, D., Attiya, H., Gafni, E., Merritt, M., Shavit, N.: Atomic snapshots of shared memory. In: Proceedings of the Ninth Annual ACM Symposium on Principles of Distributed Computing. PODC '90, New York, NY, USA, p. 1–13. Association for Computing Machinery (1990). https://doi.org/10.1145/93385.93394

2. Alford, M.W., et al.: Distributed systems: methods and tools for specification. An advanced course. Springer, Heidelberg (1985)

3. Apt, K.R., Kozen, D.: Limits for automatic verification of finite-state concurrent systems. Inf. Process. Lett. **22**(6), 307–309 (1986)

4. Attie, P.C., Arora, A., Emerson, E.A.: Synthesis of fault-tolerant concurrent programs. ACM Trans. Program. Lang. Syst. **26**(1), 125–185 (2004). https://doi.org/10.1145/963778.963782

5. Attie, P.C., Emerson, E.A.: Synthesis of concurrent systems with many similar processes. ACM Trans. Program. Lang. Syst. **20**(1), 51–115 (1998). https://doi.org/10.1145/271510.271519

6. Baier, C., Katoen, J.P.: Principles of Model Checking. The MIT Press (2008)

7. Bollig, B., Lehaut, M., Sznajder, N.: Round-bounded control of parameterized systems. In: Lahiri, S.K., Wang, C. (eds.) ATVA 2018. LNCS, vol. 11138, pp. 370–386. Springer, Cham (2018). https://doi.org/10.1007/978-3-030-01090-4_22

8. Chandy, K.M., Misra, J.: The drinking philosophers problem. ACM Trans. Program. Lang. Syst. **6**(4), 632-646 (1984). https://doi.org/10.1145/1780.1804

9. Chandy, K., Misra, J.: Parallel Program Design: A Foundation, Computer Science Series. Addison-Wesley Publishing Company (1988)

10. Ehlers, R.: Symbolic bounded synthesis. Formal Meth. Syst. Des. **40**(2), 232–262 (2012). https://doi.org/10.1007/s10703-011-0137-x

11. Ehlers, R., Finkbeiner, B.: Symmetric synthesis. In: FSTTCS. LIPIcs, vol. 93, pp. 26:1–26:13. Schloss Dagstuhl - Leibniz-Zentrum für Informatik (2017)

12. Emerson, A., Srinivasan, J.: A decidable temporal logic to reason about many processes. In: Proceedings of the Ninth Annual ACM Symposium on Principles of Distributed Computing. PODC '90, New York, NY, USA, pp. 233–246. Association for Computing Machinery (1990). https://doi.org/10.1145/93385.93425

13. Emerson, E.A., Clarke, E.M.: Using branching time temporal logic to synthesize synchronization skeletons. Sci. Comput. Program. **2**(3), 241–266 (1982). https://doi.org/10.1016/0167-6423(83)90017-5

14. Allen Emerson, E., Lei, C.-L.: Temporal reasoning under generalized fairness constraints. In: Monien, B., Vidal-Naquet, G. (eds.) STACS 1986. LNCS, vol. 210, pp. 21–36. Springer, Heidelberg (1986). https://doi.org/10.1007/3-540-16078-7_62

15. Finkbeiner, B., Schewe, S.: Uniform distributed synthesis. In: 20th Annual IEEE Symposium on Logic in Computer Science (LICS' 05), pp. 321–330 (2005). https://doi.org/10.1109/LICS.2005.53

16. Finkbeiner, B., Schewe, S.: Bounded synthesis. Int. J. Software Tools Technol. Transf. **15**(5–6), 519–539 (2013). https://doi.org/10.1007/s10009-012-0228-z

17. Fischer, M.J., Ladner, R.E.: Propositional dynamic logic of regular programs. J. Comput. Syst. Sci. **18**(2), 194–211 (1979). https://doi.org/10.1016/0022-0000(79)90046-1

18. Jacobs, S., Bloem, R.: Parameterized synthesis. Log. Methods Comput. Sci. **10**(1) (2014). https://doi.org/10.2168/LMCS-10(1:12)2014

19. Klinkhamer, A.P., Ebnenasir, A.: Synthesizing parameterized self-stabilizing rings with constant-space processes. In: Dastani, M., Sirjani, M. (eds.) Fundamentals of Software Engineering, pp. 100–115. Springer, Cham (2017)
20. Kupferman, O., Vardi, M.: Synthesizing distributed systems. In: Proceedings 16th Annual IEEE Symposium on Logic in Computer Science, pp. 389–398 (2001). https://doi.org/10.1109/LICS.2001.932514
21. Lamport, L.: Specifying Systems: The TLA+ Language and Tools for Hardware and Software Engineers. Addison-Wesley Longman Publishing Co., Inc, USA (2002)
22. Manna, Z., Wolper, P.: Synthesis of communicating processes from temporal logic specifications. ACM Trans. Program. Lang. Syst. **6**(1), 68-93 (1984). https://doi.org/10.1145/357233.357237
23. Mohalik, S., Walukiewicz, I.: Distributed games. In: Pandya, P.K., Radhakrishnan, J. (eds.) FSTTCS, pp. 338–351. Springer, Heidelberg (2003)
24. Namjoshi, K.S., Trefler, R.J.: Symmetry reduction for the local mu-calculus. In: Beyer, D., Huisman, M. (eds.) Tools and Algorithms for the Construction and Analysis of Systems, pp. 379–395. Springer, Cham (2018)
25. Pnueli, A., Rosner, R.: On the synthesis of a reactive module. In: Proceedings of the 16th ACM SIGPLAN-SIGACT Symposium on Principles of Programming Languages. POPL '89, New York, NY, USA, pp. 179–190. Association for Computing Machinery (1989). https://doi.org/10.1145/75277.75293
26. Pnueli, A., Rosner, R.: Distributed reactive systems are hard to synthesize. In: Proceedings 31st Annual Symposium on Foundations of Computer Science, vol. 2, pp. 746–757 (1990). https://doi.org/10.1109/FSCS.1990.89597
27. Pnueli, A., Rosner, R.: On the synthesis of an asynchronous reactive module. In: Proceedings of the 16th International Colloquium on Automata, Languages and Programming. ICALP '89, Berlin, Heidelberg, pp. 652–671. Springer (1989)
28. Schewe, S., Finkbeiner, B.: Bounded synthesis. In: Namjoshi, K.S., Yoneda, T., Higashino, T., Okamura, Y. (eds.) Automated Technology for Verification and Analysis, pp. 474–488. Springer, Heidelberg (2007)
29. Zhang, R., Trefler, R.J., Namjoshi, K.S.: Synthesizing locally symmetric parameterized protocols from temporal specifications. In: Griggio, A., Rungta, N. (eds.) 22nd Formal Methods in Computer-Aided Design, FMCAD 2022, Trento, Italy, October 17–21, 2022, pp. 235–244. IEEE (2022). https://doi.org/10.34727/2022/ISBN.978-3-85448-053-2_30

1–2–3–Go! Policy Synthesis for Parameterized Markov Decision Processes via Decision-Tree Learning and Generalization

Muqsit Azeem[1], Debraj Chakraborty[3], Sudeep Kanav[3], Jan Křetínský[1,3(✉)],
Mohammadsadegh Mohagheghi[2], Stefanie Mohr[1], and Maximilian Weininger[1,4]

[1] Technical University of Munich, Munich, Germany
jan.kretinsky@fi.muni.cz
[2] Vali-e-Asr University of Rafsanjan, Rafsanjan, Iran
[3] Masaryk University, Brno, Czech Republic
[4] Institute of Science and Technology Austria, Vienna, Austria

Abstract. Despite the advances in probabilistic model checking, the scalability of the verification methods remains limited. In particular, the state space often becomes extremely large when instantiating parameterized Markov decision processes (MDPs) even with moderate values. Synthesizing policies for such *huge* MDPs is beyond the reach of available tools. We propose a learning-based approach to obtain a reasonable policy for such huge MDPs.

The idea is to generalize optimal policies obtained by model-checking small instances to larger ones using decision-tree learning. Consequently, our method bypasses the need for explicit state-space exploration of large models, providing a practical solution to the state-space explosion problem. We demonstrate the efficacy of our approach by performing extensive experimentation on the relevant models from the quantitative verification benchmark set. The experimental results indicate that our policies perform well, even when the size of the model is orders of magnitude beyond the reach of state-of-the-art analysis tools.

Keywords: model checking · probabilistic verification · Markov decision process · policy synthesis

1 Introduction

Markov decision processes (MDPs) are *the* model for combining probabilistic uncertainty and non-determinism. MDPs come with a rich theory and

This research was funded in part by the DFG project 427755713 GOPro, the DFG GRK 2428 (ConVeY), the MUNI Award in Science and Humanities (MUNI/I/1757/2021) of the Grant Agency of Masaryk University, and the EU under MSCA grant agreement 101034413 (IST-BRIDGE).

algorithmics developed over several decades with mature verification tools arising 20 years ago [34] and proliferating since then [12]. Despite all this effort, the *scalability* of the methods is considerably worse than of those used for verification of non-deterministic systems with no probabilities, even for basic problems.

What to do about very large models? Researchers have made various attempts to tackle this issue, however, only with limited success. Firstly, prominent techniques which work well in *non-probabilistic* verification, such as symbolic techniques [34], abstraction [35], and symmetry reduction [20], are harder to apply efficiently in the probabilistic setting. Secondly, *"engineering"* improvements, such as the use of external storage [24] or parallelization, help by a significant, but principally very limited factor. Thirdly, there is a *relaxation of the guarantees* on the precision and/or certainty of the result, which we describe in detail below.

The result of the analysis is typically a number (called *the value*), such as the expected reward or the probability to reach a given state, maximized (or minimized) over all resolutions of non-deterministic choices (called *policies, strategies, schedulers, adversaries, or controllers* in different applications). It is generally accepted that the precise number is not needed and an approximation is sufficient in most settings. Interestingly, until a few years ago [6,9,21], typically only the *under-approximations* were computed for the fundamental (maximization) problems, with no reliable over-approximations (with dual issues for minimization).

It is worth noting that over-approximating is inherently harder since reasoning that the value cannot be greater than x involves the claim that *all* policies induce a lower value. In contrast to this universal quantification, the existential one is sufficient for under-approximating: upon providing *a* policy, its value forms automatically a lower bound, which is typically easier to compute. Consequently, many *best-effort* approaches, such as reinforcement learning (RL) [50] and lightweight statistical model checking [11] simply try to find a good policy while giving only empirically good chances to be close to the optimum.

This is sufficient in the setting of (i) *policy synthesis*, where a "good enough" (close to optimum), but not necessarily optimal, controller is sought, or (ii) *bug hunting and falsification*, where finding significant counter-examples cheaply is desirable. However, the practical quality of the results relies on certain assumptions of these methods: RL suffers when the rewards in the model are sparse (e.g., in the case of reachability) and lightweight statistical model checking suffers when near-optimal policies are not abundant.

To summarize, synthesizing **practically good policies** is sufficient in many settings and also the only way when the systems are too large. Yet, when the system is extremely large, the available techniques either run out of resources or yield policies that are far from optimum (and close to random).

Examining the structure of large MDPs in standard benchmark sets, e.g. [25], reveals that their huge sizes are typically not due to astronomically large human-written code, but rather because the MDPs are *parameterized* (e.g., by the number of participants in a protocol) and then instantiated with large values.

Accordingly, this paper proposes a new approach to scalable policy synthesis for parameterized MDPs. Namely, it produces **good policies for arbitrarily large instantiations of parameterized MDPs in particular those beyond the reach of any state-of-the-art tools**. We focus on probabilistic reachability (i) for simplicity and (ii) because it is a fundamental building block for many other problems.

Our main idea is to generalize the decisions taken by the optimal policies for the smaller instances: Instead of investigating a huge MDP, we synthesize optimal policies for the given parameterized MDP by instantiating it with *small* numbers. We then *generalize* this information and learn a policy that can be applied to any instantiation with an arbitrarily larger number. It is important to note that we generalize the corresponding *decisions* (i.e., the policy itself), not the *values* of the states across different parameterizations. Indeed, while the numeric values can differ vastly, the optimal behaviour is often similar in all instantiations. In order to capture this regularity, we thus need a *symbolic* representation of a policy, which applies to *all* instantiations. **Decision trees (DT)** can provide such a representation. Moreover, since they can represent policies explainably [2,8] and capture the essence of the decisions, not just a list of state-action pairs, they generalize well.

As an illustrative task for our "generalization", consider a buggy mutual exclusion protocol with a high number of participants. While finding the bug with many participants may be hard, an exhaustive investigation of the case with two participants may reveal a scenario violating the exclusion. A similar scenario can then also happen with many participants where the choices of the remaining participants may be irrelevant. Consequently, the key *decisions in the policy* to find bug with two participant may also be used for finding the bug with multiple participants.

Our contribution can be summarized as follows:

– We provide a *simple* and elegant way of computing practically good policies for parameterized MDPs of *any* size (as long as some instantiations exist that are small enough so that some technique can be applied), in particular also orders of magnitude beyond the reach of any other methods. The method is based on generalizing[1] policies via their decision-tree representations. The method scales *constantly* in the parameter instantiation since it applies available techniques to a fixed number of *small base instantiations*, and the large instantiation is never explicitly considered.

– We demonstrate the efficacy of the method experimentally on standard benchmarks. In particular, from the practical perspective, we observe that our policies mostly achieve values that are *close to the actual optimum*. Note that, this in principle cannot be guaranteed for instantiations too large for precise methods to apply, which are exactly of our interest. Nevertheless, the often

---

[1] The nature of our *generalization-based* policy synthesis is also portrayed by our quipping title "1–2–3–Go!": Find out what works for cases 1, 2, and 3, then "Go!" and apply it for arbitrary large values of the parameters.

consistent results on the smaller instantiations convincingly substantiate the expectation that the policies perform well also for the large instantiations.
- Finally, comparing to the benchmarks where our policies do not perform so well, we identify aspects of the models indicating where our heuristic generalizes well and where either more tailored or completely different techniques are required.

It should be emphasized that we regard the simplicity of our approach rather as an advantage, making it easy to exploit. While there is a body of work on policy representation (via post-processing them), the use of DT to compute policies is very limited (as described in the Related Work below) and, to the best of our knowledge, non-existent for computing/generalizing them for arbitrarily large systems. Altogether, this simple, yet efficient idea deserves to be finally explored.

Related Work

Symbolic approaches are widely used as for alleviating the challenges of the state explosion problem [5]. These approaches are based on data structures storing the information of a model compactly. In particular, the multi-terminal version of BDDs (MTBDDs) has been developed for probabilistic model checking [31,41,44]. In a sense, our approach is also symbolic, since we represent the policy using a decision tree. In contrast to MTBDDs, DTs are more suitable for the goal of explainability, as argued in, e.g., [2].

Reduction techniques try to reduce the state space of the model while the smaller model satisfies the same set of properties. A symmetry reduction technique for probabilistic models has been proposed in [36] for systems with several symmetric components. Probabilistic bisimulation is available for MDPs and discrete-time Markov chains (DTMCs) that reduce the original model to the smallest one that satisfies the same set of temporal logic formulae [18]. Considering a subset of temporal logic formulae, more efficient techniques have been proposed in [30] for reducing the model to a smaller one. Applying reduction on a high-level description before constructing the resulting model is available in [40,49].

Further techniques improving scalability of traditional algorithms include the following. Using *secondary storage* in an efficient way to keep a sparse representation of a large model has been studied in [24]. *Compositional techniques* have been developed for the verification of stochastic systems [16,39]. *Prioritizing computation* can reduce running times in many case studies by using topological state ordering [13,38] or learnt prioritizing [9,33,43].

All the above techniques help solving larger models, however, only up to a certain limit. In contrast, our approach synthesizes a policy that can be applied to arbitrarily large instances.

Statistical model checking (SMC) is an alternative solution for approximating the quantitative properties [3,9,26,27,32,52] by running a set of simulations on the model to approximate the requested values, while providing a confidence interval for the precision of computed values for discrete and continuous-time Markov chains (DTMCs and CTMCs). This is scalable since the number of samples does not depend on the size of the model. Still, the length of the simulations does. Using SMC for MDPs faces the difficulty of resolving non-determinism. A smart-sampling method has been proposed in [15] that considers a set of random policies, with some of them hopefully approximating the optimal one; however, this method cannot generally provide a confidence interval for the precision of computations [26].

Machine learning within formal verification of MDP has been widely studied for a decade since the seminal [27]. In particular, **reinforcement learning (RL)** has been adapted to the setting of verification with objectives such as reachability [9,22], but the sparsity of the rewards is still an issue affecting the scalability. Still, prioritizing the subset of the states that has the biggest impact on the value can allow for verifying huge models if such a subset is small and easy to find [9,33]. Unlike RL, which suggests policies for specific models through random exploration, our approach generalizes policies for various instances by computing optimal policies on smaller models and generalizing them. Pyeatt and Howe [46] propose using decision trees to approximate value function for discounted rewards in reinforcement learning. However, we learn a DT that is a valid policy for any instantiations of the parameterized MDP, whereas the DT learned in [46] is applicable only to the model under consideration.

An L^* learning approach has been developed in [51] to learn an MDP model efficiently. Neural networks and regression can be used to resolve non-determinism of large MDPs and provide the opportunity of applying SMC for this class of models [19,47].

Decision trees have been used as a data structure for representing MDP policies [1,2,8]. Interestingly, while binary decision diagrams (BDD) may appear to the verification community as a suitable candidate, it has been shown that DT are more appropriate if adequately used [2,8,43] due to their ability to handle various predicates and complex relationships, enhancing explainability. Another advantage of DTs is the ability to declare some inputs as uninteresting ("don't care" inputs), saving on size via semantics of the controller.

2   Preliminaries

We provide basic definitions in Sect. 2.1, then describe what it means for models to be parameterized and scalable in Sect. 2.2 and finally recall how decision trees can be used for representing policies in Sect. 2.3.

2.1   Markov Decision Processes with a Reachability Objective

A *probability distribution* over a discrete set X is a function $\mu : X \to [0,1]$ where $\sum_{x \in X} \mu(X) = 1$. We denote the set of all probability distributions over X by $\mathcal{D}(X)$.

Definition 1. *A (finite) Markov Decision Process (MDP) is a tuple* $\mathcal{M} = (S, A, \delta, \bar{s}, G)$ *where* S *is a finite set of states,* A *is a finite set of actions, overloading* A(s) *to denote the (non-empty) set of enabled actions for every state* $s \in S$, $\delta : S \times A \to \mathcal{D}(S)$ *is a probabilistic transition function mapping each state* $s \in S$ *and enabled action* $a \in A(s)$ *to a probability distribution over successor states,* $\bar{s} \in S$ *is the initial state, and* $G \subseteq S$ *is the set of goal states.*

A *Markov chain* (MC) can be seen as an MDP where $|A(s)| = 1$ for all $s \in S$, i.e. a system exhibiting only probabilistic behavior, but no non-determinism.

The **semantics** of an MDP are defined in the usual way by means of policies and paths in the induced Markov chain. An infinite *path* $\rho = s_1 s_2 \ldots \in S^\omega$ is an infinite sequence of states. A *policy* is a function $\sigma : S \to \mathcal{D}(A)$ that, intuitively, prescribes for every state which action to play. The policy is called deterministic if in every state it selects a single action surely, otherwise it is randomized. Note that we limit our definition of policies w.l.o.g. to those that are memoryless (history-independent), i.e. those that depend only on the current state, not on a whole path. We denote the i-th state on a path as ρ_i, the set of all paths as Paths, and the set of all policies as Σ. By using a policy σ to resolve all nondeterministic choices in an MDP $\mathcal{M}$, we obtain a Markov chain $\mathcal{M}^\sigma$ [5, Definition 10.92]. This Markov chain induces a unique probability measure $\mathbb{P}_s^\sigma$ over paths starting in state s [5, Definition 10.10].

The **reachability objective** is included in our definition of MDP in the form of the initial state $\bar{s}$ and the set of goal states G. Intuitively, the *value* V of a reachability objective is the optimal probability to reach some goal state when starting in the initial state; formally $V(\mathcal{M}) = \mathsf{opt}_{\sigma \in \Sigma} \mathbb{P}_{\bar{s}}^\sigma[\Diamond G]$, where $\mathsf{opt} \in \{\max, \min\}$ indicates whether we are trying to reach or avoid the set of goal states and $\Diamond G = \{\rho \in \mathsf{Paths} \mid \exists i. \rho_i \in G\}$ denotes the set of all paths that reach a goal state. One can restrict this optimum to the deterministic memoryless policies [45, Proposition 6.2.1].

2.2   State Space Structure and Scalable Models

For learning (e.g. of DTs) to be effective, it is important that the state space of MDP is *structured*, i.e. every state is a tuple of values of state variables. In other words, the state space of the system is not monolithic (e.g. states defined by a simple numbering), but in fact, there are multiple factors defining it, for example, time or protocol state. Each of these factors is represented by a state variable v_i with domain $\mathbb{D}_i$. Thus, every state $s \in S$ is in fact a tuple $(v_1, v_2, \ldots, v_n)$, where each $v_i \in \mathbb{D}_i$ is the value of a state variable.

A *parameterized MDP* can be described as a variant of standard MDP where certain parameters are not fixed constants but instead can take different values within a *parameter space*. These parameters can be associated to the state-space of the system (for example, lower or upper bound of a state variable) or transition dynamics (the probabilities can be functions of the parameter). We provide the formal definition below.

Definition 2. *A* parameterized MDP *is a tuple* $M = (S_\theta, A_\theta, \delta_\theta, \bar{s}_\theta, G_\theta, \Theta)$, *where:*

- Θ *is the parameter space,*
- *For each* $\theta \in \Theta$, *the tuple* $(S_\theta, A_\theta, \delta_\theta, \bar{s}_\theta, G_\theta)$ *defines an MDP instance, where:*
  - S_θ *is the set of states,*
  - A_θ *is the set of actions,*
  - $\delta_\theta : S_\theta \times A_\theta \to \mathcal{D}(S_\theta)$ *is the probabilistic transition function,*
  - $\bar{s}_\theta \in S_\theta$ *is the initial state,*
  - $G_\theta \subseteq S_\theta$ *is the set of goal states.*

In this framework, different parameter values $\theta \in \Theta$ *yield different instances of the MDP, and the parameterization can affect the state space, transition dynamics, or both.*

Intuitively, a parameterized MDP can be seen as a family of MDPs where different value of parameter gives different *instance* of the MDP. In particular, this typically makes the models *scalable*: by increasing the values of parameters in the model description, one can scale up the size of the state space of the model. MDPs in the PRISM benchmark suite [37] and the quantitative verification benchmark set [25] have these properties of being structured and scalable.

Example 1. Consider the MDP in Fig. 1. Every state is a tuple (m, x) of two state variables m and x with domains $\mathbb{D}_m = \{0, 1, \ldots, k\}$ and $\mathbb{D}_x = \{0, 1, 2\}$. The state variable m indicates which of the $k + 1$ blocks we are in, while the state variable x indicates the position inside a block.

The block with $m = 0$ is special: it contains the initial state $(0,0)$, the goal state $(0,2)$, and a sink state $(0,1)$ which cannot reach the goal. All other blocks look as follows: for every $m \in [1, k]$, the $x = 0$ state can choose to continue to $x = 1$ (action a) or self-loop (action b). For every $m \in [1, k-1]$, the $x = 1$ state can go back to $x = 0$ in the same block (action a) or leave the block (action b). When using b, there is a 50% chance of going to the sink state $(0, 1)$ and a 50% chance to continue to $x = 0$ in the $(m+1)$-th block. In the k-th block, the action leaving the block progresses to the goal state.

The model is scalable, since the number of blocks k can be increased arbitrarily. This affects both the size of the state space, which is $2k + 3$, as well as the maximum reachability probability, which is 0.5^{k-1}.

We assume the MDPs are defined using high-level modeling languages such as Probmela [5], PRISM [37] or MODEST [23]. In such modeling languages, MDPs are often represented as a composition of multiple identical components, called *modules*. For example, in a distributed system where multiple processes are interacting or sharing common resources, each process can be described as a separate module. In the context of this paper, we also consider the number of modules as a parameter.

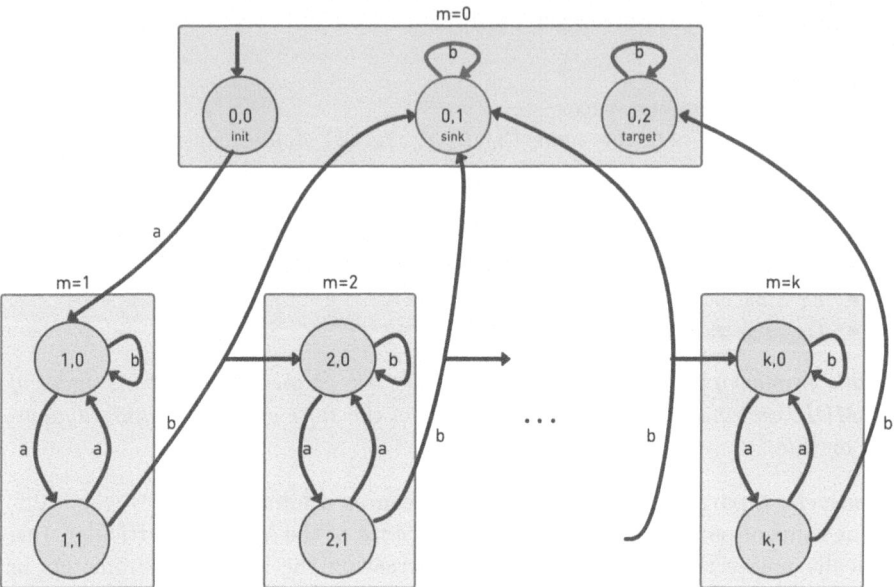

Fig. 1. A parameterized, scalable MDP with $k + 1$ blocks, described in Example 1.

Example 2 (Dining philosophers). The dining philosophers problem involves a number of philosophers, seated around a circular table, blessed with infinite availability of food and things to ponder about. There is a fork placed between each pair of neighbouring philosophers and a philosopher must have both forks in order to eat. They may only pick up one fork at a time and once they finish eating, they place the forks back at the table and return to thinking.

This can be described as an MDP where each philosopher is modeled as a separate module. Thus, the number of philosophers is a parameter.

2.3   Decision Trees for Policy Representation

Knowing that the state space is a product of state-variables, a deterministic policy is a mapping $\prod_i \mathbb{D}_i \to \mathsf{A}$ from tuples of state variables to actions. By viewing the state variables as features and the actions as labels, we can employ machine learning classification techniques such as decision trees, see e.g. [42, Chapter 3], to represent a policy concisely. We refer to [2] for an extensive description of the approach and its advantages. Here, we shortly recall the most relevant definitions in order to formally state our results.

Definition 3. *A decision tree (DT) T is defined as follows:*

- *T is a rooted full binary tree, meaning every node either is an inner node and has exactly two children or is a leaf node and has no children.*
- *Every inner node v is associated with a decision predicate α_v which is a boolean function $\mathsf{S} \to \{false, true\}$ (or equivalently $\prod_i \mathbb{D}_i \to \{false, true\}$).*

– *Every leaf ℓ is associated with an action $a_\ell \in A$.*

For a given state s, we use the following recursive procedure to obtain the action $\sigma(s) = a$ that a DT prescribes: Start at the root. At an inner node, evaluate the decision predicate on the given state s. Depending on whether it evaluates to false or true, recursively continue evaluating on the left or right child, respectively. At a leaf node, return the associated action a.

Example 3. Consider again the MDP given in Fig. 1. The optimal policy needs to continue towards the goal and not be stuck in any loops. This can be achieved by playing action a in states where $x = 0$ and action b in all other states. Traditionally, this policy would be represented as a lookup table, storing $2k + 3$ state-action pairs explicitly. Instead, we can condense the policy to the DT given in Fig. 2b, mimicking the intuitive description of the policy: If $x > 0$, we play b, otherwise, we play a.

Constructing an optimal binary decision tree is an NP-complete problem [29]. Consequently, practical decision tree learning algorithms are based on heuristics. But they tend to work reasonably well. Here, we briefly recall a general framework of learning the decision tree representation of a policy σ as described in [2].

If the policy suggests same action a for all states (i.e., for all states s, we have $\sigma(s) = a$), the tree is just a single leaf node with label a. Otherwise, we split the policy. A predicate ρ, defined on state variables, is chosen, and an inner node labeled ρ is created. Then we partition the policy by evaluating the predicate on the state space, and recursively construct one DT for the policy restricted to the states $\{s \in S | \rho(s)\}$ where the predicate is true, and one for the policy restricted to the states $\{s \in S | \neg\rho(s)\}$ where ρ is false. These policies become the children of the inner node with label ρ and the process repeats recursively.

The selection of "best" predicate is done by selecting the one which is able to split the policy as homogeneous as possible. This can be determined by optimizing some impurity measure such as Gini impurity [10, Chapter 4] or entropy [48]. As we want the learnt DT to exactly represent the policy, unlike other ML algorithms, we do not stop the learning early based on a stopping criterion. Instead, the iteration stops when every state in the leaf node of the tree has the same labeling.

3   Generalizing Policies from Small Problem Instances

In this section, we develop an approach for obtaining good policies for MDPs that are practically beyond the reach of *any available* rigorous analysis.

Our approach exploits the regularity in structure of the MDPs, therefore, we focus on parameterized MDPs where we expect regularity in the state space. Intuitively, we solve a few small instances (colloquially speaking "1, 2, and 3") where an optimal policy is easy to compute. Then we generalize these policies by

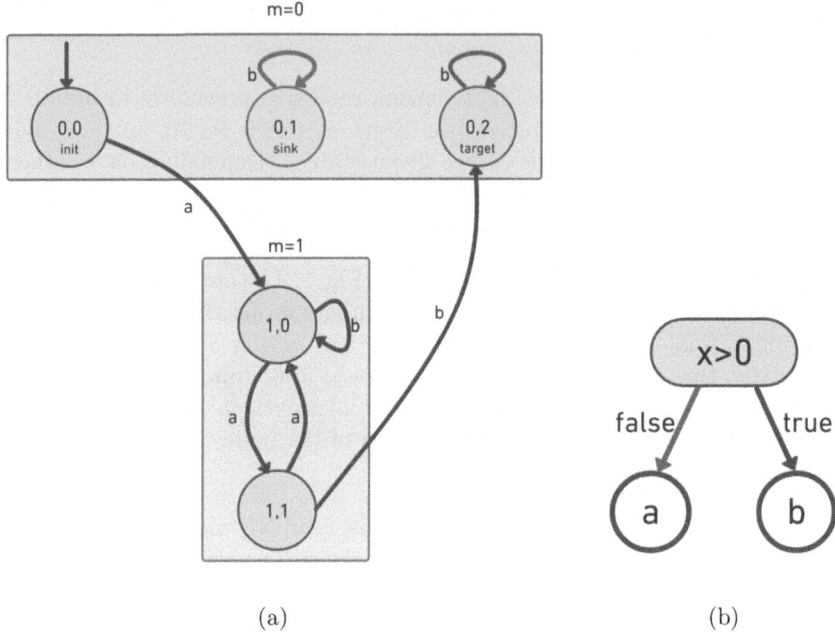

(a)   (b)

Fig. 2. (a) MDP instance for $m = 1$ from Fig. 1. (b) The optimal policy represented as a DT as learned by our approach.

learning a DT from the combined information. The policy represented by this DT is applicable to any instantiation of the parameterized MDP, even the ones that are infeasibly large for any state-of-the-art solver.

More concretely, our approach proceeds in three phases, each of which is described in detail in the following subsections.

1. Select the parameters for the small instances (*base* instances) to learn on.
2. Collect the optimal policies on the base instances.
3. Generalize these policies by learning a DT.

Finally, we discuss how to apply the DT to the MDP and evaluate the policy in order to judge its performance. We illustrate every phase on our running example, the MDP in Fig. 1.

3.1   Parameter Selection

In principle, one can use any of the solvable instances as the set of base instances. However, too small instances may not contain enough information to learn a good generalization. Therefore, we select a small set of instances $\mathcal{B} = \{b_1, ..., b_n\}$, such that the computed policies of these instances are rich enough to learn a generalized DT (see Sect. 4.1 for the details how we practically choose this set). This process can be seen as hyper-parameter search. Domain knowledge can be

Algorithm 1. Computing input for the DT learning from small problem instances.

Input: A parameterized MDP $\mathcal{M}$ and base instances $\mathcal{B} = \{b_1, ..., b_n\}$
Output: A data set $D \subseteq \mathsf{S}_\mathcal{B} \times 2^{\mathsf{A}_\mathcal{B}}$

$D \leftarrow \emptyset$
for all $b_i \in \mathcal{B}$ **do**
    Solve $\mathcal{M}_{b_i}$ and get optimal policy σ_{b_i}
    for all $s \in \mathsf{S}_{b_i}$ **do**
        if s is reachable in $\mathcal{M}_{b_i}^{\sigma_{b_i}}$ and $s \notin \mathsf{G}_{b_i}$ **then**
            add $(s, \sigma_{b_i}(s))$ to D ▷ Add optimal actions
    return D;

---

very useful to obtain this small set of instances. If not available, we can choose a time budget and solve as many instances as possible in that time.

For the remainder of this section, we assume that we are given a set $\mathcal{B} = \{b_1, ..., b_n\}$ of base instances to learn on. We shall interchangeably use b_i for ith instance of the MDP, as well as for the parameters of the ith instance. Hence each b_i can be seen as a vector of parameters with its corresponding values. Formally, $b_i := \langle p_1 = v_1, ..., p_m = v_m \rangle$, where $p_1, ..., p_m$ are the parameters and $v_1, ...v_m$ are the values that are assigned to each of the respective parameters to obtain the instance.

Example 4. Our running example has one parameter k, the number of modules. In fact, we will see that it suffices to consider $\mathcal{B} = \{b_1\}$ where $b_1 := \langle k=1 \rangle$, i.e. only learn on the simplest instance of the MDP as depicted in Fig. 2a.

3.2 Collecting Policies

We collect the optimal decisions from the optimal policies of each of the base instances into a single dataset, later to be used for learning their generalization. The input of the learning algorithms is a data set (possibly a multiset) of samples of (input, output)-pairs. In our case, it is a set of pairs of the structured state and the chosen action.

Algorithm 1 describes how to obtain the input function that can be used for the DT learning from a parameterized MDP and a set of parameterizations. Let $\mathcal{M}_b = (\mathsf{S}_b, \mathsf{A}_b, \delta_b, \bar{s}_b, \mathsf{G}_b)$ be a concrete instance of a parameterized MDP with $b := \langle p_1 = v_1, ..., p_m = v_m \rangle$. For a set of base instances $\mathcal{B} = \{b_1, ..., b_n\}$, we denote the union of all state spaces as $\mathsf{S}_\mathcal{B} := \bigcup_{i=1}^{n} \mathsf{S}_{b_i}$, and similarly define $\mathsf{A}_\mathcal{B}$ as the union of all action spaces. Note that when aggregating the information, we exclude states that are not reachable in the Markov Chain (MC) induced by the computed optimal policy, as well as goal states. This reduces the input size for the DT learning which has two advantages: firstly, it speeds up the computation and secondly, it allows the DT to focus on the relevant states.

Example 5. For our running example in Example 1, we only consider one base instance. Thus, our data is given by the function σ_1 for all reachable non-goal states in the MDP. Concretely, we have pairs of state $(0,0)$ with action a, then $(1,0)$ also with a, and $(1,1)$ with b.

Note that the algorithms are even able to deal with data where the output is non-deterministic (different outputs for the same input within the data set). This corresponds to accommodating *permissive* policies, i.e. ones that allow multiple actions per state. Thus, we can include all optimal actions from a base instance in the data set. Similarly, we can aggregate the information over multiple base instances by a simple union of the recommended actions over all policies. In such cases, the derived DT would also predict multiple actions. Various determinization approaches (see [1]) exist to get one action in these cases, e.g., determinization by voting picks the action that appears most often in the base instances.

Action Labels. Usually, actions are internally represented as integer values in the model-checkers, e.g. in STORM [28] and PRISM [34]. A natural choice would be to use these integers as labels for the actions during decision tree learning. However, this creates a problem: When the set of actions varies among different instances, "identical" action choices are denoted by different integers in each instance.

Example 6. Consider the dining philosophers problem in Example 2. For three philosophers, the initial state would have 6 actions. The first 3 are labeled with $0, 1, 2$, where an action i represents that the $(i + 1)^{th}$ philosopher thinks. The second 3 actions are labeled with $3, 4, 5$, where an action i represents that the $(i - 2)^{th}$ philosopher eats. If we try to generalize this to an MDP with $n > 3$ philosophers, the label 3 is now interpreted as the action that the fourth philosopher thinks, not the first philosopher eats as it was in the case of three philosophers. Thus, representing actions only by the index in which they appear can be sufficient if the number of modules does not change, but is problematic in the opposite case.

To overcome this issue, we take advantage of the action-labeling feature in the PRISM language. An action in PRISM is described by a command of the following form: `[label] guard -> prob_1 : update_1 +...+ prob_n : update_n`. This means that when the condition in the `guard` is true, `update_i` happens with probability `prob_i`. The label of the action is optional (except when the action is a synchronizing action). But, as a simple preprocessing step, we always define the label in each command in the PRISM file. The DT learning then can use these labels instead of the action indices. These labels need to be unique: assigning same label to two actions in two different modules would force the modules to take these actions simultaneously (i.e. to synchronize) changing the structure of the MDP. Also, we only need to define labels for non-synchronizing actions as synchronizing actions already have labels defined that we can use.

For example, the problem in Example 6 can be avoided by giving the unique label `phil_i_line_j` to the action for $(i + 1)^{th}$ philosopher defined by the command at line j in the PRISM file.

3.3   Decision Tree Learning

We use standard DT learning algorithms to learn a DT from the dataset constructed in the Algorithm 1. For predicates to be used, we consider standard, axis-aligned predicates[2], selected by Gini impurity, the default impurity measure of MLPACK [14]. The resulting DT generalizes the policy in two ways:

1. The DT is trained using smaller base instances. The same state variables in the DT's inner node predicates are present in the larger MDP instances, but they can have a larger domain. Despite this difference, the DT would still partition the state space of the larger MDP instances and still recommend actions corresponding to each state.
2. As we are aggregating multiple policies, in our dataset, unlike the learning algorithm described in Sect. 2.3, we can have a state with more than one suggested actions. The learning algorithm considers them as distinct data-points sharing the same value but different labels. Since the values are the same, there are no predicates that can distinguish them. So these data-points traverse the same path in the tree until they reach a leaf node. The classification at the leaf node is determined by 'majority voting', the label that appears most often is assigned to the leaf node. This approach helps filter out actions suggested by only a few less generalizing base instances.

Example 7. For our example data set D constructed in Example 5, the result of the DT learning is the DT depicted in Fig. 2b. This policy is in fact optimal for all k; see Example 3 for an explanation of this. In addition to being optimal, it is also small and perfectly explainable.

   In contrast, if we are interested in a huge instance of this model, e.g., setting $k = 10^{15}$, already storing the resulting MDP in the memory in order to compute an optimal policy is challenging or even infeasible for a large enough k. Additionally, the policy produced by state-of-the-art model checkers is represented as a lookup table with as many rows as there are states.

3.4   Applying and Evaluating the Resulting Policy

Once we have a decision-tree representation of a policy, we can apply it to MDP instances of arbitrary size. To evaluate a policy, we simply need to compute the value of the MC induced by applying the DT. Since solving MCs is computationally easier than solving MDPs, we can explicitly compute values for larger MDPs (which we could not do otherwise). Nonetheless, one can still scale the parameter to such an extent that the construction of the corresponding MC requires too much time or memory. In such cases, we can use statistical model checking (SMC) [52], whose runtime and memory used are independent of the size of the system.

---

[2] Axis-aligned predicates are of the form $x > c$ where x is a state variable and $c \in \mathbb{R}$. One can also consider DTs with richer predicates in the decision nodes.

The resulting value is not only a measure for the performance of the DT policy, but also a guaranteed lower bound on the value of the MDP (or an upper bound in the case of minimization).

Since our approach is based on generalization, the learned DT may, in principle, recommend an action that is not available in that state. In such cases our implementation would choose an action uniformly from the available actions. While this may occur in principle, we have not encountered this situation in any of the experiments we conducted for evaluation.

4   Evaluation

4.1   Experiment Setup

Benchmark Selection. We selected parameterized MDPs with reachability objective from the quantitative verification benchmark set (QVBS) [25]. Models with reward-bounded reachability (*e.g.*, eajs and resource-gathering) were excluded. We also identified *trivial* model and property combinations where the equation $\min_\sigma \mathbb{P}^\sigma_{\bar{s}}[\lozenge G] = \max_\sigma \mathbb{P}^\sigma_{\bar{s}}[\lozenge G]$ holds for the set of goal states G and the initial state $\bar{s}$. In such cases, any valid policy would act as an optimal policy. We have excluded these from our benchmark set. We extended the benchmark set with the Mars Exploration Rovers (mer) case study, which was introduced in [17] and appears frequently in recent literature. This model is interesting because the probability of its property is non-trivial and it is scalable to large parameter values without degenerating into a trivial model.

Choice of Base Instances. We conducted experiments to observe the effect of the set of base instances on the value produced by the learned policy. We synthesized decision trees from different sets of base instances, increasing the parameter(s) linearly as well as exponentially, and evaluated them on models larger than the base instances. We observed that one or two instances are often already enough to generalize the policy in the considered benchmark set (See Appendix A of the extended version [4] for the chosen set of base instances used in our experiments).

System Configuration. The experiments were executed on an AMD EPYC^{TM} 7443 server with 48 physical cores, 192 GB RAM, running Ubuntu 22.04.2 LTS operating system with Linux kernel version 5.15.0-83-generic. This powerful server was used to execute many runs in parallel. We assigned 2 cores and 8 GB RAM to each run. For all experiments, we used BenchExec [7], a state of the art benchmarking tool, to isolate the executions and enforce the resource limitations.

Implementation Details. We implemented our approach as an extension of the probabilistic model checker Storm [28].

Method of Comparison. Our aim is to provide a method for policy synthesis for arbitrarily large instances of parameterized MDPs, in particular for MDPs beyond the reach of any available rigorous analysis. Consequently, the optimal value for such an MDP is by definition *unknown* and **optimality becomes not only uncheckable, but also unexpectable**—rather, one can hope for values close to the range where the unknown optimum is expected to lie.

Hence a straightforward evaluation is thus beyond reach, and we devise the following ancillary evaluation process. First, we also compare on *small* benchmarks, although our approach is by no means meant as a competitor of STORM on them. Nonetheless, it gives us the following two details: (i) optimal values for various parameter instantiations, often allowing for a simple *extrapolation*, and (ii) our relative error for these various parameter instantiations, allowing for another extrapolation. *While the performance on small models is irrelevant (of course, exact methods are to be used when feasible), the resulting extrapolations give us some idea how our approach performs in the area of interest.* In addition to comparing to the theoretical optimum (obtained by extrapolation), we compare to *SMC*, which is the key state-of-the-art technique for too large systems, and to randomly chosen policies as a baseline.

Technical Description. First, to obtain *optimal* values for each of the MDP instances, we executed all the engines of STORM (sparse, dd, hybrid, dd-to-sparse). We executed each run with a CPU time limit of 1 h and memory limit of 8GB. We considered the CPU time taken by the fastest engine for each instance.

Second, we obtain values by using the state-of-the-art statistical model checker MODES [11], part of the MODEST toolset [23]. The approaches for picking the policies are (i) smart lightweight scheduler sampling (Smart LSS) [15], executed in the default configuration, producing the policy value with confidence bound 0.99 and error bound 0.01; (ii) the uniform policy, which resolves each non-deterministic choice by picking an action uniformly at random, again with confidence bound 0.99 and error bound 0.01; and (iii) an aggregate of 1000 randomly generated deterministic policies, i.e., non-randomizing policies where each non-deterministic choice is resolved by a single action, sampled independently according to the uniform distribution, each evaluated with 1000 simulation runs.

Finally, we evaluate our approach by computing the value of the MCs resulting from applying our generalizing DTs. In most of the cases (except 4), we were able to evaluate our policy precisely. In the 4 remaining cases, we used our own implementation of SMC to evaluate the learned DT. For 3 out of these 4, we were able to produce a value with with confidence bound 0.99 and error bound 0.01, and in the remaining one (csma+some_before, $N = 8$) we had to use the confidence bound 0.95 and error bound 0.05. The key idea of the evaluation is to show how the values (optimal / for our approach / for different random schedulers) evolve with the parameter.

We executed all the tools on the MDPs obtained by scaling the value of the parameter. In case of MDPs with a single parameter, we start with the smallest parameter values suggested in the QVBS and then increase it. In cases where there was more than one parameter, we scaled each parameter while fixing the values of the other ones. The values chosen to be fixed were the smallest values for these parameters taken from the QVBS website. Since we could not run experiments for all the parameter values due to resource constraints, we sampled the parameters. (Please see Table 5 in Appendix B of the extended version [4] for the concrete parameter values used in our experiments.)

We present the results for the parameters that STORM was able to solve within a minute of CPU time, within one hour of CPU time, and an instance that even STORM was not able to solve within an hour of CPU time.

Sometimes, parameter scaling does not increase the time required to solve the given MDP. In such cases, we still present several parameter valuations and the corresponding values to assess how the values evolve.

4.2   Results

Table 1 and Table 2 show the results of our evaluation for the minimizing and maximizing properties, respectively. Each *instance* refers to a combination of model, property and parameter. The tables show, for each model+property combination, the parameter value and CPU time taken by STORM to solve it (<1 min, <60 min, and Beyond in case STORM could not solve the instance in an hour), the values produced by STORM, our approach and the sampling based SMC. The tables report OOR when running out of resource (time or memory). Also, a few MODES runs resulted in a *run length exceeded* (RLE) error.

Some models converge to triviality (*i.e.*, max = min) as we scale the parameter. *Zeroconf_dl+deadline_min* becomes trivial for higher values of the parameter K than 3, *firewire+deadline* becomes trivial for deadline > 1300, and *csma+some_before* becomes trivial when the value of K is more than twice the value of N. Pacman also approaches closer to triviality for higher values of the parameter MAX_STEPS (the horizon).

The results show that our approach gives near optimal values for 13 out of 21 cases (the upper halves of Table 1 and Table 2), better than Smart LSS for 2 out of remaining 8 cases, and generally better than random and uniform in remaining cases. There are two instances where random performs better than our approach (pacman for the MAX_STEPS 25, and csma+all_before_max for $N = 3$), see the discussion below.

4.3   Discussion

Although our approach is simple, it performs well in a number of cases. We often generalize from a single instance or two, yielding satisfactory solutions for arbitrarily large instantiations. In a number of cases, we can justifiably extrapolate that the policies are (nearly) optimal for all instances. For instance, consider the two benchmarks of Fig. 3. No matter how much the model is scaled up, the value of our policy seems to remain stable. While its (near-)optimality can be proven only up to a certain point (beyond which no ground truth can be known), the apparent stability suggests it is true onwards, too. Note that MODES returns low values as the optimal policies are rather rare.

In the sequel, we discuss the scope and the limitations of our approach in details. As discussed earlier, we can divide the parameters in two types.

Table 1. The results table for *minimizing* properties. MODES sometimes gives smaller value than optimal value (marked by ♯) as it uses SMC which does not report an exact value, but reports an approximate value (with 0.01 error bound) with high (99%) confidence. For 1-2-3-Go, the values marked by † were approximated using SMC. We shorten the parameters delay to d, deadline to dl, and MAX_STEPS to MS. OOR means out-of-resources (both time and memory) and RLE means *run length exceeded*.

Model+property (values of parameters)	Scale		Values				
					MODES		
	Variable	Time	STORM	1 2 3-Go	Smart LSS	Uniform	Random
zeroconf_dl+deadline_min (N = 1000, K = 1)	dl = 200	<1 min	5.02×10^{-207}	2.62×10^{-43}	$0^{\sharp}$	0.00	0.00
	dl = 1600	<60 min	0	6.81×10^{-86}	0	0.00	0.00
	dl = 3200	Beyond	OOR	8.71×10^{-135}	0	0	0.00
zeroconf_dl+deadline_min (N = 1000, deadline = 10)	K = 2	<1 min	0.34	0.34	0.34	$0.34^{\sharp}$	0.34
	K = 8	<1 min	1	1	1	1	1
firewire+deadline (deadline = 200)	delay = 5	<1 min	0.50	0.50	0.56	0.99	0.99
	delay = 21	<1 min	0.50	0.50	1	1	1
	delay = 34	<1 min	0	0	1	1	1
	delay = 89	<1 min	0	0	1	1	1
firewire+deadline (delay = 3)	dl = 200	<1 min	0.50	0.50	$0.50^{\sharp}$	0.98	0.97
	dl = 500	<1 min	0.85	0.85	1	1	1
	dl = 1300	<1 min	1.00	1.00	1	1	1
csma+some_before (N = 2)	K = 2	<1 min	0.50	0.50	$0.50^{\sharp}$	$0.50^{\sharp}$	0.50
	K = 3	<1 min	0.88	0.88	0.88	$0.87^{\sharp}$	$0.87^{\sharp}$
csma+some_before (fix K = 2)	N = 3	<1 min	0.59	0.59	$0.58^{\sharp}$	0.89	0.90
	N = 5	<60 min	0.21	0.21	0.39	0.79	0.78
	N = 8	<60 min	0.04	0.03 †	0.51	0.75	0.76
	N = 13	Beyond	OOR	OOR	0.56	0.75	0.76
consensus+c2 (N = 2)	K = 2	<1 min	0.38	0.47	0.42	0.49	0.49
	K = 55	<1 min	0.49	0.50	RLE	RLE	RLE
	K = 144	<1 min	0.49	0.50	RLE	RLE	RLE
consensus+c2 (K = 2)	N = 6	<1 min	0.29	0.45	0.49	0.48	0.48
	N = 7	<60 min	0.29	0.46	0.48	0.49	0.48
	N = 13	Beyond	OOR	0.46	RLE	0.49	RLE
zeroconf_dl+deadline_min (deadline = 10, K = 1)	N = 1000	<1 min	0.00	0.00	0.00	$0.01^{\sharp}$	0.01
	N = 8000	<1 min	0.01	0.04	0.04	0.07	0.06
	N = 32000	<1 min	0.11	0.22	0.21	0.35	0.32
pacman+crash	MS = 5	<1 min	0.55	0.55	$0.55^{\sharp}$	0.55	$0.55^{\sharp}$
	MS = 25	<1 min	0.55	0.87	0.73	0.93	0.92
	MS = 200	<60 min	0.55	1.00	1	1	1
	MS = 300	Beyond	OOR	1.00	1	1	1

Type 1: Parameters that dictate the number of PRISM-modules. This type of parameter not only changes the structure of the MDP, but also increases the number of state variables. Note that when we train a DT from the policies from smaller base instances, the predicates in decision tree would not use the state variables present only in the bigger instances.

Table 2. The results table for *maximizing* properties. The values marked by † were approximated using SMC. We shorten the parameter deadline to *dl*. OOR means out-of-resources (both time and memory) and RLE means *run length exceeded*.

Model+property (values of parameters)	Scale Variable	Time	STORM	1-2-3-Go	Smart LSS	Uniform	Random
zeroconf_dl+deadline_max (N = 1000, K = 1)	dl = 200	<1 min	0.01	0.01	0.00	0.00	0.00
	dl = 1600	<60 min	0.01	0.01	0.00	0.00	0.00
	dl = 3200	Beyond	OOR	0.01	0.00	0	0.00
zeroconf_dl+deadline_max (N = 1000, deadline = 10)	K = 2	<1 min	0.34	0.34	0.34	0.34	0.34
	K = 8	<1 min	1	1	1	1	1
	K = 32	<1 min	1	1	1	1	1
zeroconf_dl+deadline_max (deadline = 10, K = 1)	N = 1000	<1 min	0.02	0.02	0.01	0.01	0.01
	N = 8000	<1 min	0.12	0.12	0.10	0.07	0.06
	N = 32000	<1 min	0.49	0.49	0.47	0.35	0.32
philosophers+eat	N = 5	<1 min	1	1	1	1	0.04
	N = 21	<60 min	1	1	0.51	1	0.00
	N = 34	Beyond	OOR	1	0.49	1	0
pnueli-zuck+live	N = 5	<1 min	1	1	1	1	0.04
	N = 21	<60 min	1	1	1	1	0.00
	N = 34	Beyond	OOR	1	1	1	0.00
csma+all_before_max (N = 2)	K = 8	<1 min	1	1	1	1	1
	K = 11	<60 min	1	1 †	1	1	1
	K = 13	Beyond	OOR	1 †	1	1	OOR
mer+p1 (x = 0.01)	n = 1000	<1 min	0.20	0.20	RLE	RLE	0.00
	n = 21000	<60 min	0.20	0.20	RLE	RLE	0
	n = 55000	Beyond	OOR	0.20	RLE	RLE	0.00
mer+p1 (n = 10)	x = 0.01	<1 min	0.20	0.20	RLE	RLE	0.00
	x = 0.08	<1 min	0.21	0.20	RLE	RLE	0.00
	x = 0.55	<1 min	0.36	0.20	RLE	RLE	0.00
	x = 0.89	<1 min	0.67	0.20	RLE	RLE	0.00
consensus+disagree (K = 2)	N = 5	<1 min	0.34	0.10	0.05	0.03	0.04
	N = 7	<60 min	0.38	0.07	0.03	0.04	0.03
	N = 13	Beyond	OOR	0.05	RLE	0.03	RLE
consensus+disagree (N = 2)	K = 2	<1 min	0.11	0.06	0.08	0.03	0.03
	K = 55	<1 min	0.00	0.00	RLE	RLE	RLE
	K = 144	<1 min	0.00	0.00	RLE	RLE	RLE
csma+all_before_max (K = 2)	N = 3	<1 min	0.86	0.52	0.85	0.03	0.68
	N = 5	<60 min	0.70	0.05	0.41	0.04	0.24
	N = 6	Beyond	OOR	0.01 †	0.21	0.04	0.11

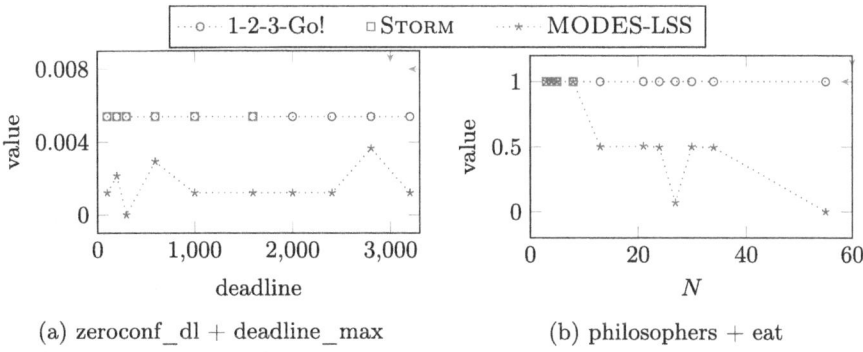

(a) zeroconf_dl + deadline_max   (b) philosophers + eat

Fig. 3. Robustness of the policy given by 1-2-3-Go!, STORM and MODES for different instances of zeroconf_dl + deadline_max and philosophers + eat. For larger models, while STORM times out and MODES gives sub-optimal policies, 1-2-3-Go! consistently gives better results irrespective of N.

Even then, as the system is a product of these isomorphic modules, we can think the smaller base instances as *projections* of the larger instance to the first few state variables. Then, for some interesting properties, our method still gives good policies: cases where there is a generalizing optimal policy that does not depend on the additional modules.

For example, consider the *csma* model describing the CSMA/CD consensus protocol when N stations use a network with a single channel. Each station is represented by a module in the PRISM file. Now consider two different properties "all_before_max" and "some_before". The first one checks the maximum probability that all stations send the message successfully avoiding a data collision. A policy maximizing successful transmission for all stations, needs to take account of the state variables corresponding to all modules in the PRISM file. Our approach fails with this kind of property. But on the other hand, the second property checks the minimum probability that some station eventually sends the message successfully. Thus, a decision tree generated from the base instances gives a policy that minimizes the collision probability for some station among the first three stations. This would act as an optimal policy (as it optimizes the collision probability for some station) in the larger instances even though the predicates in the DT does not contain state variables related to stations with larger index. Thus, for this property, our approach succeeds in generalizing the optimal policy.

Type 2: Other parameters which can be changed by setting the value externally. Often, this would mean expanding the domain of the state variables (as in the case of Example 1), which increases the size of the state space linearly. Then our approach can still work as we can have an optimal policy that is independent of that specific state variable, or if the newly added states are not relevant.

However, this is not always the case: The parameter MAX_STEPS in *pacman + crash* denotes the number of steps Pac-Man needs to stay safe from the ghosts.

Our algorithm fails for this model as the policies across instances cannot be generalized. Indeed, a policy that minimized the probability of crash for K steps does not provide any information about how to stay safe for $K' > K$ steps.

In the case of *mer + p1*, if we fix n and vary the parameter x, our approach fails to generalize a policy. Changing the value of x does not change the state-space or the structure of the model, but changes the probability values of the transitions, which in turn would change the optimal policy across different instances. For that reason, we cannot construct a generalizing DT as the decision predicates are defined on the state variables and not on the probability value of the transition.

Time. The approach typically needs less than 5s to generate a policy, except for zeroconf_dl (see Table 8 and Table 9 in Appendix D of the extended version [4]). In the case of zeroconf_dl, the time taken is a couple of minutes because we learned from a larger instance, with non-trivial values for all three involved parameters. However, this instance is then so informative that we could later use it for all instances derived by varying different parameters (i.e., varying N and fixing others, varying delay and fixing other, and varying deadline and fixing others, not having to take care of these three families separately).

5   Conclusion

We have seen that *practically good* policies (with values close to the unknown optimum in the sense of the "Method of Comparison" above) can be generated in a lightweight way even for very large parameterized models, beyond reach of any other methods. In order to synthesize policies for arbitrarily large models, we generalize the policies computed for the smaller instances using (more explainable and thus more generalizable) decision trees, coining the "generalizability by explainability".

The generalization is an example of *unreliable* reasoning, which can contribute to better scalability. On the one hand, the unreliability results in *no guarantees* that the produced policies are anywhere close to optimum, which, however, often cannot be computed anyway. On the other hand, the values of the policies can be *reliably* approximated: either numerically with absolute guarantees if the resulting Markov chain is still analyzable (e.g., with partial-exploration methods [33]) or with statistical guarantees by SMC on the Markov chain. Consequently, although optimal control policies might be out of reach, we can still produce what we thus coin here as *provably good enough* policies. Moreover, the consistency of the values over the different instantiations often suggests practical proximity to optimum.

A possibly surprising point is the conclusion of our experiments that very few base instances need to be analyzed. Such *robustness* (together with the robustness across the target instances as seen in Fig. 3) suggests that this generalizability is a deeply inherent property of many models, and thus deserves further investigation and exploitation. In particular, our approach is only the first, generic try to exploit this property, opening the new paradigm. As suggested by our experimental results, more *specific heuristics* for certain types of systems where parameters play different roles, such as number of modules, number of repetitions, time-outs, etc., offer a desirable direction of future work.

References

1. Ashok, P., Jackermeier, M., Jagtap, P., Kretínský, J., Weininger, M., Zamani, M.: dtControl: decision tree learning algorithms for controller representation. In: Ames, A.D., Seshia, S.A., Deshmukh, J. (eds.) HSCC '20: 23rd ACM International Conference on Hybrid Systems: Computation and Control, Sydney, New South Wales, Australia, April 21–24, 2020, pp. 30:1–30:2. ACM (2020). https://doi.org/10.1145/3365365.3383468

2. Ashok, P., Jackermeier, M., Kretínský, J., Weinhuber, C., Weininger, M., Yadav, M.: dtControl 2.0: explainable strategy representation via decision tree learning steered by experts. In: TACAS (2). Lecture Notes in Computer Science, vol. 12652, pp. 326–345. Springer (2021)

3. Ashok, P., Křetínský, J., Weininger, M.: PAC statistical model checking for Markov decision processes and stochastic games. In: Dillig, I., Tasiran, S. (eds.) CAV 2019. LNCS, vol. 11561, pp. 497–519. Springer, Cham (2019). https://doi.org/10.1007/978-3-030-25540-4_29

4. Azeem, M., Chakraborty, D., Kanav, S., Kretínský, J., Mohagheghi, M., Mohr, S., Weininger, M.: 1-2-3-Go! policy synthesis for parameterized Markov decision processes via decision-tree learning and generalization. CoRR abs/2410.18293 (2024). https://arxiv.org/abs/2410.18293

5. Baier, C., Katoen, J.: Principles of Model Checking. MIT Press (2008)

6. Baier, C., Klein, J., Leuschner, L., Parker, D., Wunderlich, S.: Ensuring the reliability of your model checker: interval iteration for Markov decision processes. In: Majumdar, R., Kunčak, V. (eds.) CAV 2017. LNCS, vol. 10426, pp. 160–180. Springer, Cham (2017). https://doi.org/10.1007/978-3-319-63387-9_8

7. Beyer, D., Löwe, S., Wendler, P.: Reliable benchmarking: requirements and solutions. Int. J. Softw. Tools Technol. Transf. **21**(1), 1–29 (2019). https://doi.org/10.1007/s10009-017-0469-y

8. Brázdil, T., Chatterjee, K., Chmelík, M., Fellner, A., Křetínský, J.: Counterexample explanation by learning small strategies in Markov decision processes. In: Kroening, D., Păsăreanu, C.S. (eds.) CAV 2015. LNCS, vol. 9206, pp. 158–177. Springer, Cham (2015). https://doi.org/10.1007/978-3-319-21690-4_10

9. Brázdil, T., et al.: Verification of Markov decision processes using learning algorithms. In: Cassez, F., Raskin, J.-F. (eds.) ATVA 2014. LNCS, vol. 8837, pp. 98–114. Springer, Cham (2014). https://doi.org/10.1007/978-3-319-11936-6_8

10. Breiman, L.: Classification and Regression Trees. The Wadsworth statistics / probability series), Wadsworth International Group (1984)

11. Budde, C.E., D'Argenio, P.R., Hartmanns, A., Sedwards, S.: A statistical model checker for nondeterminism and rare events. In: Beyer, D., Huisman, M. (eds.) TACAS 2018. LNCS, vol. 10806, pp. 340–358. Springer, Cham (2018). https://doi.org/10.1007/978-3-319-89963-3_20

12. Budde, C.E., et al.: On correctness, precision, and performance in quantitative verification - QComp 2020 competition report. In: ISoLA (4). Lecture Notes in Computer Science, vol. 12479, pp. 216–241. Springer (2020)

13. Ciesinski, F., Baier, C., Größer, M., Klein, J.: Reduction techniques for model checking markov decision processes. In: QEST, pp. 45–54. IEEE Computer Society (2008)
14. Curtin, R.R., Edel, M., Lozhnikov, M., Mentekidis, Y., Ghaisas, S., Zhang, S.: mlpack 3: a fast, flexible machine learning library. J. Open Source Softw. **3**(26), 726 (2018)
15. D'Argenio, P.R., Legay, A., Sedwards, S., Traonouez, L.: Smart sampling for lightweight verification of Markov decision processes. Int. J. Softw. Tools Technol. Transf. **17**(4), 469–484 (2015)
16. Feng, L.: On learning assumptions for compositional verification of probabilistic systems. Ph.D. thesis, University of Oxford, UK (2014)
17. Feng, L., Kwiatkowska, M., Parker, D.: Automated learning of probabilistic assumptions for compositional reasoning. In: Giannakopoulou, D., Orejas, F. (eds.) Fundamental Approaches to Software Engineering, pp. 2–17. Springer, Heidelberg (2011)
18. Groote, J.F., Verduzco, J.R., de Vink, E.P.: An efficient algorithm to determine probabilistic bisimulation. Algorithms **11**(9), 131 (2018)
19. Gros, T.P., Hermanns, H., Hoffmann, J., Klauck, M., Steinmetz, M.: Deep statistical model checking. In: Gotsman, A., Sokolova, A. (eds.) FORTE 2020. LNCS, vol. 12136, pp. 96–114. Springer, Cham (2020). https://doi.org/10.1007/978-3-030-50086-3_6
20. Größer, M., Baier, C.: Partial order reduction for Markov decision processes: a survey. In: FMCO. Lecture Notes in Computer Science, vol. 4111, pp. 408–427. Springer (2005)
21. Haddad, S., Monmege, B.: Reachability in MDPs: refining convergence of value iteration. In: Ouaknine, J., Potapov, I., Worrell, J. (eds.) RP 2014. LNCS, vol. 8762, pp. 125–137. Springer, Cham (2014). https://doi.org/10.1007/978-3-319-11439-2_10
22. Hahn, E.M., Perez, M., Schewe, S., Somenzi, F., Trivedi, A., Wojtczak, D.: Omega-regular objectives in model-free reinforcement learning. In: Vojnar, T., Zhang, L. (eds.) TACAS 2019. LNCS, vol. 11427, pp. 395–412. Springer, Cham (2019). https://doi.org/10.1007/978-3-030-17462-0_27
23. Hartmanns, A.: MODEST - A unified language for quantitative models. In: FDL, pp. 44–51. IEEE (2012). https://ieeexplore.ieee.org/document/6336982/
24. Hartmanns, A., Hermanns, H.: Explicit model checking of very large MDP using partitioning and secondary storage. In: Finkbeiner, B., Pu, G., Zhang, L. (eds.) Automated Technology for Verification and Analysis - 13th International Symposium, ATVA 2015, Shanghai, China, October 12–15, 2015, Proceedings. Lecture Notes in Computer Science, vol. 9364, pp. 131–147. Springer (2015). https://doi.org/10.1007/978-3-319-24953-7_10
25. Hartmanns, A., Klauck, M., Parker, D., Quatmann, T., Ruijters, E.: The quantitative verification benchmark set. In: TACAS (1). Lecture Notes in Computer Science, vol. 11427, pp. 344–350. Springer (2019).https://doi.org/10.1007/978-3-030-17462-0_20
26. Hartmanns, A., Timmer, M.: Sound statistical model checking for MDP using partial order and confluence reduction. Int. J. Softw. Tools Technol. Transf. **17**(4), 429–456 (2015)
27. Henriques, D., Martins, J.G., Zuliani, P., Platzer, A., Clarke, E.M.: Statistical model checking for Markov decision processes. In: QEST. pp. 84–93. IEEE Computer Society (2012)

28. Hensel, C., Junges, S., Katoen, J., Quatmann, T., Volk, M.: The probabilistic model checker storm. Int. J. Softw. Tools Technol. Transf. **24**(4), 589–610 (2022). https://doi.org/10.1007/s10009-021-00633-z
29. Hyafil, L., Rivest, R.L.: Constructing optimal binary decision trees is np-complete. Inf. Process. Lett. **5**(1), 15–17 (1976). https://doi.org/10.1016/0020-0190(76)90095-8
30. Kamaleson, N.: Model reduction techniques for probabilistic verification of Markov chains. Ph.D. thesis, University of Birmingham, UK (2018)
31. Klein, J., et al.: Advances in probabilistic model checking with PRISM: variable reordering, quantiles and weak deterministic büchi automata. Int. J. Softw. Tools Technol. Transf. **20**(2), 179–194 (2018)
32. Křetínský, J.: Survey of statistical verification of linear unbounded properties: model checking and distances. In: Margaria, T., Steffen, B. (eds.) ISoLA 2016, Part I. LNCS, vol. 9952, pp. 27–45. Springer, Cham (2016). https://doi.org/10.1007/978-3-319-47166-2_3
33. Kretínský, J., Meggendorfer, T.: Of cores: a partial-exploration framework for Markov decision processes. Log. Methods Comput. Sci. **16**(4) (2020)
34. Kwiatkowska, M., Norman, G., Parker, D.: PRISM: probabilistic symbolic model checker. In: Field, T., Harrison, P.G., Bradley, J., Harder, U. (eds.) TOOLS 2002. LNCS, vol. 2324, pp. 200–204. Springer, Heidelberg (2002). https://doi.org/10.1007/3-540-46029-2_13
35. Kwiatkowska, M.Z., Norman, G., Parker, D.: Game-based abstraction for Markov decision processes. In: QEST, pp. 157–166. IEEE Computer Society (2006)
36. Kwiatkowska, M., Norman, G., Parker, D.: Symmetry reduction for probabilistic model checking. In: Ball, T., Jones, R.B. (eds.) CAV 2006. LNCS, vol. 4144, pp. 234–248. Springer, Heidelberg (2006). https://doi.org/10.1007/11817963_23
37. Kwiatkowska, M.Z., Norman, G., Parker, D.: The PRISM benchmark suite. In: QEST, pp. 203–204. IEEE Computer Society (2012). https://doi.org/10.1109/QEST.2012.14
38. Kwiatkowska, M.Z., Parker, D., Qu, H.: Incremental quantitative verification for markov decision processes. In: DSN, pp. 359–370. IEEE Compute Society (2011)
39. Li, R., Liu, Y.: Compositional stochastic model checking probabilistic automata via symmetric assume-guarantee rule. In: 2019 IEEE 17th International Conference on Software Engineering Research, Management and Applications (SERA), pp. 110–115. IEEE (2019)
40. Lomuscio, A., Pirovano, E.: A counter abstraction technique for the verification of probabilistic swarm systems. In: AAMAS, pp. 161–169. International Foundation for Autonomous Agents and Multiagent Systems (2019)
41. Maisonneuve, V.: Automatic heuristic-based generation of MTBDD variable orderings for prism models. internship report (2009)
42. Mitchell, T.: Machine Learning, vol. 1. McGraw-Hill, New York (1997)
43. Mohagheghi, M., Salehi, K.: Machine learning and disk-based methods for qualitative verification of Markov decision processes. In: ICTERI Workshops. CEUR Workshop Proceedings, vol. 2732, pp. 74–88. CEUR-WS.org (2020)
44. Parker, D.A.: Implementation of symbolic model checking for probabilistic systems. Ph.D. thesis, University of Birmingham, UK (2003)
45. Puterman, M.L.: Markov Decision Processes: Discrete Stochastic Dynamic Programming. Wiley Series in Probability and Statistics, Wiley (1994). https://doi.org/10.1002/9780470316887
46. Pyeatt, L.D., Howe, A.E.: Decision tree function approximation in reinforcement learning (1999)

47. Rataj, A., Wozna-Szczesniak, B.: Extrapolation of an optimal policy using statistical probabilistic model checking. Fundam. Informaticae **157**(4), 443–461 (2018)
48. Shannon, C.E.: A mathematical theory of communication. The Bell system technical journal **27**(3), 379–423 (1948)
49. Smolka, S., et al.: Scalable verification of probabilistic networks. In: PLDI, pp. 190–203. ACM (2019)
50. Sutton, R.S., Barto, A.G.: Introduction to Reinforcement Learning, 1st edn. Cambridge, MA, USA (1998)
51. Tappler, M., Aichernig, B.K., Bacci, G., Eichlseder, M., Larsen, K.G.: L^*-based learning of Markov decision processes. In: FM. Lecture Notes in Computer Science, vol. 11800, pp. 651–669. Springer (2019)
52. Younes, H.L.S., Simmons, R.G.: Probabilistic verification of discrete event systems using acceptance sampling. In: Brinksma, E., Larsen, K.G. (eds.) CAV 2002. LNCS, vol. 2404, pp. 223–235. Springer, Heidelberg (2002). https://doi.org/10.1007/3-540-45657-0_17

LLOR: Automated Repair of OpenMP Programs

Utpal Bora[2] , Saurabh Joshi[3] , Gautam Muduganti[1]([✉]) ,
and Ramakrishna Upadrasta[1]

[1] Indian Institute of Technology Hyderabad, Hyderabad, India
{cs17resch01003,ramakrishna}@iith.ac.in
[2] University of Cambridge, Cambridge, UK
utpal.bora@cl.cam.ac.uk
[3] Supra Research, Miami, USA
sbjoshi@iith.ac.in

Abstract. In this paper, we present a technique for repairing data race errors in parallel programs written in C/C++ and Fortran using the OpenMP API. Our technique can also remove barriers that are deemed unnecessary for correctness. We implement these ideas in our tool called *LLOR*, which takes a language-independent approach to provide appropriate placements of synchronization constructs to avoid data races. To the best of our knowledge, *LLOR* is the only tool that can repair parallel programs that use the OpenMP API. We showcase the capabilities of *LLOR* by performing extensive experiments on 415 parallel programs.

Keywords: OpenMP · Verification · Automated Repair · C · C++ · Fortran

1 Introduction

Programs that can solve problems using multiple threads in parallel are defined as parallel programs. Parallelism can be achieved through concurrent computing using threads on CPUs or GPUs. There are two common approaches for achieving parallelism: Multiple Instruction, Multiple Data (MIMD) and Single Instruction, Multiple Data (SIMD). The OpenMP API [18] provides a cross-platform abstraction for programs written in C/C++ and Fortran for both of these approaches.

A data race occurs in a parallel program when two or more threads are accessing the same memory location at the same time, and at least one of those accesses is a write. The behavior of such programs is unpredictable. Data races are one of the frequently encountered issues in concurrent computing. There is a substantial positive financial impact in identifying and repairing these errors early in the development cycle [10].

The author names are in alphabetical order.

K. Shankaranarayanan et al. (Eds.): VMCAI 2025, LNCS 15530, pp. 121–136, 2025.
https://doi.org/10.1007/978-3-031-82703-7_6

Listing 1.1 without the highlighted line illustrates an OpenMP program that contains a data race. The data race exists because of the parallel reading and writing of the shared array `data`. Placing a barrier (`#pragma omp barrier`) in the program at line 5 mitigates the data race by ensuring that all the threads reach it before any of them can proceed further.

Listing 1.2 presents another OpenMP program that has a similar data race. In this example, the data race is because of the `parallel for` loop. OpenMP does not support barrier constructs inside a `parallel for` loop. The data race can be avoided by adding the ordered clause to the `for` loop (line 2) and placing the statements causing the data race inside an ordered region (`#pragma omp ordered`), as shown in line 5. Note that there can be multiple statements in this ordered region.

```
1   int data[NUM_THREADS+1];
2   #pragma omp parallel {
3        int id = omp_get_thread_num();
4        int temp = data[id+1];
5        #pragma omp barrier
6        data[id] = temp;
7   }
```

Listing 1.1. This OpenMP program without the highlighted line contains a data race

```
1   int data[count+1];
2   #pragma omp parallel for ordered
3   for (int i=0; i<count; i++) {
4        int temp = data[i+1];
5        #pragma omp ordered
6        data[i] = temp;
7   }
```

Listing 1.2. This OpenMP program without the highlighted snippets contains a data race inside the `for` loop

In this tool paper, we make the following contributions:

– We introduce our tool, *LLOR*, which takes a language-independent approach to automatically fix data race errors in programs written using the OpenMP API.
– For OpenMP programs that contain **parallel** regions, our paper proposes a technique to provide barrier placements to avoid data races.
– For OpenMP programs that contain **parallel for** loops, our paper proposes a technique to identify the statements that need to be placed in an ordered region to avoid data races.
– In addition to the above two techniques that introduce synchronization constructs to avoid data races, our paper proposes a methodology to remove existing barriers and ordered regions inserted by the programmer if deemed unnecessary.

- Our paper showcases the practical differences between the two different solver strategies (*mhs* and *MaxSAT*) that are employed during the repair process.
- We showcase the effectiveness of our tool by performing extensive experimental evaluation on 415 parallel programs that use the OpenMP API. This benchmark set consists of 235 C/C++ programs and 180 Fortran programs, highlighting the language versatility of *LLOR*.

To the best of our knowledge, ours is the only technique and tool that can propose a fix for parallel programs written using the OpenMP API. In Sect. 2, we describe the working of *LLOR* in fine detail, discuss related work in Sect. 3, and in Sect. 4, we provide a detailed background on the experimental setup and present the results summary of running *LLOR* against the benchmark suite.

2   LLOR

2.1   LLOR Architecture and Workflow

The implementation of *LLOR* leverages *LLOV* [11], a state-of-the-art data race checker, as depicted in Fig. 1. *LLOV* identifies data races in C/C++ and Fortran programs. It is built as an analysis pass using the *LLVM* compiler infrastructure. The architecture of *LLOR* consists of two components: Instrumentation and Repair.

The instrumentation component adds metadata that indicates the possible locations of barriers in the case of `parallel` regions. In the case of `parallel for` loops, this metadata marks the possible statements that have to be a part of an ordered region.

The repair component takes an iterative approach. In each iteration, *LLOV* is called with a repair candidate to check if the program is error-free. If *LLOV* identifies any data races, constraints are generated from these errors to avoid them in subsequent iterations. The *Solver* is called with these constraints to obtain a solution that determines which synchronization constructs have to be enabled or disabled while generating the next repair candidate.

If *LLOR* can find an error-free repair candidate, it generates the LLVM intermediate representation (LLVM IR) [34] of the repaired program and a summary file. The summary file contains the necessary changes to repair the program, along with the source location details of the original C/C++ or Fortran input program. In this paper, we have used *LLOV* [11] as the verifier, but in principle, any OpenMP verifier [2,4,16,21,51,54] can be used for this technique to work.

2.2   Instrumentation

Since *LLOR* attempts to fix errors caused only due to data races, it proposes a solution that involves either adding new barriers in a `parallel` region or creating an ordered region for a subset of instructions inside a `parallel for` loop. It also tries to remove unnecessary barriers and ordered regions. Data races can only be caused when multiple threads read and write the same shared variable. Taking

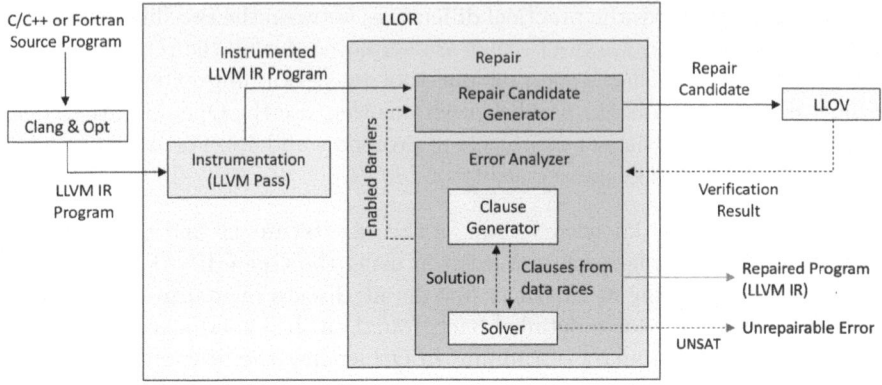

Fig. 1. Architecture of *LLOR* showcasing the various components involved in the repair process. The solid lines represent the source code, and the dashed lines represent the information flow between the components.

this into account, the instrumentation component identifies all the instructions that are either reading or writing a shared variable.

If these instructions are in a `parallel` region, *LLOR* generates possible repair candidates by inserting barriers in front of these instructions. If these instructions are in a `parallel for` loop, *LLOR* generates possible repair candidates by selecting the smallest subset of instructions that need to be placed inside an ordered region. Existing barriers and ordered regions in the program are removed to see if they are indeed required for program correctness. Using this technique significantly reduces the search space of repair candidates since the number of instructions that involve a shared variable is typically much fewer than the total number of instructions in the program.

Consider the OpenMP program presented in Listing 1.1. The instrumentation component identifies that the shared variable `data` is being accessed at Line 4 and Line 6 and adds metadata indicating that a barrier might be necessary before these instructions. Similarly, in Listing 1.2, it identifies that the shared variable `data` is being accessed at Line 4 and Line 6 and that these instructions are inside a for loop. Metadata indicating that these instructions might have to be placed inside an ordered region is added in this case. Each of these instructions is marked with a Boolean variable that indicates whether the barrier should be enabled or not in the case of `parallel` regions. In the case of `parallel for` loops, these variables indicate if the instruction should be placed inside an ordered region. These variables are referred to as *barrier variables*. By default, these barrier variables are set to `false` to check if the program can be concluded as error-free without introducing any additional synchronization constructs.

2.3   The Repair Algorithm

The repair technique used in *LLOR* is depicted in Algorithm 1. The input to this algorithm is the LLVM IR program obtained after instrumentation. We walk through this algorithm using the example presented in Listing 1.3. Without the highlighted portions, this program contains two data races: one caused by memory accesses of data_a at Line 5 and Line 8 and the other caused by memory accesses of data_b at Line 6 and Line 9. The instrumentation component identifies that these four statements either read or write a shared variable. For each statement, a barrier variable is introduced that indicates whether a synchronization construct is needed before the statement.

```
1   int data_a[NUM_THREADS+1];
2   int data_b[NUM_THREADS+1];
3   #pragma omp parallel {
4       int id = omp_get_thread_num();
5       int temp_a = data_a[id+1];    // b1=false
6       int temp_b = data_b[id+1];    // b2=false
7       #pragma omp barrier
8       data_a[id] = temp_a;          // b3=false (true)
9       data_b[id] = temp_b;          // b4=false
10  }
```

Listing 1.3. Instrumented OpenMP program that contains multiple data races in a **parallel** region. The highlighted parts represent the changes in the second iteration.

```
1   int data_a[count+1];
2   int data_b[count+1];
3   #pragma omp parallel for ordered
4   for (int i=0; i<count; i++) {
5       int temp_a = data_a[i+1];     // b1=false
6       int temp_b = data_b[i+1];     // b2=false
7       #pragma omp ordered {
8           data_a[i] = temp_a;       // b3=false (true)
9           data_b[i] = temp_b;       // b4=false
10      }
11  }
```

Listing 1.4. Instrumented OpenMP program that contains multiple data races in a **parallel for** loop. The highlighted parts represent the changes in the second iteration.

The algorithm iteratively calls the verifier (Line 4–21) until it either finds a solution or concludes that the program cannot be repaired. In every iteration, *GenerateRepairCandidate* (Line 10) generates a repair candidate using the error traces seen till then. The first iteration of this algorithm always verifies the input program without enabling any synchronization constructs since all barrier variables are set to false by default. The verifier returns the verification *result* along with an error trace π if it identifies an error in the program. If the verifier is unable to find an error (Line 12) with the proposed solution *sol*, then the

Algorithm 1. The Repair Algorithm

1: Input: Instrumented Program P
2: Output: Repaired Program P_{sol}
3: $\varphi := true$
4: **loop**
5:     $\langle res, sol \rangle := Solve(\varphi)$
6:     **if** $res = UNSAT$ **then**
7:         **print** Error: Program cannot be repaired
8:         **return** errorcode
9:     **end if**
10:    $P_{sol} := GenerateRepairCandidate(P, sol)$
11:    $\langle result, \pi \rangle := Verify(P_{sol})$
12:    **if** $result = SAFE$ **then**
13:        break
14:    **end if**
15:    **if** $result \neq RACE$ **then**
16:        **print** Error: Program cannot be repaired
17:        **return** errorcode
18:    **end if**
19:    $c := GenerateClause(\pi)$
20:    $\varphi := \varphi \cup \{c\}$
21: **end loop**
22: **return** P_{sol}

---

algorithm exits the loop, and the instrumented LLVM IR program constrained with *sol* is returned (Line 22). If the verifier returns an error that is not a data race (Line 15), then Algorithm 1 terminates with an error stating that it cannot repair the program.

The error trace provided by the verifier contains the line numbers of the statements involved in the data race. Introducing a synchronization construct between these lines will mitigate the data race. However, there could be multiple locations between these line numbers where the synchronization construct could be introduced. The algorithm aims to identify the optimum placement of these synchronization constructs since there are performance penalties for every synchronization construct that is introduced.

GenerateClause (Line 19) generates a *positive monotone clause* (a clause having only positive literals) for every data race error using the disabled barrier variables that exist between the line numbers obtained from the error trace. These clauses are added to the constraint φ (Line 20). If φ is satisfiable by *Solve* (Line 5), we proceed to generate the repair candidate. Since φ consists of only positive monotone clauses, it will always be satisfiable as such a formula can be satisfied by an assignment that sets all the literals to true. The only exception is the case when we encounter an empty clause, which is generated when the data race occurs due to a write-write conflict on the same line. In these cases (Line 6), the algorithm terminates, stating that the program cannot be repaired.

Based on the error trace information obtained from the verifier for Listing 1.3, two clauses are generated: $b_2 \lor b_3$ and $b_3 \lor b_4$. There are multiple solutions available for *Solve* to make these clauses satisfiable. Let us consider that *Solve* chooses the optimum solution, which is to enable b_3 and disable the rest. The highlighted portions of Listing 1.3 are then introduced to obtain the repair candidate for the next iteration. The verifier does not identify any data race errors with this repair candidate, so the repair algorithm terminates successfully by returning the repaired program.

In the case of `parallel for` loops, the repair candidate generator does not introduce barriers. Instead, it creates an ordered region that starts with the earliest statement associated with an enabled barrier variable to the last statement that uses a shared variable. Note that there can be only one ordered region per loop. Because of this restriction, our technique may include statements that do not even use a shared variable in the ordered region. Listing 1.4 shows an example of how the repair algorithm works for `parallel for` loops.

There are several ways to obtain *sol* from φ in *Solve* (Line 5). A basic implementation of *Solve* could involve using a *SAT* solver. However, using a *SAT* solver could cause the repair algorithm to enable several unnecessary synchronization constructs since the solver does not guarantee optimality. Instead, we propose two different strategies for implementing *Solve*. The *MaxSAT* strategy implements this as a partial *MaxSAT* problem [23] with φ as hard clauses and $\{\neg b_1, \ldots, \neg b_m\}$ as soft clauses. The *mhs* strategy computes a minimal-hitting-set (*mhs*) over φ using a polynomial time greedy algorithm [27].

3 Related Work

3.1 Verification of Parallel Programs

Identifying data races in parallel programs has been an active area of research for quite a while. Some of the early techniques proposed in this area were based on the *happens-before* relation defined by Lamport [33]. Mellor-Crummey [41] introduced a new protocol that uses Offset-Span Labeling of the nodes in a fork-join graph that represents the concurrency relationships among threads in an execution of a fork-join program.

Eraser [50] tries to address the limitations of the tools built using the happens-before relation by introducing a new Lockset algorithm. In addition to these, several other techniques [20,31,46,47] have been introduced to detect races in parallel programs. Eraser [50], Relay [53], Locksmith [48], and RacerX [22] use the Lockset algorithm to detect races in the pthreads execution model of C/C++ programs. RacerD [8] is a static analysis tool for Java programs. Tools like GPU-Verify [5,6], ESBMC-GPU [43], VerCors [1,9], PUG [35], and GKLEE [36] propose techniques to detect data races in GPU programs.

Tools like ompVerify [4], DRACO [54], PolyOMP [15,16], SWORD [2], OMPT [21], Helgrind [51], and LLOV [11] have been introduced to identify data races in programs using the OpenMP API. Some of these tools [4,11,15,16,54] use only static techniques, while some [2,21,51] employ both static and dynamic techniques. *LLOV* is a static data race detection tool built on top of the *LLVM* [34] compiler infrastructure, which uses *RDG (Reduced Dependence Graph)* to model parallel regions of an LLVM IR program and infer race conditions based on the presence of data dependencies.

3.2 Automatic Program Repair

Using automatic techniques to repair bugs in programs has been another area of research that aligns closely with our work. The research works in this area can be broadly classified into techniques that attempt to repair sequential programs and techniques that attempt to repair concurrent programs.

The approaches that focus on sequential programs [14,24,26,40] take a set of assertions that need to be met and an input program that fails one or more of these assertions. The technique then modifies a small subset of statements in the input program to ensure that all the provided assertions pass.

The approaches that focus on concurrent programs [13,19,25,28–30,45,52] work on an input program that passes all the assertions when run sequentially but has inconsistent behavior when run in parallel using multiple threads. This inconsistent behavior is because of the interleaving of threads. Repairing such programs involves introducing critical regions, locks, and problem-specific synchronization constructs.

In this paper, we instrument the given concurrent input program with the possible locations of the synchronization constructs and generate repair candidates based on the error traces obtained from the verifier. Similar techniques [28–30] have been used in the past to repair concurrent programs. To the best of our knowledge, *LLOR* is the only tool that can fix data races in OpenMP programs written in C/C++ and Fortran.

3.3 Comparison with GPURepair

The counter-example driven repair technique used in this paper is similar to one of our earlier works [30] that focuses on the automated repair of GPU kernels and the implementation of the technique in our tool, *GPURepair*. However, the motivations behind *GPURepair* and *LLOR* are fundamentally different, causing the repair algorithm to differ in a non-trivial fashion.

GPU kernels work on the principle of Single Instruction Multiple Data (SIMD). The threads created in the kernel are organized as blocks. Blocks consist of warps. The threads within a warp generally execute in a lock-step manner. There could be data races within the threads of a block. Data races in GPU kernels are avoided using barrier constructs. However, there might be a scenario where some of the threads in a block reach the barrier, and some do not, resulting in a deadlock. This problem is called barrier divergence. *GPURepair* aims to fix data race and barrier divergence errors in CUDA and OpenCL kernels.

LLOR aims at fixing data races in OpenMP programs that have `omp parallel` regions and `omp parallel for` loops. Since OpenMP programs with these constructs work on the principle of Multiple Instruction Multiple Data (MIMD), the problem of barrier divergence does not exist here. Besides proposing barrier placements for `omp parallel` regions, our technique also identifies the statements that need to be placed within ordered regions for `omp parallel for` loops.

Since *GPURepair* has to repair errors due to data race and barrier divergence errors, the repair algorithm has to evaluate which barriers need to be removed and which barriers need to be added. The algorithm achieves this by generating positive monotone clauses for data race errors and negative monotone clauses for barrier divergence errors. However, in the case of *LLOR*, the repair algorithm generates only positive monotone clauses.

The instrumentation component of *LLOR* instruments the LLVM IR generated from the OpenMP programs. On the other hand, *GPURepair* uses Boogie [3] as an intermediate representation which is generated from GPU kernels. *GPURepair* adds the instrumentation also in Boogie. These differences give rise to an entirely different implementation stack, with *LLOR* using C++ while *GPURepair* uses C#. Another notable difference in *LLOR* is that we do not add conditional barriers but instead add metadata on possible locations of barriers.

4   Experiments

In this section, we present the experimental results of running *LLOR* on various C/C++ and Fortran benchmarks. The source code of *LLOR* is available at [38]. The artifacts used to reproduce the results of this paper are available at [39].

4.1   Experimental Setup

Several tools are involved in the pipeline of *LLOR*, as introduced in Sect. 2.1. The instrumentation component is developed as an LLVM pass and is therefore built in C++. The repair component is outside the LLVM infrastructure and is built using the .NET Framework with C# as the programming language. We use the *Z3* solver [7,44] to solve the constraints generated during the repair phase. The tools used in *LLOR* and their versions are: LLVM 12.0.0, Clang 12.0.0, Flang 12.0.1, and LLOV 0.3.

The experiments were performed on Standard_F2s_v2 Azure® virtual machine, which has 2 vCPUs and 4 GiB of memory. More details on the virtual machine can be found at [42]. A total of 415 programs (235 C/C++ and 180 Fortran) were considered for the evaluation of *LLOR*. This benchmark set consists of programs from the DataRaceBench [37] test suite, Exascale [49] project, Rodinia [17] test suite, and Parallel Research Kernels [32], along with additional benchmarks introduced while developing *LLOR*. Table 1 summarizes the distribution of the benchmark set.

The benchmarks from DataRaceBench and *LLOR* test suites have one source file per benchmark, and the remaining have multiple source files per benchmark.

For benchmarks that have a single source file, the experiments were performed with a timeout of 300 seconds for each benchmark. For benchmarks that have multiple source files, the experiments were performed with a timeout of 60 seconds for each source file and an overall timeout of 1800 seconds for the benchmark. Each benchmark has been executed 3 times, and the average time of these 3 runs is taken into consideration. We used the average since there was a negligible difference between the median and the average.

Table 1. Benchmark Summary

Source	Programs		
	C/C++	Fortran	Total
DataRaceBench	181	168	349
Exascale Project	8	0	8
Rodinia	18	0	18
Parallel Research Kernels	11	0	11
Other Large Benchmarks	5	0	5
LLOR Test Suite	12	12	24

4.2   Results

The results obtained from running *LLOR* against the benchmark suite are summarized in Table 2. The table categorizes the results into three categories based on the output of *LLOV*. The first category includes all the programs for which *LLOV* concluded that there were no errors. On manual inspection, we found out that for some of the Fortran programs (e.g., `baseline_fortran/B01_simple_race.f95`, `dataracebench_fortran/DRB002-anti dep1-var-yes.f95`), *LLOV* was not detecting a data race, even if one existed. Since the soundness and completeness guarantees of *LLOR* are modulo the completeness of the verifier, *LLOR* could not propose a fix for these programs. On the other hand, because of this behavior, *LLOR* suggests removing existing barriers or ordered regions in some Fortran programs even though they are necessary. Note that this behavior of *LLOV* is limited to Fortran programs, and we did not come across any C/C++ programs that were impacted by this.

For 11 programs in this category, *LLOR* recommended changes. 3 of these were Fortran programs (`baseline_fortran/B05_incorrect_barrier.f95`, `baseline_fortran/B07_racefree.f95`, `dataracebench_fortran/DRB110-ordered-orig-no.f95`) that *LLOV* had incorrectly declared as race-free when the barrier or ordered region was removed. For the remaining 8 programs, *LLOR* was correct in suggesting the removal of barriers and ordered regions.

The second category includes the programs for which *LLOV* had identified data races. Out of the 147 programs in this category, *LLOR* was able to fix 107 programs correctly. *LLOR* was unable to repair 16 programs since it could not find any assignment that satisfied the clauses generated during the repair phase. 5 of these were C/C++ programs, and 11 were Fortran programs. Upon manual inspection, we found that 11 of these were similar, where an integer value was incremented through a static pointer variable. In the LLVM IR code, the address of the static pointer was copied to local variables, and the value was updated through these local variables. The instrumentation component of *LLOR* could not track these operations, causing it to not instrument instructions that should have been instrumented. 2 programs could not be fixed since multiple threads were writing different values to the same shared variable in parallel. No placement of barriers can fix such programs. The remaining 3 programs had complex OpenMP constructs like *teams* and *target*, because of which simple barrier placements could not fix the data races. 15 programs were using OpenMP constructs like *sections* and *simd*. These programs are beyond the scope of *LLOR* since barriers and ordered regions cannot fix data races in these programs. *LLOR* clearly states that these programs are unsupported when provided as input. 9 programs timed out.

Table 2. Count of programs grouped by category

Source	C/C++	Fortran
Total Benchmarks	235	180
I. No data races identified by *LLOV*	75	129
No changes made by *LLOR*	71	122
Changes recommended by *LLOR*	4	7
II. Data races identified by *LLOV*	117	30
Repaired by *LLOR*	92	15
Could not be repaired by *LLOR*	5	11
Timeouts	9	0
Unsupported	11	4
III. Unsupported by *LLOV*	43	21
Unsupported by *LLOR*	37	11
Compilation errors	0	10
Verification errors	6	0

The final category includes the programs that are either unsupported by *LLOV* or failed compilation. There are 64 programs in this category. The flang version that we used for experimentation threw compilation errors for

10 Fortran programs and did not generate *LLVM* IR. Because of that, we could not attempt to repair them. *LLOV* does not support the verification of OpenMP programs that use task-based parallelism. Since *LLOR* depends on *LLOV* in the repair process, it could not generate any repair candidates for 48 programs. *LLOV* either timed out or threw a runtime error for 6 programs.

4.3   Solver Comparison

As mentioned in Sect. 2.3, *LLOR* offers two different ways of solving the clauses generated from the error traces. The default approach uses the minimal-hitting-set (*mhs*) strategy. The user can switch to the *MaxSAT* strategy if they wish to. The *MaxSAT* strategy is computationally heavier than the *mhs* strategy since the *mhs* solver uses a polynomial time algorithm, whereas the *MaxSAT* solver makes multiple queries to the SAT solver. In most practical cases, the *mhs* strategy would provide a similar solution to the *MaxSAT* strategy. Because of this, the *mhs* strategy has a performance advantage over the *MaxSAT* strategy.

The behavior of the solver also impacts which strategy performs better. Consider the clause $a \lor b$. The *mhs* solver could choose a to be the solution, and the *MaxSAT* solver could choose b to be the solution. Both of the solutions are valid for the clause, but choosing b could fix the program, and choosing a may not, thus forcing more iterations. Because of these reasons, we notice that for some benchmarks, the *mhs* strategy performs better, and the *MaxSAT* strategy performs better for others.

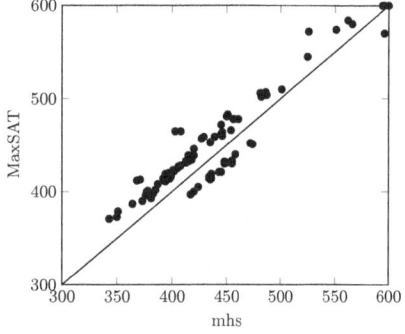

Fig. 2. mhs vs. MaxSAT Runtime in milliseconds

Out of the 415 benchmarks, only 133 benchmarks did not result in a timeout and needed the use of a solver strategy (either *mhs* or *MaxSAT*). In 93 of these benchmarks, the *mhs* strategy performed better than the *MaxSAT* strategy. Figure 2 shows the comparison of these benchmarks. Most of the benchmarks executed within a second, so Fig. 2 is limited to 600 milliseconds on the x-axis and y-axis to showcase those benchmarks. Any benchmark that took more than 600 milliseconds has been plotted on the top-right corner of the graph.

Additional experiments and analysis are provided in the extended manuscript [12].

5   Conclusion and Future Work

In this tool paper, we introduce *LLOR*, which can fix data race errors in OpenMP programs written in C/C++ and Fortran. *LLOR* can also remove unnecessary barriers and ordered regions in the programs while preserving correctness. We have affirmed the effectiveness of our work by attempting repair on multiple benchmark suites (consisting of 415 C/C++ and Fortran programs) using our tool. *LLOR* was able to repair more than 80% of the programs that had a valid data race error in them.

The technique behind *LLOR* can support any OpenMP verifier. Leveraging other static and dynamic verifiers might help repair some of the programs that the combination of *LLOR* and *LLOV* could not. More details on these verifiers have been provided in [12].

Acknowledgements. We thank the Ministry of Education, India, for their financial support.

References

1. Amighi, A., Darabi, S., Blom, S., Huisman, M.: Specification and verification of atomic operations in GPGPU programs. In: SEFM 2015, pp. 69–83. Springer (2015)
2. Atzeni, S., Gopalakrishnan, G., Rakamaric, Z., Laguna, I., Lee, G.L., Ahn, D.H.: SWORD: a bounded memory-overhead detector of OpenMP data races in production runs. In: IPDPS 2018, pp. 845–854. IEEE Computer Society (2018)
3. Barnett, M., Chang, B.E., DeLine, R., Jacobs, B., Leino, K.R.M.: Boogie: a modular reusable verifier for object-oriented programs. In: FMCO 2005, pp. 364–387. Springer (2005)
4. Basupalli, V., et al.: ompVerify: polyhedral analysis for the OpenMP programmer. In: Chapman, B.M., Gropp, W.D., Kumaran, K., Müller, M.S. (eds.) IWOMP 2011. LNCS, vol. 6665, pp. 37–53. Springer, Heidelberg (2011). https://doi.org/10.1007/978-3-642-21487-5_4
5. Betts, A., et al.: The design and implementation of a verification technique for GPU kernels. TOPLAS **37**(3), 10:1–10:49 (2015)
6. Betts, A., Chong, N., Donaldson, A.F., Qadeer, S., Thomson, P.: GPUVerify: a verifier for GPU kernels. In: OOPSLA 2012, pp. 113–132. ACM (2012)
7. Bjørner, N., Phan, A., Fleckenstein, L.: νZ - an optimizing SMT solver. In: TACAS 2015, pp. 194–199. Springer (2015)
8. Blackshear, S., Gorogiannis, N., O'Hearn, P.W., Sergey, I.: RacerD: compositional static race detection. Proc. ACM Program. Lang. **2**(OOPSLA), 144:1–144:28 (2018)
9. Blom, S., Huisman, M., Mihelcic, M.: Specification and verification of GPGPU programs. Sci. Comput. Program. **95**, 376–388 (2014)
10. Boehm, B.W., Papaccio, P.N.: Understanding and controlling software costs. IEEE Trans. Software Eng. **14**(10), 1462–1477 (1988)
11. Bora, U., Das, S., Kukreja, P., Joshi, S., Upadrasta, R., Rajopadhye, S.V.: LLOV: a fast static data-race checker for OpenMP programs. ACM Trans. Archit. Code Optim. **17**(4), 35:1–35:26 (2020)

12. Bora, U., Joshi, S., Muduganti, G., Upadrasta, R.: LLOR: automated repair of OpenMP programs (2024). https://arxiv.org/abs/2411.14590
13. Černý, P., Chatterjee, K., Henzinger, T.A., Radhakrishna, A., Singh, R.: Quantitative synthesis for concurrent programs. In: Gopalakrishnan, G., Qadeer, S. (eds.) CAV 2011. LNCS, vol. 6806, pp. 243–259. Springer, Heidelberg (2011). https://doi.org/10.1007/978-3-642-22110-1_20
14. Chandra, S., Torlak, E., Barman, S., Bodík, R.: Angelic debugging. In: ICSE 2011, pp. 121–130. ACM (2011)
15. Chatarasi, P., Shirako, J., Kong, M., Sarkar, V.: An extended polyhedral model for SPMD programs and its use in static data race detection. In: LCPC 2016, pp. 106–120. Springer (2016)
16. Chatarasi, P., Shirako, J., Sarkar, V.: Static data race detection for SPMD programs via an extended polyhedral representation. In: IMPACT 2016, vol. 16 (2016)
17. Che, S., Boyer, M., Meng, J., Tarjan, D., Sheaffer, J.W., Lee, S., Skadron, K.: Rodinia: a benchmark suite for heterogeneous computing. In: IISWC 2009, pp. 44–54. IEEE Computer Society (2009)
18. Dagum, L., Menon, R.: OpenMP: an industry-standard API for shared-memory programming. IEEE Comput. Sci. Eng. 5(1), 46–55 (1998)
19. Deshmukh, J., Ramalingam, G., Ranganath, V.-P., Vaswani, K.: Logical concurrency control from sequential proofs. In: Gordon, A.D. (ed.) ESOP 2010. LNCS, vol. 6012, pp. 226–245. Springer, Heidelberg (2010). https://doi.org/10.1007/978-3-642-11957-6_13
20. Dinning, A., Schonberg, E.: Detecting access anomalies in programs with critical sections. In: PADD 1991, pp. 85–96. ACM (1991)
21. Eichenberger, A.E., et al.: OMPT: an OpenMP tools application programming interface for performance analysis. In: Rendell, A.P., Chapman, B.M., Müller, M.S. (eds.) IWOMP 2013. LNCS, vol. 8122, pp. 171–185. Springer, Heidelberg (2013). https://doi.org/10.1007/978-3-642-40698-0_13
22. Engler, D.R., Ashcraft, K.: RacerX: effective, static detection of race conditions and deadlocks. In: SOSP 2003, pp. 237–252. ACM (2003)
23. Fu, Z., Malik, S.: On solving the partial MAX-SAT problem. In: Biere, A., Gomes, C.P. (eds.) SAT 2006. LNCS, vol. 4121, pp. 252–265. Springer, Heidelberg (2006). https://doi.org/10.1007/11814948_25
24. Griesmayer, A., Bloem, R., Cook, B.: Repair of Boolean programs with an application to C. In: Ball, T., Jones, R.B. (eds.) CAV 2006. LNCS, vol. 4144, pp. 358–371. Springer, Heidelberg (2006). https://doi.org/10.1007/11817963_33
25. Jin, G., Song, L., Zhang, W., Lu, S., Liblit, B.: Automated atomicity-violation fixing. In: PLDI 2011, pp. 389–400. ACM (2011)
26. Jobstmann, B., Griesmayer, A., Bloem, R.: Program repair as a game. In: Etessami, K., Rajamani, S.K. (eds.) CAV 2005. LNCS, vol. 3576, pp. 226–238. Springer, Heidelberg (2005). https://doi.org/10.1007/11513988_23
27. Johnson, D.S.: Approximation algorithms for combinatorial problems. J. Comput. Syst. Sci. 9(3), 256–278 (1974)
28. Joshi, S., Kroening, D.: Property-driven fence insertion using reorder bounded model checking. In: Bjørner, N., de Boer, F. (eds.) FM 2015. LNCS, vol. 9109, pp. 291–307. Springer, Cham (2015). https://doi.org/10.1007/978-3-319-19249-9_19
29. Joshi, S., Lal, A.: Automatically finding atomic regions for fixing bugs in Concurrent programs. CoRR abs/1403.1749 (2014)
30. Joshi, S., Muduganti, G.: GPURepair: automated repair of GPU kernels. In: Henglein, F., Shoham, S., Vizel, Y. (eds.) VMCAI 2021. LNCS, vol. 12597, pp. 401–414. Springer, Cham (2021). https://doi.org/10.1007/978-3-030-67067-2_18

31. Joshi, S., Shyamasundar, R.K., Aggarwal, S.K.: A new method of MHP analysis for languages with dynamic barriers. In: IPDPS 2012, pp. 519–528. IEEE Computer Society (2012)
32. Kernels, P.R.: Parallel Research Kernels. https://github.com/ParRes/Kernels. Accessed 30 Sept 2024
33. Lamport, L.: Time, clocks, and the ordering of events in a distributed system. Commun. ACM **21**(7), 558–565 (1978)
34. Lattner, C., Adve, V.S.: LLVM: a compilation framework for lifelong program analysis & transformation. In: CGO 2004, pp. 75–88. IEEE Computer Society (2004)
35. Li, G., Gopalakrishnan, G.: Scalable SMT-based verification of GPU kernel functions. In: FSE 2010, pp. 187–196. ACM (2010)
36. Li, G., Li, P., Sawaya, G., Gopalakrishnan, G., Ghosh, I., Rajan, S.P.: GKLEE: concolic verification and test generation for GPUs. In: PPOPP 2012, pp. 215–224. ACM (2012)
37. Liao, C., Lin, P., Asplund, J., Schordan, M., Karlin, I.: DataRaceBench: a benchmark suite for systematic evaluation of data race detection tools. In: SC 2017, p. 11. ACM (2017)
38. LLOR: LLOR Github Repository. https://github.com/cs17resch01003/llor. Accessed 30 Sept 2024
39. LLOR: LLOR VMCAI 2025 Artifacts. https://doi.org/10.5281/zenodo.13886253. Accessed 30 Sept 2024
40. Malik, M.Z., Siddiqui, J.H., Khurshid, S.: Constraint-based program debugging using data structure repair. In: ICST 2011, pp. 190–199. IEEE Computer Society (2011)
41. Mellor-Crummey, J.M.: On-the-fly detection of data races for programs with nested fork-join parallelism. In: SC 1991, pp. 24–33. ACM (1991)
42. Microsoft: Microsoft Azure Fsv2-Series Virtual Machine Sizes. https://docs.microsoft.com/en-us/azure/virtual-machines/fsv2-series. Accessed 30 Sept 2024
43. Monteiro, F.R., da S. Alves, E.H., da Silva, I., Ismail, H., Cordeiro, L.C., de Lima Filho, E.B.: ESBMC-GPU a context-bounded model checking tool to verify CUDA programs. Sci. Comput. Program. **152**, 63–69 (2018)
44. de Moura, L., Bjørner, N.: Z3: an efficient SMT solver. In: Ramakrishnan, C.R., Rehof, J. (eds.) TACAS 2008. LNCS, vol. 4963, pp. 337–340. Springer, Heidelberg (2008). https://doi.org/10.1007/978-3-540-78800-3_24
45. Muzahid, A., Otsuki, N., Torrellas, J.: AtomTracker: a comprehensive approach to atomic region inference and violation detection. In: MICRO 2010, pp. 287–297. IEEE Computer Society (2010)
46. Netzer, R.: Race condition detection for debugging shared-memory parallel programs. Ph.D. thesis, University of Wisconsin Madison (1991)
47. Perkovic, D., Keleher, P.J.: Online data-race detection via coherency guarantees. In: OSDI 1996, pp. 47–57. ACM (1996)
48. Pratikakis, P., Foster, J.S., Hicks, M.: LOCKSMITH: practical static race detection for C. TOPLAS **33**(1), 3:1–3:55 (2011)
49. Project, E.C.: ECP Proxy Applications. https://proxyapps.exascaleproject.org/. Accessed 30 Sept 2024
50. Savage, S., Burrows, M., Nelson, G., Sobalvarro, P., Anderson, T.E.: Eraser: a dynamic data race detector for multithreaded programs. ACM Trans. Comput. Syst. **15**(4), 391–411 (1997)
51. Valgrind-project: Helgrind: a thread error detector. http://valgrind.org/docs/manual/hg-manual.html. Accessed 30 Sept 2024

52. Vechev, M.T., Yahav, E., Yorsh, G.: Abstraction-guided synthesis of synchronization. In: POPL 2010, pp. 327–338. ACM (2010)
53. Voung, J.W., Jhala, R., Lerner, S.: RELAY: static race detection on millions of lines of code. In: FSE 2007, pp. 205–214. ACM (2007)
54. Ye, F., Schordan, M., Liao, C., Lin, P., Karlin, I., Sarkar, V.: Using polyhedral analysis to verify openmp applications are data race free. In: CORRECTNESS 2018, pp. 42–50. IEEE (2018)

Synthesis of Controllers for Continuous Blackbox Systems

Benedikt Maderbacher[1]([✉]) [iD], Felix Windisch[1] [iD], Alberto Larrauri[2] [iD], and Roderick Bloem[1] [iD]

[1] Graz University of Technology, Graz, Austria
{benedikt.maderbacher,roderick.bloem}@iaik.tugraz.at,
felix.windisch@tugraz.at
[2] University of Oxford, Oxford, UK
alberto.larrauri@cs.ox.ac.uk

Abstract. Feed-forward controllers compute control outputs to adapt to changes in environmental parameters in a cyber-physical system. When synthesizing control functions it can be difficult to give an analytical description of the controlled plant or to decide the expected control output ahead of time. These systems must, however, adhere to strict safety requirements, which makes it hard to write correct controllers. In this paper, we propose a novel blackbox synthesis approach to construct a continuous control function while dynamically sampling a limited number of test cases. The controller is guaranteed to be correct for a given Lipschitz bound. It can be adapted to work for increasingly conservative estimates of the bound based on observed behavior, iteratively providing increasing confidence in its correctness. Our algorithm employs a linear interpolation model, based on a Delaunay triangulation, to identify candidate control functions. It then generates additional test cases to either confirm a candidate or to improve the model. We evaluate our approach on random benchmarks and CPS examples to show its effectiveness.

1 Introduction

In this paper, we focus on the synthesis of a type of feed-forward control functions that interact with a blackbox environment that can only be sampled but have strong requirements to be correct. A feed-forward controller measures an environmental parameter, such as temperature, and produces control outputs to ensure the system behaves correctly. This architecture is well suited to handle sudden large changes in the environment, because it can preemptively change its control outputs. In practice feed-forward and feedback controllers are often used

We would like to thank Dejan Ničković for helpful discussions on synthesis for blackbox systems. This work has received funding from the Austrian research promotion agency FFG under project FATE (№ 894789). This work was supported partially by UKRI EP/X024431/1.

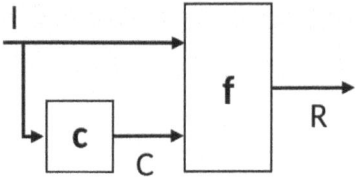

Fig. 1. Schematic

together where the feed-forward controller handles large disturbances with measurable causes and the feedback controller handles smaller disturbances. This paper focuses on synthesizing feed-forward controllers that control a black box system.

Figure 1 show the high-level structure of the synthesis problem. The system and its specification is modeled as function f that takes as inputs the uncontrollable environment input I (e.g. temperature) and the controllable input C (e.g. valve setting) that is controlled by the controller c. The output of f is a real-valued *robustness score* quantifying how robustly the plant satisfies the specification, where a positive (negative) value signifies a satisfied specification (a violated specification, resp.). We want to automatically synthesize a controller c for cyber-physical systems (CPS) where both the uncontrollable (I) and controllable inputs C are real-valued. The controllable inputs C have to be chosen such that for every uncontrollable input I the system satisfies the specification.

We consider a setting where the system is only available as a blackbox function. Cyber-physical systems often depend on a physical environment for which no accurate model is available. Even for applications where source code is available, such as analog systems, analytical reasoning about the system is often impossible. Synthesis is performed in a test environment (a laboratory or simulator) where both the controllable and uncontrollable inputs can be selected for a simulation. This setup makes it impossible to use traditional synthesis methods that require a white-box model of the environment. When a white-box model is available traditional methods such as reactive synthesis [5] or program synthesis [3,18] can be used. Other methods use a set of input-output examples [21] which requires the user to know the desired output.

As a further complication, the evaluation of a cyber-physical system may be very time-consuming. For instance, executing a robotic system in physical environments can take a long time. Performing gate-level hardware simulations is also much slower than high-level software models. For example, a single simulation using the circuit simulator in our industrial case study takes between one and two minutes. We thus have a setting in which the behavior of the system can only be approximated by testing it, and the number of test runs has to be minimized as much as possible for the procedure to be time efficient.

Formal correctness guarantees are not available when using blackbox optimization and machine learning methods that can be used to handle blackbox systems, these also require a large number of simulations. We use robustness

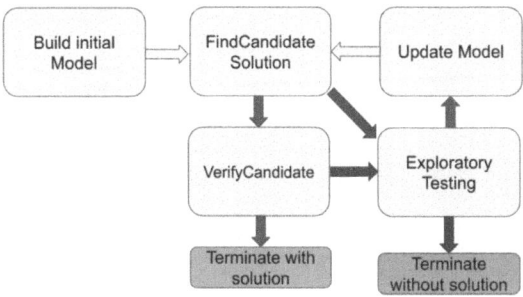

Fig. 2. Blackbox synthesis overview.

measures [10] along with the assumption of Lipschitz continuity to show formal correctness in a blackbox setting. The assumption of Lipschitz continuity is well established in the control community [30] and also applied in the verification of continuous systems [17]. Lipschitz continuity together with a known Lipschitz constant means that for every sample that satisfies the specification we can conclude that all inputs within a certain distance around the sample are guaranteed to satisfy the specification without having to sample them. This makes it possible to verify a function over a continuous state space using a finite number of samples.

Note that if the plant is discontinuous in arbitrary ways, we cannot draw conclusions from any finite number of samples. We thus adopt the assumption that the plant is Lipschitz continuous with constant k.

Our Approach. We present a novel approach that combines blackbox testing with functional synthesis to find control functions for blackbox environments. Our algorithm uses a linear interpolation model of the CPS based on a Delaunay triangulation of samples of the CPS. To build an initial model, we run a small number of random tests consisting of uncontrollable and controllable inputs.

We then proceed in three steps. (See Fig. 2 for an overview.) First, we construct a candidate control function that is correct according to the model by performing a search for a path starting at the minimum input value and ending at the maximum input value that goes only through vertices with positive values. Second, if we find a candidate that is correct according to the model, we run additional tests with the aim to falsify or verify it (for the given Lipschitz constant). The testing effort is guided by an uncertainty metric that quantifies the likelihood that the model is wrong at any given point by comparing the model's prediction with the upper and lower bounds provided by the Lipschitz constant. If the candidate is falsified by finding an input for which it does not meet the specification, we use the executed tests to improve the model and identify a different candidate based on the new model. Third, if we cannot find a candidate, we perform exploratory testing to reduce the uncertainty of the overall model. This helps to identify promising regions for candidate functions that might otherwise be overlooked.

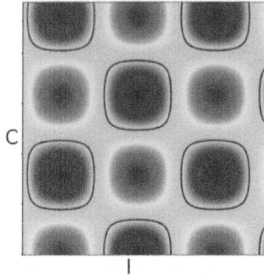 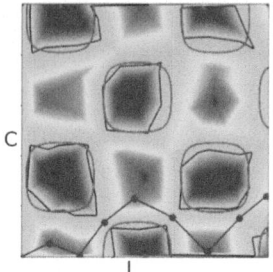

Fig. 3. Blackbox function of running example (left). Final model and solution (right).

The tests that we run during exploration and validation lead to an iteratively-improving model. If a solution with a good enough robustness exists, the model will eventually be good enough to find it. If no solution exists the model improves until it can guarantee that no solution exists (Proposition 8).

The Lipschitz constant may be known, or it can be estimated using the samples obtained during the synthesis of the controller. In the latter case, we can estimate the actual value of k using one of several known means such as the Strongin estimator. This allows us to generate a controller for an arbitrarily conservative estimate of k, or even a sequence of controllers that is guaranteed to be correct for increasingly conservative estimates.

We use the simple synthetic function $f(i,c) = \sin(i)*\cos(c)+0.2$ as a running example. Figure 3(left) shows the black box function f, where the function values are shown as a color gradient between red for high values (correct behavior) and blue for low values (incorrect behavior). The zero-level curves are shown in black. On the right, we show the interpolation model including the solution in blue, the zero-level lines of the model in black, and the zero-level lines of f in red. The right part shows the constructed model and the found solution.

Contributions. Our main results can be summarized as follows.

- We provide a novel approach to synthesize continuous control functions for blackbox CPSs that uses guided testing and counterexample-guided model improvement to find promising solutions.
- We provide an uncertainty measure to estimate the quality of a model that improves iteratively.
- We evaluate our method on multiple case studies and demonstrate that it can find provably correct control functions using a small number of test cases.

2   Related Work

2.1   Simulation Based Validation

Simulation-based validation [7,8,16] or falsification testing can be used to falsify cyber-physical systems. Given a blackbox function representing a plant, the goal is to find a controller input that violates a given formal specification. These robustness specifications can be used to guide an optimizer to find inputs that violate the specification [11,13,25].

These methods focus on testing and falsifying a plant in combination with a given controller. We use similar robustness specifications to quantify the satisfaction of a controller. In contrast to these falsification approaches, we consider a synthesis problem. We are not given the controller, but we try to find one while executing tests. The given Lipschitz bound also allows us to verify a system by obtaining a test suite that guarantees correctness.

2.2 Min-max Blackbox Optimization

A variation of the blackbox synthesis problem can be formulated as a multi-objective optimization problem. These are often referred to as min-max blackbox optimization or robust blackbox optimization [2,12,22,24]. Given a function with two sets of parameters, in blackbox optimization, we want to find a value for the first parameter that maximizes the minimal output of the function over all possible values of the second parameter. This approach could be used to perform a parameter synthesis version of our problem: it can find a *fixed* controller output that maximizes the worst plant output for any of the uncontrollable input values. If this output is positive, the controller output satisfies the specification for any uncontrollable input.

We propose a more general approach in which the controller output is not fixed but rather depends on its input. Of course, in the case that the structure of the control function is known, we can use blackbox optimization to search for a set of template parameters. However, often such a template is not available. Furthermore, blackbox optimization methods provide no guarantee that the found minimum is correct or greater than zero (especially for functions that are not convex). Our method, on the other hand, can prove that the proposed solution is correct and uses fewer samples as we are not interested in an optimal solution but only in one that satisfies our specification.

2.3 Models for Blackbox Systems

Models are widely used in blackbox optimization settings, or even in situations where the objective function is known but too complex, so performing optimization on a simpler approximate model is more desirable than doing so on the original function.

A popular approach to blackbox optimization is Bayesian Optimization, which relies on statistical models. These models provide an estimate of the objective function at each point together with a measure of uncertainty, given by the standard deviations of some underlying probability distributions. This measure of uncertainty is useful for balancing the exploration of low information areas and exploitation of the information accrued so far. Out of these models, Gaussian Processes are a popular choice across multiple domains of application [27].

In particular, they have been used in the verification of cyber-physical systems [9,34].

Another important family of models is those obtained via interpolation of different kinds of basis functions. It is possible to equip such models with tailored uncertainty measures and use them in blackbox optimization, achieving similar results to Bayesian optimization [4].

In our proposed method we opt for piece-wise linear interpolations as models. These are simpler than the models described above, and hence it is easier to perform tasks like finding level curves on them. This can be done exactly and efficiently on our models, whereas one would need to resort to inexact and more inefficient numerical methods to find level curves on Gaussian Processes or other more involved kinds of interpolations. A potential drawback of our simpler models is its lower precision, but whereas in traditional blackbox optimization precision plays a more important role (i.e., one needs to distinguish between potentially close and high values when looking for a global maximum), we only need to ensure that the output of the plant remains positive, which often allows us to use less precise models.

3   Preliminaries

3.1   Formal Problem Statement

We consider blackbox functions $f : I \times C \to \mathbb{R}$ with two types of inputs: the uncontrollable input I and the controllable inputs C. We assume that I and C are box-shaped real-valued domains with dimensions 1 and n. We assume without loss of generality that every dimension has a lower bound of min and an upper bound of max. The input domains are thus $I = [min, max]$ and $C = [min, max]^n$. We use s^I to refer to the uncontrollable value and s^C to the controllable values of s from $I \times C$. The controllable inputs to the plant are computed by a control function $c \in (I \to C)$, a continuous function that selects a controllable input for each uncontrollable input.

The goal of our approach is to find a control function c for a given black box plant and specification f that always satisfies the specification, i.e., $\forall i \in I : f(i, c(i)) \geq 0$. This should be done by collecting as few samples as possible. We define a sample s as a point in the domain $I \times C$ for which we evaluate the blackbox function f and we refer to the set of all collected samples as $S \subset I \times C$.

Many systems do not have a single output where positive values correspond to satisfying the specification. In that case the expected behavior of the system is provided in the form of a quantitative specification. This evaluates the quality of the produced output and assigns it a real-valued score. Positive values indicate that the specification is satisfied whereas negative values signify a violated specification. Similar robustness measures are often used for temporal specifications [10,14,23]. This quantitative specification is integrated into the blackbox function $f/$

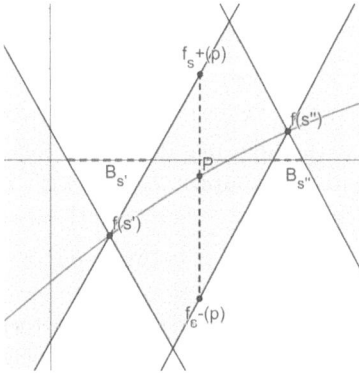

Fig. 4. Lipschitz bound and derived quantities.

3.2 Lipschitz Continuity

We assume the blackbox function f consisting of plant and specification is Lipschitz continuous with constant k, that is, $\forall x_1, x_2 \in I \times C : |f(x_1) - f(x_2)| \leq k\|x_1 - x_2\|$ where $\|\cdot\|$ refers to the Euclidean norm.

Given a set of samples $S \in I \times C$, we can compute the maximum and minimum value that f may take at a point p in $I \times C$ as $\hat{f}_S^+(p) = \min_{s \in S} f(s) + k\,\|s - p\|$ and $\hat{f}_S^-(p) = \max_{s \in S} f(s) - k\,\|s - p\|$. Lipschitz continuity provides such a guarantee for the neighborhood of any positive or negative samples. We define the *Lipschitz ball* around a sample s with radius $|f(s)|/k$ as $B_k(s) = \{p \in I \times C \mid \|p - s\| \leq \frac{|f(s)|}{k}\}$. For all $p \in B_k(s)$ for a sample s, Lipschitz continuity with a Lipschitz constant less than or equal to k guarantees that $f(p) \geq 0$ if $f(s) \geq 0$ and that $f(p) \leq 0$ if $f(s) \leq 0$.

Figure 4 depicts these concepts using a function f with the one-dimensional input s on the x-axis and the function value $f(s)$ on the y-axis. Based on the samples s' and s'', the function f is guaranteed to lie inside the shaded area created by the cones around the samples. The balls (intervals in the one-dimensional case) around each s are denoted by $B(s)$. For one point p the figure shows $\hat{f}_S^+(p)$ and $\hat{f}_S^-(p)$.

If the Lipschitz constant is not known, it can be approximated using the collected samples. A classic approach is the Strongin estimator [29], defined as $\hat{k} = r \cdot \max_{i \neq j} |f(s_i) - f(s_j)|/\|s_i - s_j\|$ for $r > 1$. There are also more precise stochastic estimations [15,19,33].

3.3 Delaunay Triangulation and Interpolation

Let $S \subseteq I \times C$ be a finite set of points. A *Delaunay triangulation* $\mathcal{T}$ of S is a subdivision of the convex hull of S into d-simplices (d-dimensional generalizations of triangles) that are "compact" in the sense that no point lies strictly within the circum-hypersphere (the generalization of the circumcircle of a triangle) of any simplex T.

Given a function $f : I \times C \to \mathbb{R}$, a set of samples $\{(s_1, f(s_1)), \ldots, (s_n, f(s_n))\}$, and a triangulation $\mathcal{T}$ of the set $S = \{s_1, \ldots, s_n\}$, the piece-wise linear interpolation of f given by $\mathcal{T}$ is the map $m_{\mathcal{T}}$ defined as follows: The domain of $m_{\mathcal{T}}$ is the convex hull of S. Given a simplex $T \in \mathcal{T}$ with vertices $s_{r_1}, \ldots, s_{r_d}$ and a point $x \in T$, there are unique $\lambda_1, \ldots \lambda_d \in [0, 1]$ satisfying $\sum_{t=1}^{d} \lambda_t = 1$ and $x = \sum_{t=1}^{d} \lambda_t s_{r_t}$ where the second sum is between vectors in the Euclidean space. Then the value $m_{\mathcal{T}}(x)$ is defined as $\sum_{t=1}^{d} \lambda_t f(s_{r_t})$.

Delaunay triangulations are well-suited for the interpolation of spatial data [32], as they avoid long thin triangles and induce a neighborhood structure on the vertices that relates well to their distances on the space [28].

4   Blackbox Synthesis Method

4.1   Exploration and Model

Our method uses samples collected from simulations to identify new candidate solutions. We use $m_{\mathcal{T}} : I \times C \to \mathbb{R}$ to denote the value predicted by the linear approximation model for an arbitrary point in the domain. In combination with a known Lipschitz bound a precise model of f can be obtained.

Proposition 1. *Suppose f's Lipschitz constant is $k \geq 0$, and that the maximum length of an edge inside $\mathcal{T}$ is d. Then $|m_{\mathcal{T}}(x) - f(x)| \leq kd$ for all $x \in I \times C$.*

Proof. Consider a simplex T in $\mathcal{T}$ with vertices $v_1, \ldots, v_m$. For any point $x \in T$ it holds that

$$\min\{f(v_h) \mid 1 \leq h \leq m\} \leq m_{\mathcal{T}}(x) \leq \max\{f(v_h) \mid 1 \leq h \leq m\}.$$

The maximal distance between any vertex of T to any point inside T is bounded by the length of the longest edge d.

As such, $\forall v \in \{v_1, \ldots, v_m\} \forall x \in T : f(v) - kd < f(x) < f(v) + kd$. Symmetrically, this also implies $\forall v \in \{v_1, \ldots, v_m\} \forall x \in T : f(x) - kd < f(v) < f(x) + kd$. As $m_{\mathcal{T}}(x)$ is a weighted average of $f(v_i)$ for $i \in \{1, \ldots, m\}$, $m_{\mathcal{T}}(x)$ is contained in the interval $[f(x) - kd, f(x) + kd]$, which directly implies $|m_{\mathcal{T}}(x) - f(x)| \leq kd$.

A corollary is the following:

Proposition 2. *Let k be f's Lipschitz constant, and let d be the smallest edge length in the triangulation $\mathcal{T}$. Suppose that $kd \leq \epsilon$ for some $\epsilon > 0$. Then the following hold:*

- *If $c : I \to C$ is such that $m_{\mathcal{T}}(x, c(x)) \geq \epsilon$ for all x, then c is a solution for f.*
- *If $c : I \to C$ is such that $f(x, c(x)) \geq \epsilon$, then c is a solution for $m_{\mathcal{T}}$.*

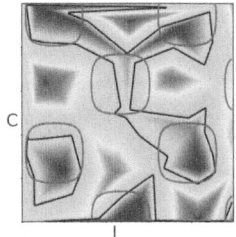 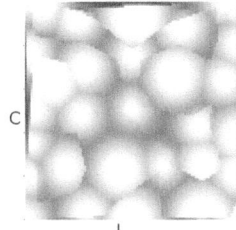

Fig. 5. Delaunay interpolation model for the running example (left). Uncertainty scores $\hat{u}$ for the same model (right), white denotes zero uncertainty and the darker the color the higher the uncertainty.

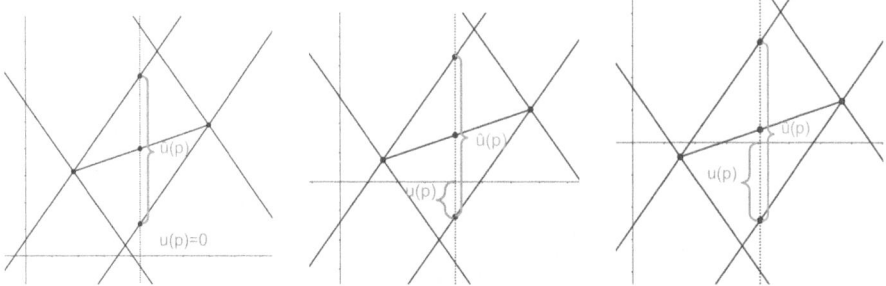

Fig. 6. Uncertainty estimation.

To guide the selection of samples we define a heuristic to select interesting points. An *uncertainty estimator* assigns a non-negative value to each point in the domain $I \times C$, where zero means that the model is correct at this point and a higher score indicates less confidence in the predicted value. We provide two uncertainty estimates, each using the Lipschitz property. The first estimate, $\hat{u}$, compares the upper and lower bounds: $\hat{u}_S(p) = \hat{f}_S^+(p) - \hat{f}_S^-(p)$. This assigns the highest uncertainty to points with the largest range of possible values.

We propose a novel uncertainty estimator u that takes into account that we are looking for a model that correctly predicts the sign of a function, but might be less certain about the exact value.

$$u_S(p) = \begin{cases} \max(\hat{f}_S^+(p), 0) & \text{if } m_T(p) \leq 0, \\ \max(-\hat{f}_S^-(p), 0) & \text{otherwise.} \end{cases}$$

Figure 6 shows three cases that are assigned the same uncertainty score using $\hat{u}$, but have different uncertainty scores using u based on how far the prediction is from zero. Figure 5 shows the Delaunay interpolation model of an early iteration and the values of the uncertainty measure u for the same model.

To improve the global quality of our model we use exploratory tests. These help to find new candidates in under-explored areas. Algorithm 1 select points using three heuristics. Two of them are based on the uncertainty estimators u

---

Algorithm 1. Exploratory Testing
___

1: **function** EXPLORATORYTESTING(f, m_T)
2:    **for** *number of explorations* **do**
3:        $x_1 \leftarrow \arg\max_{x \in I \times C} \hat{u}_V(x)$
4:        $x_2 \leftarrow \arg\max_{x \in I \times C} u_V(x)$
5:        $x_3 \leftarrow (s_0 l, \ldots, s_n l)$ for unused $s_0 \ldots s_n \in \{0, 1, \ldots, max/l\}$ ▷ Terminate if
    the grid is complete.
6:        $m_T \leftarrow$ DELAUNAYTRIANGULATION($S \cup \{x_1, x_2, x_3\}$)

---

and $\hat{u}$, for which we use the shorthand $\hat{u}_V(x)$ and $u_V(x)$ where the set V is implicitly selected as the vertices of the simplex in T that contains x. The third heuristic eventually selects all points on a grid with spacing $l = \epsilon/(2k\sqrt{n+1})$.

Proposition 3. *Let k be f's Lipschitz constant, and let d be the longest edge length in the triangulation T. Exploratory testing will eventually result in a model such that $k \cdot d \leq \epsilon$ for some $\epsilon > 0$.*

Every call to exploratory testing includes one sample from a grid with spacing $l = \epsilon/(2k\sqrt{n+1})$, where n is the number of controllable parameters. The factor $\sqrt{n+1}$ is the length of the diagonal in such a grid cell. The domain is bounded so this grid contains only a finite number of points. After all the points from this grid have been sampled, any Delaunay triangulation contains no edges longer than ϵ/k.

4.2   Searching for Candidate Solutions

The candidate search procedure takes a model in the form of a triangulation graph T and its associated interpolation function m_T. We want to identify candidate solutions c for the controller. The main criterion for a candidate controller is that it satisfies the specification according to our model. This leaves us with a very large and unstructured search space. To make the problem more tractable we limit the candidates to piece-wise linear functions. This is a reasonable restriction as with enough segments we can represent a solution to every realistic problem for which a solution exists.

Most models will allow for multiple solutions. In principle, any of them would serve as a candidate solution. However, candidates with a higher minimal value seem to result in faster overall convergence. We rank candidates c by the lowest point according to the interpolation model i.e. $\min_{i \in I}(m_T(i, c(i)))$.

We present two algorithms to search for candidates based on the triangulation model. The first uses the bottleneck shortest path algorithm along the edges of the triangulation to find candidates with high minimal value. This works well in practice, but it might miss some candidates. For completeness, we also present a second algorithm which is guaranteed to find a candidate if one exists.

Algorithm 2. Path search for finding candidates.

1: **function** FINDCANDIDATESOLUTION(m_T)
2:     $S, E \leftarrow$ vertices and edges from T
3:     $E \leftarrow \{((s_1^I, s_1^C), (s_2^I, s_2^C)) \in E \mid s_1^I < s_2^I\}$
4:     $E' \leftarrow \{(s_1, s_2) \in E \mid m_T(s_1) > 0 \wedge m_T(s_2) > 0\}$
5:     $S' \leftarrow \{s \in S \mid m_T(s) > 0\}$
6:     $S'' \leftarrow S' \cup \{start, end\}$
7:     $E'' \leftarrow E' \cup \{(start, s) \mid s^I = min\} \cup \{(s, end) \mid s^I = max\}$
8:     $w(s_1, s_2) = \begin{cases} \min(m_T(s_1), m_T(s_2)) & (s_1, s_2) \in E' \\ 0 & \text{otherwise} \end{cases}$
9:     **return** BOTTLENECKSHORTESTPATH$((S'', E''), start, end, w)$

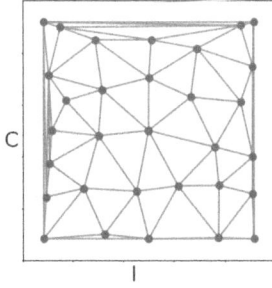

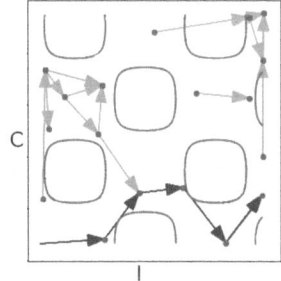

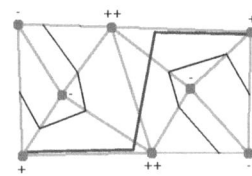

Fig. 7. Delaunay graph.

Fig. 8. Directed positive subgraph with candidate.

Fig. 9. Configuration where Algorithm 2 does not find an existing candidate solution.

Using Bottleneck Shortest Paths. A solution corresponds to a path that connects a point at the lower bound of the input domain to another point at the upper bound of the input domain. Instead of searching all possible piecewise linear functions, we limit the search to only segments that are part of the Delaunay triangulation graph. This allows us to use standard algorithms for finite graphs, but it might miss some candidates. The limited number of considered paths is alleviated by improving the model which increases the number of edges.

Algorithm 2 shows how to translate our synthesis problem into a bottleneck shortest path problem on a directed subgraph of the Delaunay triangulation consisting of the edges that are positive and in the direction of increasing I. We add *source* and *sink* nodes and edges to allow any vertices with $s^I = min$ as start points and vertices with $s^I = max$ as end points. The bottleneck shortest path on the graph from *source* to *sink* is the path where the minimal edge weight is maximized. This is efficiently computable using a variant of Dijkstra's algorithm or more specialized implementations. See [20] for an overview. Figure 7 shows the graph resulting from the triangulation of the running example and Fig. 8 the preprocessed graph including the bottleneck shortest path.

Proposition 4 (Soundness of the candidate search). *Any candidate found by Algorithm 2 is a solution to the synthesis problem on the model.*

The search is performed on a graph with only positive edges that point in the positive I direction. A candidate solution is a continuous path from the minimum input to the maximum input. Any found path is thus a valid candidate solution.

Search Algorithm with Completeness Guarantee. In some cases, searching only along the edges of the triangulation might result in no candidate solution being found, even when one exists (an example of this situation can be seen in Fig. 9).

To solve this problem we add some auxiliary vertices and edges to the graph of the triangulation and search for a path in this extended graph. However, in this new graph vertical edges are included, so a left-to-right path in the graph does not necessarily correspond to a candidate solution c. Hence, one needs to be careful and consider only paths whose vertical segments can be "perturbed" slightly and remain non-negative. We give an outline of the procedure below.

We consider a triangulation $\mathcal{T}$ of $I \times C$. Recall that $\mathcal{T}$ is a simplicial complex. A facet $F \in \mathcal{T}$ is a d-dimensional simplex in the complex. Similarly to Algorithm 2, we build a graph $G_{\mathcal{T}}$ from $\mathcal{T}$ in such a way that paths in the graph correspond to solutions in $m_{\mathcal{T}}$. Initially, the vertex set V corresponds to the vertices $v \in \mathcal{T}$ with $m_{\mathcal{T}}(v) \geq 0$. Afterwards, for each edge $e \in \mathcal{T}$ that has at least one endpoint with a non-zero value, we consider the only point (if any) $v_e \in e$ that achieves $m_{\mathcal{T}}(v_e) = 0$. We add to V all those new points of the zero-level curve of $m_{\mathcal{T}}$. At this stage we consider for each $v \in V$ the vertical hyper-plane given by $h_v = \{x \in I \times C \mid x^I = v^I\}$. Given a non-vertical edge $e \in \mathcal{T}$, we add to V all the intersections of e with the hyperplanes h_v where the value of $m_{\mathcal{T}}$ is non-negative. Now we define a set E of directed edges on V, given by $(u, v) \in E$ if $u^I < v^I$ and u, v share a common facet in $\mathcal{T}$. The final graph $G_{\mathcal{T}}$ has vertex set $V(G_{\mathcal{T}}) = V \cup E$ (that is, we have a new vertex representing each edge in E), and its set of directed edges is given by

$$
\begin{aligned}
E(G_{\mathcal{T}}) =& E \cup \{(v, (u_1, u_2)) \in V \times E \mid v = u_1\} \\
& \cup \{((u_1, u_2), v) \in E \times V \mid v = u_2\} \\
& \cup \{((v_1, v_2), (u_1, u_2)) \in E \times E \mid v_1^I = u_1^I, \text{ and} \\
& \quad v_1, v_2, u_1, u_2 \text{ share a facet in } \mathcal{T}\}.
\end{aligned}
$$

In summary $G_{\mathcal{T}}$ has two types of vertices: points $x \in I \times C$ and ones that correspond to "segments" $(x, y) \in (I \times C)^2$. Two points are joined by a directed edge when they lie on a shared facet and the I-coordinate of the first is smaller than the I-coordinate of the second. Vertices representing segments are joined to both their endpoints via directed edges. Finally, two vertices that represent segments are joined by an edge if the left endpoints of the segments have the same I-coordinates and the endpoints of the segments lie in some common facet.

A left-to-right path in $G_{\mathcal{T}}$ is just a directed path $v_0, \ldots, v_k$ in the graph where $v_0, v_k \in I \times C$ and $v_0^I = min$, $v_k^I = max$. The **extended search algorithm**

Algorithm 3. Verification of a candidate solution.

1: **function** VERIFYCANDIDATE(c,f,m)
2:     $edges \leftarrow$ edges of the candidate c
3:     **while** $edges \neq \emptyset$ **do**
4:         $e \leftarrow$ get and remove the edge with the maximum u from $edges$.
5:         $v_1, v_2 \leftarrow$ vertices of e
6:         **if** $\exists p \in e : u_{\{v_1,v_2\}}(p) > 0$ **then**   ▷ equivalent to $e \not\subset B_k(v_1) \cup B_k(v_2)$
7:             $z \leftarrow \arg\max_{z \in e} u_{\{v_1,v_2\}}(z)$
8:             $S \leftarrow S \cup \{z\}$
9:             **if** $f(z) < \epsilon$ **then**   ▷ Evaluate blackbox function
10:                 **return** *False*, S
11:             $edges \leftarrow edges \cup \{(v_1, z), (z, v_2)\}$
12:     **return** *True*, S

---

builds the graph $G_\mathcal{T}$ and searches for a left-to-right path inside. The following two results state that such a path exists if and only if there is some candidate solution for $m_\mathcal{T}$.

Proposition 5 (Soundness of the Extended Search Algorithm). *Suppose that there is a left-to-right path in $G_\mathcal{T}$. Then there is a continuous map $c : I \to C$ with $m_\mathcal{T}(i, c(i)) \geq 0$ for all $i \in I$.*

It is illustrative to see how a left-to-right path $v_0, \dots, v_k$ in $G_\mathcal{T}$ corresponds to a solution $c : I \to C$ in last proposition. The idea is that from this path we obtain a sequence of points $x_0, \dots, x_k \in I \times C$ whose I-coordinates are increasing. Then the curve $p(i) = (i, c(i))$ is just the union of the segments $[x_h, x_{h+1}]$. If the vertex v_h is just a point in $I \times C$, then we set $x_h = v_h$. Alternatively, suppose that $v_h, v_{h+1}, \dots, v_{h+\ell}$ is a maximal sub-path where all vertices represent segments. I.e., for each $j = 0, \dots, \ell$, $v_{h+j} = (u_j, w_j) \in (I \times C)^2$. Observe that by the definition of $G_\mathcal{T}$, u_j^I remains constant for all j, and $w_j^I > u_j^I$. Then for each $j = 0, \dots, \ell$ we pick x_{h+j} as an arbitrary point in the interval $[u_j, w_j]$ in such a way that the I-coordinates of the points are increasing. This allows us to complete the choice of the points $x_0, \dots, x_k$ and define the curve $(i, c(i))$.

Proposition 6 (Completeness of the Extended Search Algorithm). *Suppose there is a continuous map $c : I \to C$ with $m_\mathcal{T}(i, c(i)) \geq 0$ for all $i \in I$. Then there is a left-to-right path in $G_\mathcal{T}$.*

4.3 Verification of Candidate Solutions

When a candidate solution is found we need to verify whether $\forall i \in I : f(i, c(i)) \geq \epsilon$ by sampling new values and using previous samples of f. This is done via a guided testing procedure shown in Algorithm 3. It aims to verify a candidate solution by finding samples S such that the Lipschitz balls induced by the samples completely cover the candidate.

A candidate solution consists of a set of edges. The verification is done one edge at a time and either confirms that it is covered by the Lipschitz balls induced by its vertices or adds a new sample on the edge. The new sample might be negative which falsifies the candidate, or it is used to split the edge into two parts that need to be verified.

The edges are verified in the order of their uncertainty u, based on their vertices. We define the uncertainty of an edge $e = (v_1, v_2)$ as $u(e) = \max_{x \in e} u_{\{v_1, v_2\}}(x)$. Selecting edges in this order makes it more likely to quickly reject spurious candidates.

If an edge e is not covered, the point on the edge with the highest uncertainty is added as a new sample. This either falsifies the candidate or splits the edge into two smaller edges to be tested. The algorithm rejects all candidates with values less than ϵ, as the verification might otherwise get stuck for a candidate that asymptotically approaches zero.

Proposition 7 (Soundness and ϵ-Completeness of the Verification).
Given a Lipschitz constant k and an $\epsilon > 0$, for all candidates c, the verification will terminate with either a counterexample $p \in I$ such that $f(p, c(p)) < \epsilon$, or by showing that $f(i, c(i)) \geq 0$ for all $i \in I$.

Proof. In case a candidate is verified every point in the candidate solution is included in a Lipschitz-ball of a positive sample. If a counter-example is found this is witnessed by a point on the candidate with a function value less than ϵ. To show termination we observe that the highest uncertainty of any edge is decreasing in every iteration of the verification loop and it is bounded below by 0. Given an edge $e = (v_1, v_2)$ and a point $z \in e$ with $f(z) \geq \epsilon$ and $u_{\{v_1, v_2\}}(z) > 0$. We have that $\|v_i - z\| \geq \epsilon/k$ for $i \in \{1, 2\}$, because any point closer to one of the vertices has a uncertainty of 0. Splitting e into two edges $e_1 = (v_1, z)$ and $e_2 = (z, v_2)$ results in the uncertainties $u_{\{v_1, z\}}(e_1)$ and $u_{\{z, v_2\}}(e_2)$ which are both less than $u_{\{v_1, v_2\}}(e) - \epsilon$. Every iteration of the loop reduces the highest uncertainty of all edges by at least ϵ, or it decreases the finite number of edges that share the highest uncertainty. Therefore, the verification loop terminates with a correct result.

4.4   Blackbox Synthesis

Based on the components defined in the previous sections we can define our method for blackbox function synthesis as shown in Algorithm 4. The main loop starts by performing some guided exploration tests as described in Sect. 4.1 and constructs a triangulation model from all samples. Then, one of the search algorithms from Sect. 4.2 is used to search for a candidate solution. If no candidate can be found, more exploration tests are performed. Once a candidate has been found, verification as described in Sect. 4.3 is attempted. When a counterexample is found, it and all other samples collected during verification are incorporated into the model. The algorithm terminates once a candidate has been verified successfully.

Algorithm 4. Blackbox Functional Synthesis

1: **function** BLACKBOXSYNTHESIS(f)
2:     Using blackbox function f, set of samples S, model m, and candidate solution c.
3:     **loop**
4:         $S \leftarrow S \cup \text{EXPLORATORYTESTING}(f, m)$
5:         $m \leftarrow \text{DELAUNAYTRIANGULATION}(S)$
6:         $c \leftarrow \text{FINDCANDIDATESOLUTION}(m)$
7:         **if** $c = $ **null then**
8:             **continue**
9:         $Correct, S' \leftarrow \text{VERIFYCANDIDATE}(c, f)$
10:        **if** $Correct$ **then**
11:            **return** c
12:        **else**
13:            $S \leftarrow S \cup S'$

Proposition 8 (Soundness and ϵ-Completeness of the Synthesis Algorithm). *For a known Lipschitz constant k and an $\epsilon > 0$. Given there exists a $c' \in I \rightarrow C$ with $f(i, c'(i)) \geq 2\epsilon$ for all $i \in I$. The Algorithm 4 finds a control function $c \in I \rightarrow C$ such that $f(i, c(i)) \geq 0$ for all $i \in I$.*

Proof. As long as no solution has been found, the algorithm performs repeated calls to EXPLORATORYTESTING, because all functions called in Algorithm 4 terminate. EXPLORATORYTESTING performs a finite number of simulations. DELAUNAYTRIANGULATIONS and the path search in FINDCANDIDATESOLUTION are graph algorithms known to terminate. The termination of FINDCANDIDATESOLUTION is shown in Proposition 7. Given that we perform arbitrarily many calls to EXPLORATORYTESTING eventually all grid points are sampled. So the length d of the longest edge in the triangulation satisfies $kd \leq \epsilon$ (Proposition 3).

Proposition 2 shows for the given c with $\forall i : f(i, c(i)) \geq 2\epsilon$, there is a c' in the model such that $m_T(i, c'(i)) \geq \epsilon$. By Proposition 6 the search finds a candidate solution c such that $m_T(i, c(i)) \geq \epsilon$. By Proposition 7 the verification is guaranteed to succeed for the candidate c as no valid counterexample exists. As such, VERIFYCANDIDATE will show that $f(i, c(i)) \geq 0$ for all $i \in I$.

The output upon termination of the algorithm depends on the quality of possible solutions in the blackbox function. We distinguish three cases based on the bottleneck value $b := \max_c(\min_{i \in I} f(i, c(i)))$:

$b > 2\epsilon$ The algorithm is guaranteed to find a solution c such that
    $\forall i f(i, c(i)) > 0$.
$0 < b < 2\epsilon$ The algorithm may find a valid solution c such that
    $\forall i f(i, c(i)) > 0$ or it will terminate without a solution.
$b < 0$ The algorithm is guaranteed to terminate without a solution.

The ϵ-completeness guarantees rely on an extensive exploration. Even though a negative result can only be shown for very small problems, this property shows that our algorithm is guaranteed to progress towards a solution.

4.5   Dynamic Lipschitz Bound Estimation

The previous section describes a sound and ϵ-complete synthesis algorithm for the case where the Lipschitz bound for the blackbox function is known. For some systems only the fact that they are Lipschitz continuous is know without also knowing the value of the Lipschitz constant. In that case no soundness or completeness guarantees are possible. However, Lipschitz bound estimation can be incorporated into our method by computing an estimate based on the collected samples after a solution has been found. The algorithm can then be restarted with a more precise or conservative Lipschitz bound and using all previous samples to create the initial model. This way solutions for ever-improving Lipschitz bound estimates can be obtained. Algorithm 5 describes how to combine our method with Strongin's method for Lipschitz estimation. First a small number of random samples are used to compute a first estimate for the Lipschitz bound. Then the blackbox synthesis algorithm is run with the estimated Lipschitz bounds and the existing samples are used to build the initial model. If the Lipschitz bound estimate did change due to the samples collected during synthesis the synthesis algorithm is rerun with the updated Lipschitz bound. Otherwise a result is reported together with the confidence parameter for which the result is correct. The loop can then either be terminated if the confidence is good enough or repeated with a more conservative Lipschitz estimation. To improve the confidence in a found solution the candidate verification algorithm can also be run with more conservative estimates computed only from samples close to the candidate. This is justified, because rapid changes that are far away from the solution do not affect the validity of that solution.

Algorithm 5. Synthesis with dynamic Lipschitz bound estimation.

1: $S \leftarrow \textsc{ExploratoryTesting}(f)$ ▷ Initialization with some random samples
2: select $r \geq 1.0$
3: **loop**
4:      $k \leftarrow \textsc{LipschitzEstimate}(S, r)$
5:      $c, S \leftarrow \textsc{BlackBoxSynthesis}(f, S, k)$
6:      $k' \leftarrow \textsc{LipschitzEstimate}(S, r)$
7:      **if** $k' \leq k$ **then**   ▷ estimate did not change
8:          **if** c is **null** **then**
9:              **yield** (**null**, r)   ▷ Return null and confidence and continue.
10:             increase r
11:         **else**
12:             **yield** (c, r) ▷ Return solution and confidence and continue.
13:             increase r

---

5   Evaluation

We implemented a prototype implementation of our approach in Python. The implementation uses QHull[1] to compute Delaunay triangulations. SciPy [31] is used for its graph implementations. Our implementation uses the incomplete candidate search described in Algorithm 2 with a known Lipschitz constant. For exploratory testing one grid point and two of each uncertainty sample (u and $\hat{u}$) are added. Our prototype implementation and experiments are available at https://doi.org/10.5281/zenodo.13880504.

The focus of our algorithm is to find solutions with few evaluations of the blackbox function. We perform our evaluation on randomly generated problem instances, six CPS examples, and an industrial case study. To provide a baseline for our implementation we compare with a simple DFS search and a state-of-the-art blackbox min-max optimization tool [1,2]. Blackbox optimization has some differences compared to our approach, it requires a template and tries to maximize the function value, but does not provide a correctness guarantee. To obtain a comparable setting we use a piece-wise linear function defined by four points as the template, combine it with our verification function, and report the number of samples where the first verifiably correct function was found.

5.1   Random Perlin Noise Functions

Perlin noise [26] is a type of random gradient noise that includes a smoothing step. By varying the scale parameter, random functions with specific Lipschitz constants may be sampled for arbitrary dimensions. Instances of Perlin noise also tend to include features like dead ends and low valleys that make the problem more difficult. Setting f to an instance of Perlin noise can then supply us with an infinite number of random test cases. Not all instances of Perlin noise, particularly in low dimensions contain a valid solution c. By adding a constant value $o \in \mathbb{R}^+$ to the Perlin noise, the difficulty of the problem can be varied. We use three configurations with a different number of controllables for Perlin noise generation and generate 100 random instances of each. The results for the three evaluated methods are shown as cactus plots in Fig. 10. Our method consistently finds solutions with the lowest amount of samples. In the case of one controllable about half the instances are only solved by our method.

5.2   Cyber-Physical Systems

We also evaluated our approach on CPS problems three of which are illustrated in Fig. 11. The first problem is a robot arm controller where the gripper at the top of the arm has to remain in a target zone and the arm must avoid collisions with obstacles. The input is the position of the target and the controllables are the two joints and the length of the upper segment. The robustness specification is based on the distance to the obstacles and to the target. The next set of problems are

---

[1] http://www.qhull.org/.

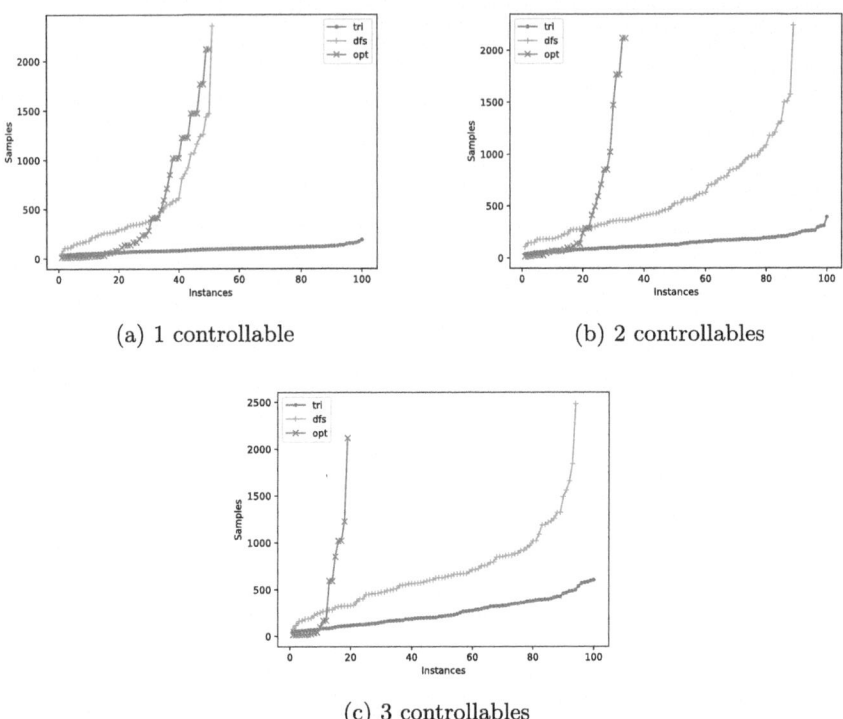

(a) 1 controllable

(b) 2 controllables

(c) 3 controllables

Fig. 10. Cactus plots for solving random instances of Perlin noise problems. It shows how many instances can be solved with a certain amount of samples. The number of simulations is capped at 2500.

motion planning tasks where one or two agents have to avoid moving obstacles and each other. The controller has to select a position for each agent based on the current time. Another problem is a controller for the lights in an office space with four desks, four lamps, and windows. Based on the sunlight intensity the lamps have to be dimmed such that the amount of light at each desk is within a given tolerance.

Our experiments show that our triangulation-based method outperforms the two other approaches on most benchmarks. The optimization-based method performs well on some of the simpler benchmarks where it is slightly better than our method. Our method can solve the harder examples using only a few hundred samples whereas the other methods exceeded the maximum of 10000 samples.

5.3   Industrial Case Study

We applied our method to an industrial bandgap device [6] provided by an industry partner. It is used to provide a reference for voltage and current that is as close as possible to a target value. Figure 12 shows a schematic of the case study.

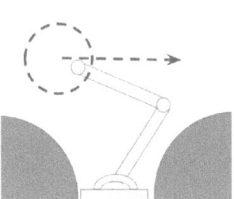
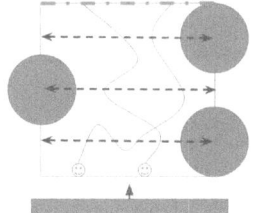

(a) Robot arm control with red obstacles and blue moving target zone.

(b) Multi agent motion planning.

(c) Light control with 4 desks, 4 lamps, and windows on the south side.

Fig. 11. Illustrations of selected CPS problems.

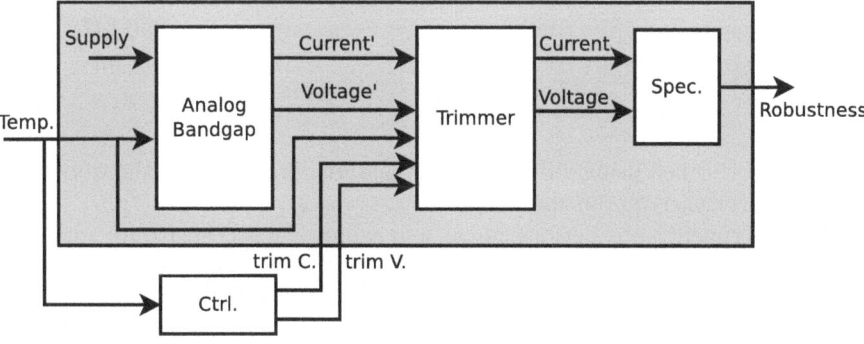

Fig. 12. Schematic of the industrial bandgap case study.

The device consists of two main parts an analog bandgap and a digital trimmer. The analog part produces intermediate current and voltage from a power supply. The trimmer is then used to adjust these to be as close as possible to their respective reference values. Both components show different behavior based on the environment temperature. To compensate for these effects a small controller is used to adjust the trimmer based on the current temperature. In this case study we synthesize this temperature-trimming controller. The relationship between temperature, trimming parameters, and output voltage and current is too complex to be analyzed as a whitebox system. Therefore, we treat it as blackbox system on which simulations can be performed. The blackbox function for our method is marked by a grey box, it consists of the analog bandgap, the trimmer and the specification. This compares the output current and voltages to their reference values and an allowed tolerance. The robustness value shows the distance to the tolerance boundary with positive values inside the tolerance and negative values outside. For every call to the blackbox function, nine simu-

Table 1. Samples required to solve the synthesis problems. DFS is the greedy baseline, OPT is the min-max optimizer from [2], and TRI is the triangulation-based approach presented in this paper. The table reports the median of 5 runs.

Problem	DFS	OPT	**TRI**
Robot arm control	717	—(>2500)	**310**
Multi-agent multi-obstacle motion planning	—(>2500)	—(>2500)	**295**
Multi-obstacle motion planning	—(>2500)	1020	**50**
Multi-agent single obstacle motion planning	—(>2500)	—(>2500)	**346**
Light control	1578	**26**	68
Beach line	—(>2500)	—(>2500)	**34**
Bandgap temperature trimming v1	—(>1000)	—(>1000)	**159**
Bandgap temperature trimming v2	—(>1000)	—(>1000)	**268**
Bandgap temperature trimming v3	—(>1000)	—(>1000)	**152**
Bandgap temperature trimming v4	—(>1000)	—(>1000)	**147**
Bandgap temperature trimming v5	—(>1000)	—(>1000)	**99**

lations are performed using different supply parameters to obtain the worst-case robustness over variation in the supply.

The experiments are performed on a machine-learned surrogate model of the system that was trained from a highly accurate analog electronic circuit simulator model. One evaluation of the blackbox function takes about one second on the ML model and 10 min on the circuit simulator. Table 1 includes the median results of our experiments for five variants of the bandgap device. Out of the 25 runs (5 random seeds for each of the 5 variants) performed for the case study the optimization approach was unable find any solution, the DFS method succeeded 4 times with a best case performance of 632 samples. Our triangulation method succeeded on every run. We can conclude that our triangulation-based approach is the only one able to consistently find a solution in less than 1000 simulations.

Due to the long simulation times, this use case greatly benefits from our method which uses more resources to process the samples but can find verified solutions with a small number of samples.

6   Conclusion

We have presented a novel blackbox synthesis approach to construct a continuous control function for CPS. Our approach works for systems that have to be treated as black boxes because code is either not available or cannot easily be explored analytically. We focus on systems that are expensive to execute, as may be the case for physical or analog systems. Thus, we attempt to minimize the number of executions of the system needed to synthesize a correct controller.

Our algorithm employs a linear interpolation model based on a Delaunay triangulation to identify candidate control functions. It then generates additional test cases to either confirm a candidate or to improve the model. In case good candidates cannot be found, exploratory testing yields more global information about the CPS.

Our algorithm constructs a controller that guarantees correctness for a given Lipschitz bound on the system. This bound may be either known or may be learned during synthesis with a given confidence. Thus, given enough time, we can construct controllers that are correct with arbitrary confidence. The Lipschitz bound limits the amount of testing that needs to be done. The algorithm is guaranteed to terminate with a solution if there is a control function with a quality of at least 2ϵ. We have shown the effectiveness of our approach on random benchmarks and CPS examples.

References

1. Akimoto, Y.: Saddle point optimization with approximate minimization oracle. In: Chicano, F., Krawiec, K. (eds.) GECCO 2021: Genetic and Evolutionary Computation Conference, Lille, France, 10-14 July 2021, pp. 493–501. ACM (2021). https://doi.org/10.1145/3449639.3459266
2. Akimoto, Y., Miyauchi, Y., Maki, A.: Saddle point optimization with approximate minimization oracle and its application to robust berthing control. ACM Trans. Evol. Learn. Optim. **2**(1), 2:1–2:32 (2022). https://doi.org/10.1145/3510425
3. Alur, R., et al.: Syntax-guided synthesis. In: Formal Methods in Computer-Aided Design, FMCAD 2013, Portland, OR, USA, 20-23 October 2013, pp. 1–8. IEEE (2013)
4. Bemporad, A.: Global optimization via inverse distance weighting and radial basis functions. Comput. Optim. Appl. **77**(2), 571–595 (2020). https://doi.org/10.1007/S10589-020-00215-W
5. Bloem, R., Chatterjee, K., Jobstmann, B.: Graph games and reactive synthesis. In: Clarke, E.M., Henzinger, T.A., Veith, H., Bloem, R. (eds.) Handbook of Model Checking, pp. 921–962. Springer (2018). https://doi.org/10.1007/978-3-319-10575-8_27
6. Bloem, R., Larrauri, A., Lengfeldner, R., Mateis, C., Nickovic, D., Ziegler, B.: Industry paper: surrogate models for testing analog designs under limited budget - a bandgap case study. In: International Conference on Hardware/Software Codesign and System Synthesis, CODES+ISSS 2022, Shanghai, China, 7-14 October 2022, pp. 21–24. IEEE (2022). https://doi.org/10.1109/CODES-ISSS55005.2022.00016
7. Corso, A., Moss, R.J., Koren, M., Lee, R., Kochenderfer, M.J.: A survey of algorithms for black-box safety validation of cyber-physical systems. J. Artif. Intell. Res. **72**, 377–428 (2021). https://doi.org/10.1613/JAIR.1.12716
8. Dang, T., Donzé, A., Maler, O., Shalev, N.: Sensitive state-space exploration. In: CDC, pp. 4049–4054. IEEE (2008)
9. Deshmukh, J., Horvat, M., Jin, X., Majumdar, R., Prabhu, V.S.: Testing cyberphysical systems through bayesian optimization. ACM Trans. Embedded Comput. Syst. (TECS) **16**(5s), 1–18 (2017). https://doi.org/10.1145/3126521
10. Donzé, A., Maler, O.: Robust satisfaction of temporal logic over real-valued signals. In: Chatterjee, K., Henzinger, T.A. (eds.) FORMATS 2010. LNCS, vol. 6246, pp. 92–106. Springer, Heidelberg (2010). https://doi.org/10.1007/978-3-642-15297-9_9

11. Eddeland, J.L., Donzé, A., Åkesson, K.: Multi-requirement testing using focused falsification. In: Proc. of HSCC 2022: 25th ACM International Conference on Hybrid Systems: Computation and Control. pp. 4:1–4:11. ACM (2022). https://doi.org/10.1145/3501710

12. Edo, H., Miyauchi, Y., Maki, A., Akimoto, Y.: Trade-off between robustness and worst-case performance in min-max optimization. In: GECCO, pp. 1339–1347. ACM (2023). https://doi.org/10.1145/3583131.3590362

13. Ernst, G., Sedwards, S., Zhang, Z., Hasuo, I.: Falsification of hybrid systems using adaptive probabilistic search. ACM Trans. Model. Comput. Simul. **31**(3), 18:1–18:22 (2021). https://doi.org/10.1145/3459605

14. Fainekos, G.E., Pappas, G.J.: Robustness of temporal logic specifications for continuous-time signals. Theor. Comput. Sci. **410**(42), 4262–4291 (2009). https://doi.org/10.1016/J.TCS.2009.06.021

15. Fazlyab, M., Robey, A., Hassani, H., Morari, M., Pappas, G.J.: Efficient and accurate estimation of lipschitz constants for deep neural networks. In: NeurIPS, pp. 11423–11434 (2019)

16. Girard, A., Pappas, G.J.: Verification using simulation. In: Hespanha, J.P., Tiwari, A. (eds.) HSCC 2006. LNCS, vol. 3927, pp. 272–286. Springer, Heidelberg (2006). https://doi.org/10.1007/11730637_22

17. Gruenbacher, S.A., Lechner, M., Hasani, R.M., Rus, D., Henzinger, T.A., Smolka, S.A., Grosu, R.: Gotube: Scalable statistical verification of continuous-depth models. In: Thirty-Sixth AAAI Conference on Artificial Intelligence, AAAI 2022, Thirty-Fourth Conference on Innovative Applications of Artificial Intelligence, IAAI 2022, The Twelveth Symposium on Educational Advances in Artificial Intelligence, EAAI 2022 Virtual Event, 22 February -1 March, 2022, pp. 6755–6764. AAAI Press (2022). https://doi.org/10.1609/AAAI.V36I6.20631

18. Gulwani, S., Polozov, O., Singh, R., et al.: Program synthesis. Foundati. Trends® Program. Lang. **4**(1-2), 1–119 (2017)

19. Huang, J.W., Roberts, S.J., Calliess, J.P.: On the sample complexity of lipschitz constant estimation. Trans. Mach. Learn. Res. (2023)

20. Kaible, V., Peinhardt, M.A.: On the bottleneck shortest path problem. ZIB-Report (May 2006)

21. Kitzelmann, E.: Inductive programming: a survey of program synthesis techniques. In: Schmid, U., Kitzelmann, E., Plasmeijer, R. (eds.) AAIP 2009. LNCS, vol. 5812, pp. 50–73. Springer, Heidelberg (2010). https://doi.org/10.1007/978-3-642-11931-6_3

22. Liu, S., et al.: Min-max optimization without gradients: Convergence and applications to black-box evasion and poisoning attacks. In: ICML. Proceedings of Machine Learning Research, vol. 119, pp. 6282–6293. PMLR (2020)

23. Maler, O., Nickovic, D.: Monitoring properties of analog and mixed-signal circuits. Int. J. Softw. Tools Technol. Transf. **15**(3), 247–268 (2013). https://doi.org/10.1007/S10009-012-0247-9

24. Marzat, J., Walter, E., Piet-Lahanier, H.: Worst-case global optimization of black-box functions through kriging and relaxation. J. Glob. Optim. **55**(4), 707–727 (2013). https://doi.org/10.1007/S10898-012-9899-Y

25. Nghiem, T., Sankaranarayanan, S., Fainekos, G., Ivancic, F., Gupta, A., Pappas, G.J.: Monte-carlo techniques for falsification of temporal properties of non-linear hybrid systems. In: Proc. of HSCC 2010: the 13th ACM International Conference on Hybrid Systems: Computation and Control, pp. 211–220. ACM (2010). https://doi.org/10.1145/1755952.1755983

26. Perlin, K.: An image synthesizer. ACM SIGGRAPH Comput. Graph. **19**(3), 287–296 (1985). https://doi.org/10.1145/325165.325247
27. Seeger, M.: Gaussian processes for machine learning. Int. J. Neural Syst. **14**(02), 69–106 (2004). https://doi.org/10.1142/S0129065704001899
28. Sibson, R.: Locally equiangular triangulations. Comput. J. **21**(3), 243–245 (1978). https://doi.org/10.1093/comjnl/21.3.243
29. Strongin, R.: On the convergence of an algorithm for finding a global extremum. Eng. Cybernetics **11**, 549–555 (1973)
30. Teel, A., Praly, L.: Tools for semiglobal stabilization by partial state and output feedback. Siam J. Control Optim. (1995)
31. Virtanen, P., et al.: SciPy 1.0 Contributors: SciPy 1.0: fundamental algorithms for scientific computing in python. Nat. Methods **17**, 261–272 (2020). https://doi.org/10.1038/s41592-019-0686-2
32. Watson, D., Philip, G.: Systematic triangulations. Comput. Vis. Graph. Image Process. **26**(2), 217–223 (1984). https://doi.org/10.1016/0734-189X(84)90184-1
33. Wood, G.R., Zhang, B.P.: Estimation of the lipschitz constant of a function. J. Glob. Optim. **8**(1), 91–103 (1996). https://doi.org/10.1007/BF00229304
34. Zhang, Z., Arcaini, P.: Gaussian process-based confidence estimation for hybrid system falsification. In: Huisman, M., Păsăreanu, C., Zhan, N. (eds.) FM 2021. LNCS, vol. 13047, pp. 330–348. Springer, Cham (2021). https://doi.org/10.1007/978-3-030-90870-6_18

Applications

Automated Flaw Detection for Industrial Robot RESTful Service

Yuncheng Wang[1,2], Puzhuo Liu[3], Yaowen Zheng[1,2], Dongliang Fang[1,2], Shuaizong Si[1,2(✉)], Zhiwen Pan[1,2], Weidong Zhang[1,2], and Limin Sun[1,2(✉)]

[1] Beijing Key Laboratory of IOT Information Security Technology, Institute of Information Engineering, Chinese Academy of Sciences, Beijing, China
{wangyuncheng,zhengyaowen,fangdongliang,sishuaizong,panzhiwen,
zhangweidong,sunlimin}@iie.ac.cn
[2] School of Cyber Security, University of Chinese Academy of Sciences, Beijing, China
[3] Tsinghua University, Beijing, China
liupz@mail.tsinghua.edu.cn

Abstract. As industrial robots become an integral part of Industry 4.0 in the manufacturing sector, their interconnection and interoperability introduce significant security challenges. RESTful Web services have emerged as the preferred method for network communication due to their simplicity and ease of use. However, the effective detection of security flaws in RESTful services for industrial robots still faces three key challenges: high-quality test case generation, high-throughput testing, and anomaly detection. Unlike traditional applications deployed within cloud services, limited computational resources, unique controller states, and unclear API specifications in robots further complicate the resolution of these challenges. Consequently, a large number of security flaws persist in real and deployed devices, with some flaws even posing the risk of physical damage.

To address these challenges, we propose a novel testing technique named RobRest specifically designed for emerging RESTful services in the context of robotic systems. In test case generation, RobRest analyzes description fields extracted from the OpenAPI specification, ensuring the generation of high-quality test cases. During abnormality observation, RobRest combines both cyber and physical space states to identify anomalies in the target service. Additionally, RobRest automatically customizes each testing request to the service, minimizing resource usage within the robot controller and bypassing the quantity restrictions present in the controller. Applying RobRest to industrial robots, we identified a total of 19 system flaws (4 vulnerabilities and 15 bugs), and 2 of them have been assigned CVE IDs. Exploiting them can affect a multitude of industrial robots in the physical world.

1 Introduction

With the advancement of industry automation and intelligent manufacturing, an increasing number of industrial robots are being deployed in factory assembly

K. Shankaranarayanan et al. (Eds.): VMCAI 2025, LNCS 15530, pp. 163–184, 2025.
https://doi.org/10.1007/978-3-031-82703-7_8

processes, and this trend is expected to accelerate in the coming years [19,21,35]. Industrial robots are now seamlessly integrated into corporate networks and connected to the internet [34], enhancing communication between operators and robots. Representational State Transfer (REST, or RESTful) services, known for their popularity and flexibility in managing and accessing web resources, have gradually emerged as a preferred choice in industrial robotics, facilitating such communication [17,26,37].

However, this increased connectivity of industrial robots brings along cybersecurity risks. If vulnerabilities exist in the network interactions, attackers can exploit them to launch attacks that can cause severe damage. This may include damaging manufacturing equipment, reducing production quality and efficiency, posing risks to human operators, and even causing the complete unavailability of industrial robots. In the industrial sector, the disruption of availability is particularly severe [8] as such disruptions may cause the production line to stagnate, hinder factory production, and cause serious economic losses. For example, the average cost of robot downtime among manufacturers exceeds \$10,000 per minute [36]. Moreover, when a working robot suddenly shuts down, the robot's built-in mechanical reset setting may be triggered, stretching and moving from the working position to the reset position. This process may cause harm to on-site workers [10]. Therefore, it is urgent to detect vulnerabilities and bugs in the built-in RESTful service of industrial robots.

Research Scope. This paper aims to detect vulnerabilities and bugs in industrial robot RESTful service that affect the availability of robots. Specifically, we focus on (1) uncovering bugs that can trigger internal errors in the industrial robot RESTful service and (2) identifying vulnerabilities that can cause denial of service of industrial robots. We refer to them collectively as **flaws**.

Existing Work. Traditional RESTful API testing has made some progress. Restler [5] and Morest [27] are two cutting-edge black-box testing techniques for RESTful APIs. They take the OpenAPI [29] specification (also referred to as Swagger [30]) as input to learn API dependencies, enabling the generation of test sequences, each sequence consisting of multiple requests. NAUTILUS [11] and VoAPI [12] explore the detection of injection vulnerabilities in REST APIs by extending capture flags and employing various complementary strategies. Moreover, scanning tools such as ZAP [31] can be used to discover known vulnerabilities. However, existing work cannot effectively test robots' REST APIs. The specific reasons are described in the following paragraph.

Challenges. To the best of our knowledge, there is no RESTful API testing effort geared towards industrial robotics. Effectively testing industrial robot REST APIs requires addressing the following unique challenges compared to traditional RESTful API testing:

❶ **How to Generate High-Quality Test Cases with Correct Input Parameters?** When generating RESTful API test cases, it is crucial to generate the correct parameters for the API request, as this will help reach deeper states of the program for testing purposes. Existing techniques [5,11,12,27,31] can only generate request parameters from the response and parameter examples in the API specification, or they rely on inefficient random mutation for parameter generation. Their request construction strategy does not effectively utilize the potential parameter constraint information in the specification to enhance the quality of parameter construction during the request process. For those parameters that cannot be obtained through dependencies, their correct values cannot be obtained directly, and it is difficult to quickly mutate into valid values, resulting in insufficient testing efficiency.

❷ **How to Efficiently Launch High-Throughput Testing?** In RESTful API testing, the test space is extremely large, and the testing process consumes significant resources. However, existing methods do not consider the resource constraints present within industrial robot controllers, which can result in the inability to handle a large volume of requests within a short period, leading to a significant reduction in throughput. For traditional RESTful service testing, these services run in cloud environments with abundant resources; they can withstand a large number of testing requests. In contrast, industrial robot controllers have a strict session management mechanism for external requests. The number of sessions must be maintained within a specific limit; otherwise, new requests will be rejected to prevent resource exhaustion. Therefore, it is challenging to execute a large number of requests for robot service within a short period of time.

❸ **How to Effectively Detect Abnormal States in Robot Services?** Visibility of flaws is very important. Ignoring or omitting certain abnormal indicators can result in some flaws going uncaptured. Restler and Morest rely on identifying the status of cyberspace (e.g., "500 Internal Server Error") as a sign of capturing bugs in the API. Moreover, approaches [11,12,31] utilized additional cyber information to capture vulnerabilities within the RESTful service. However, these techniques are blind to physical device anomalies. Robots are typical cyber-physical systems, so solely relying on cyberspace states to assess whether a service has entered an abnormal state is insufficient. Therefore, it is essential to integrate both cyber and physical space states to detect anomalies in robot web services effectively.

Our Solution. To overcome the aforementioned three challenges, this paper introduces RobRest, a testing framework designed specifically for industrial robot web services. To generate high-quality test cases with correct API parameters, RobRest utilizes a Large Language Model (LLM) to extract candidate values for request parameters from the description fields of the OpenAPI specification in a one-time effort. This LLM effort establishes a parameter dictionary to store these parameter-related values, which are then heuristically filled into individual test requests throughout the API testing process. For launching high-throughput testing on the robot service, RobRest automatically customizes each testing request to the service, minimizing resource usage within the robot controller and bypassing the quantity restrictions present in the controller. To detect abnormal states in the robot service during testing, RobRest employs a comprehensive approach by utilizing both the robot message response codes and controller status codes to determine if the robot has entered an abnormal state.

Evaluation. We selected the Robot Web Service (RWS) from ABB, which represents a leading industrial robot manufacturer, as our evaluation target. Moreover, ABB's RWS REST API allows interaction with different controller resources, ensuring a thorough evaluation. We evaluate both flaw detection capability and API operation coverage of RobRest. The results demonstrate that RobRest outperforms the state-of-the-art techniques Morest and ZAP, uncovering 90% more flaws and achieving an average increase of 30.94% in API operation coverage. In total, RobRest identified 19 flaws confirmed by the developers, including 4 vulnerabilities (2 previously unknown and 2 known) and 15 bugs. According to feedback from the manufacturer, the vulnerabilities we discovered are present in two mainstream robot controllers, IRC5 and OmniCore. Since nearly all ABB robots currently use these two controllers, malicious exploitation of these vulnerabilities could result in many physical robots experiencing denial of service, severely disrupting production processes.

Contribution. We make the following contributions:

- To the best of our knowledge, we are the first to investigate system flaws affecting the availability of industrial robot RESTful services and propose an efficient detection approach for them.
- We implement an automated testing tool RobRest, capable of detecting bugs and vulnerabilities in the RESTful service of industrial robot controllers (we uploaded our artifacts).
- We apply RobRest to robot web service from the leading industrial robot manufacturer, and identify 4 vulnerabilities and 15 bugs. The developers confirmed the findings, and 2 unknown vulnerabilities were assigned CVE IDs due to their critical threat.

2   Background and Motivation

2.1   Industrial Robot Network Architecture

An industrial robot is "an automatically controlled, reprogrammable multipurpose manipulator, programmable in three or more axes, which can be either fixed in place or fixed to a mobile platform for use in automation applications in an industrial environment" [25]. It consists of manipulators and actuators, controllers, and the software and hardware for teaching or programming the robots (including communication interfaces). The industrial robot is controlled through the controller, and any interaction with the robot is actually completed through the controller. With the development of Industry 4.0 and the Industrial Internet of Things (IIOT), industrial robots can be controlled locally via a Local Area Network (LAN) or programmed and maintained remotely over a remote network [24]. Teach pendants and supervisory computers (PCs) enable on-site control and offline programming of industrial robots through a LAN connection. Remote users can communicate with the controller through industrial routers (or service boxes) and industrial gateways for remote programming and maintenance, as illustrated in Fig. 1. In consideration of the integration with external

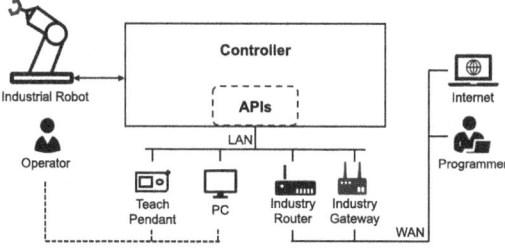

Fig. 1. Overview of Industrial Robot Network Architecture.

systems, the robot controller has incorporated complex APIs for handling external network requests.

2.2 RESTful API

REST is an architectural style and communication method for web services development [14]. Services that provide RESTful APIs are called RESTful services. RESTful APIs communicate via HTTP requests and provide a standardized set of interfaces that define external requests for creating, reading, updating and deleting (CRUD) operations on service resources. A RESTful service contains multiple endpoints, each representing a unique URL path within the service. Each URL path supports specific HTTP methods to perform predefined functions, called API operations.

OpenAPI Specification. The OpenAPI specification is typically stored as a YAML or JSON file, containing information about API endpoints, available CRUD operations, input parameters, responses, descriptions, etc. [29]. Figure 2 presents a fragment of the OpenAPI specification for RWS, where the shaded areas indicate API endpoints and the oval highlights indicate the CRUD operations supported at each endpoint.

RESTful API Testing. The core idea of RESTful API testing is to automatically generate and send test request sequences, and record the request sequences that can trigger internal errors in the RESTful service. To delve into errors existing in the reachable execution states of RESTful services, existing works [5,27] use black-box techniques to test the service through its RESTful API. They generate request sequences that can reach different states of the service under test, using the OpenAPI specification as input and guiding the testing based on the responses to these requests. The presence of 50X status codes in the response is taken as an indication of service anomalies. In addition, [11,12] have introduced indicators for injection-type vulnerabilities, such as data structure and semantic relation anomalies.

RESTful-service Property Graph. MOREST [27] proposed using the RESTful Service Property Graph (RPG) to represent the relationship between REST APIs and schemas. Utilizing RPG allows for a more detailed description of

```
swagger: '2.0'
paths:
  /rw/rapid/uiinstr/active/param/{stackurl}/{uiparam}:
    get:
      parameters:
      - name: uiparam
        in: path
        description: '{uiparam}'
        required: true
        type: string
    post:
      description: >
        ...
        Example of {uiparam} are TPFK1, TPFK2, TPFK3, TPCompleted etc.
        ...
      parameters:
      - name: uiparam
        in: path
        description: '{uiparam}'
        required: true
        type: string
      - name: mastership
        in: query
        description: '{implicit | explicit} by default mastership is explicit'
        required: false
        type: string
      - name: request-body
        in: body
        description: Data parameter(s)
        schema:
          type: object
          properties:
            value:
              description: >-
                {value} For example, TPFK3 can accept a value like `0` and
                TPCompleted can accept a value like TRUE
              type: string
              example: value=string
  /rw/panel/opmode:          /rw/panel/opmode/acknowledge:
    get:                       post:
      responses:                 parameters:
        '200':                   - name: request-body
          schema:                  schema:
            properties:              properties:
              opmode                   opmode:
                                         description: '{auto | manf | coldet}'
                                       example: opmode=string
```

Fig. 2. The OpenAPI specification of ABB RWS*. * For clarity, we omit some content in the specification.

the producer-consumer dependency relationships among APIs, as well as the attribute equivalence relationships between schemas. RPG is a multigraph $G = (V, E, \lambda, \mu)$ where $V = S \cup O$ is a set of nodes, S is the set of schemas and O is the set of operations. $E = E_{OS} \cup E_{SO} \cup E_{SS} \cup E_{OO}$ is the set of edges between any two nodes in V. λ indicates that there is a connection between two schemas or between a schema and an operation, while μ assigns properties to nodes in V. RPG can encode the following three types of dependency relationships: (1) Data-flow among schemas and operations: E_{OS} and E_{SO} contain the edges which represent an operation node in O respectively producing or consuming a schema object in S. (2) Schemas relations: if a property p_1 in schema s_1 is equivalent to a property p_2 in another schema s_2, it indicates that there is an association between these two schemas. (3) CRUD relations of operations: the order of occurrence between CRUD operations.

Industrial Robot RESTful API. RESTful APIs have been popular in traditional web services for a long time, but they have only recently begun to gain attention in industrial robots as part of the digital transformation proposed by

Industry 4.0 [15]. Many industrial robot manufacturers, such as leading companies ABB [18] and KUKA [2], have already opened or are planning to open their RESTful APIs. For instance, the Robot Web Service [7] built into ABB robot controllers offers a set of RESTful APIs. Through these interfaces, external requests or third-party applications can directly access a wealth of resources within the robot controller. This includes accessing and modifying any file, reading and setting IO signals, reading and modifying RAPID programs, and reading internal controller status, among other functionalities. With the increasing openness of robots, directly controlling robots through RESTful API has become a standardized and practical method that combines simplicity and versatility.

2.3   Motivation

Given the current limitations and inability of conventional REST API testing tools for general cloud services to be directly applicable in the context of industrial robots (as discussed in Sect. 1), this paper aims to provide a high-throughput black-box testing tool for industrial robot RESTful APIs to discover flaws in controller services. Our work is motivated by the following three aspects.

Parameter Examples Provided in the Openapi Specification May Be Inaccurate. Existing work [5,11,12,27] mainly leverages the OpenAPI specification to extract example values as input parameters to assist testing. However, example values are often incorrect or missing, which leads to inefficient testing. For instance, both POST operations in Fig. 2 have example fields, but the example values for the parameters value and opmode are incorrect. This is because some specification example fields only describe the parameter types without focusing on actual values. In contrast, description fields may contain rich and meaningful parameter information within these specifications, such as correct sample values for parameters and even unstructured natural language descriptions of certain parameter value ranges. As shown in Fig. 2, some description fields may contain the correct values for parameters. These description fields can help improve the correctness of input parameters.

Industrial Robots have Inherent Resource Limitations. Traditional application scenarios, such as cloud services, are designed to handle a large number of requests without considering communication resource consumption. In contrast, industrial robot controllers impose strict limitations on the resources and the number of network communication interfaces they can occupy. Existing research has not addressed these limitations due to varying testing objectives, resulting in inefficient testing. When communication resources become saturated, robots are unable to accept additional test requests. To address this issue, we implement resource reuse strategies within the robot to reduce communication resource consumption while maintaining compliance with request requirements.

Relying Solely on Cyber Status Monitoring May Fail to Detect Certain Flaws. Current REST testing uses changes in HTTP status codes [5,27], data structures, and semantic relationships [11,12] to observe bugs and vulnerabilities in services, which is sufficient for conventional cyber systems. However,

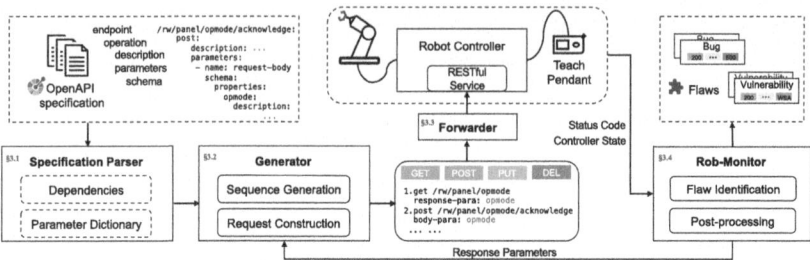

Fig. 3. The Architecture of RobRest.

for cyber-physical systems, where network information (RESTful API requests) impacts physical entities (robot controllers), solely observing anomalies from the network perspective is insufficient and may result in overlooking certain flaws. Nevertheless, we can infer potential flaws triggered by the requests by observing changes in the physical entities of the robot. Specifically, after sending test requests, monitoring the state changes of the robot controller (e.g., power on/off) can directly indicate whether the request leads to any abnormal conditions.

3 Approach

Figure 3 shows the overview of RobRest. The core process involves iteratively constructing API operation sequences and sequentially dispatching individual requests to evaluate the RESTful service within the industrial robot controller via the REST API. RobRest generates API requests using a Large Language Model (LLM) and extracts valid parameter candidate values from the OpenAPI specification (Sect. 3.1), filling in potentially valid parameters for requests within the sequence (Sect. 3.2). Then, the generated requests are sent to the industrial robot RESTful service through a forwarder (Sect. 3.3). Finally, to discover flaws in the industrial robot RESTful service, we construct a new monitor to analyze and detect the feedback from the service and the status of the industrial robot controller (Sect. 3.4).

3.1 Specification Parser

Parameter Candidate Value Extraction. RobRest aims to extract valid information from the OpenAPI specification to help generate correct parameters that can be accepted by the service, thereby improving the quality of the constructed requests. As indicated by the underline in Fig. 2, we observed that the OpenAPI specification for industrial robots contains information helpful for constructing high-quality requests. There are a large number of natural language descriptions in the description field, which include the value range and precise examples for input parameters. Furthermore, we observed that an operation within the same endpoint may contain descriptions of input parameters for other operations within that API endpoint. As shown in Fig. 2, the first endpoint contains two operations, within which the description for the POST operation includes the value for the parameter uiparam, which can be used in the GET operation.

Based on the scope of these descriptions, we categorize the extracted parameters into two types: (1) parameter-based candidate dictionary (D_P), which involves extracting possible values for a parameter from the description of that single parameter, and (2) operation-based candidate dictionary (D_O), which involves extracting possible values for parameters across all API endpoints from the description of the operations.

Extracting precise constraints from natural language descriptions in specifications is a challenging task. These specifications can span hundreds of pages, requiring significant manual effort to carefully read and interpret the documents. This process is prone to errors. Regular expression matching-based solutions are not suitable due to the flexible nature of natural language descriptions, which may use different terms or sentences to convey the same concept. Fortunately, large language models have demonstrated their capability in understanding natural language, particularly the semantic information conveyed in the description field. They excel at extracting useful information and organizing it in the desired formats, such as JSON, through the use of carefully constructed prompts. Therefore, we propose leveraging the power of LLMs to automatically obtain the potential values for each input parameter.

We first extract these description fields and corresponding parameters from the specification. Then, we perform LLM-based knowledge extraction on the descriptions to generate a candidate dictionary for parameters. Note that this process is a one-time effort. Furthermore, through our actual testing, the average cost of retrieving parameter values from each description field was only \$0.0052. Compared with the saved labor, time and other costs, we think this is acceptable. Algorithm 1 describes how RoBREST extracts D_P and D_O from the entire OpenAPI specification and uses them to generate the candidate dictionary (S) for parameters. Firstly, for D_P, if input parameters exist under the current API operation C (Line 12), we extract the candidate value d_p for each parameter p from its corresponding description field and add d_p to D_P. For example, the first endpoint in Fig. 2 contains the parameter mastership and its description "{implicit | explicit} by default mastership is explicit" within the POST operation. Secondly, for D_O, extraction is performed at the level of the API endpoint. Specifically, we extract the description fields of each operation within an endpoint (P) as *Descrips* (Line 11) and extract all input parameters under P as *Paras* (Line 13). Then, using $<Descrips, Paras>$ as input, we utilize the LLM to extract potential initial values for parameters that could be used in the context of parameter names (Line 17). After that, D_P and D_O are merged and deduplicated to obtain the candidate dictionary (S_P) of the endpoint P (Line 18–19). Finally, by continuously traversing all endpoints in the API specification, a comprehensive candidate dictionary S for all parameters is obtained (Line 2–4).

Taking Fig. 2 as an example, by utilizing the parameters and operation description field in POST /rw/rapid/uiins-tr/active/param/{stackurl}/{uiparam} and

---

Algorithm 1. Request Parameter Value Generation

Input: Ps: Specification of RESTful API endpoints
Output: S: the candidate dictionary of input parameters

1: **def** *get_dictionary*(Ps) :
2:    **for** P ∈ Ps **do**
3:        S ← S ∪ *parse_endpoint*(P);
4:    **return** S;
5: **def** *parse_endpoint*(P) :
6:    D_P ← ∅;
7:    D_O ← ∅;
8:    Paras ← [];
9:    Descrips ← [];
10:    **for** C ∈ P.operation **do**
11:        Descrips.*add*(C.descrip);
12:        **if** C.has_parameter **then**
13:            Paras.*add*(C.parameter);
14:            **for** p ∈ C.parameter **do**
15:                d_p ← *extract_value*(p.name, p.descrip);
16:                D_P ← d_p ∪ D_P;
17:    D_O ← *extract_value*(Paras, Descrips);
18:    S_P ← D_P ∪ D_O;
19:    **return** S_P;
20: **def** *extract_value*(paraname, descrip) :
21:    *value* ← ∅;
22:    **for** text ∈ descrip **do**
23:        value ← value ∪ *LLM*(prompt, paraname, text);
24:    **return** value;

---

extracting candidate values with the LLM, we can obtain the values for the parameters (S_{P1}): ① uiparam with values {'TPFK1', 'TPFK2', 'TPFK3', 'TPCompleted'}, ② mastership with values {'implicit', 'explicit'}, and ③ value with values {'0', 'TRUE'}. Among these, ① is obtained using D_O, while ② and ③ are obtained through D_P. Similarly, from the other two endpoints we can get S_{P2}={""} and S_{P3}={"opmode": ["auto", "manf", "coldet"]}. Finally, we can obtain the comprehensive candidate dictionary for all parameters, $S = S_{P1} ∪ S_{P2} ∪ S_{P3} ∪ ... ∪ S_{Pn}$.

By utilizing this parameter dictionary (S), we can pre-construct generic initial values for all parameters. Specifically, the specification writer may describe the definition of some parameters in the previous operation, but omit the description of the same parameters later. Therefore, when an input parameter in a certain operation lacks a specific description at that endpoint, it can directly obtain the description of the same parameter from other operations within the dictionary, thereby enhancing the quality of parameter generation.

Dependency Extraction. ROBREST parses the specification to extract various dependency relationships, including CRUD relations, schemas relations, and the data-flow among the schemas and operations. It constructs a RESTful-service Property Graph as described in Sect. 2.2.

3.2 Generator

RobRest divides the process of generating test cases into two steps. First, it traverses the RESTful-service Property Graph (RPG) to create API request call sequences. Subsequently, for each request within these sequences, we use (1) the candidate values in the parameter dictionary, (2) the parameter values in the response, and (3) random values of a specific type to construct the required parameter values and send them sequentially to the targeted RESTful service under testing.

Sequence Generation. The basic approach involves accessing each schema within the RPG, gathering any API operations related to that schema, and generating request call sequences. If operation o_1 connected to the current schema (s_1) is associated with another schema (s_2), those operations are linked to the current sequence to expand it. Once the call sequence is generated, CRUD relations are utilized to filter out edges that do not satisfy CRUD relationships.

For example, in Fig. 2, there exists a dependency relationship between the schema of GET /rw/panel/opmode and POST /rw/panel/opmode/acknowledge, thus generating the request sequence <GET /rw/panel/opmode, POST /rw/panel/opmode/acknowledge>.

Request Construction. After generating the request sequence, it is essential to supplement the input parameters for each request to ensure proper reception by the service. RobRest assigns values to the input parameters of the request based on the following conditions: Firstly, if the current request has been successfully executed previously, there is a high probability of continuing to use the parameters from the last successful run. Secondly, if the parameters to be generated are present in the parameter dictionary (obtained in the previous section), a value is randomly selected from the dictionary for the given parameter name. Additionally, we dynamically generate potential parameters during the testing process. Thirdly, if an operation's parameters are based on the response of a previous operation, RobRest waits for the completion of the preceding operation to obtain the required parameter values. This means that if there is a producer-consumer relationship between the parameters of the two operations, then the consumer parameters depend on the producer parameters that exist in the response of the previous request. The obtained values from the response are then used to construct the input parameters for the current request. Otherwise, based on the data type, RobRest randomly generates specific values as input parameters. If no specific data type is indicated, a randomly sized string is used as the input value.

3.3 Forwarder

The forwarder module is responsible for facilitating high-throughput testing of the industrial robot REST API. REST API testing involves generating a large number of requests in a short period to continuously test the RESTful service. However, industrial robots limit the number of connections for RESTful services.

Specifically, the industrial robot controller contains a session table with limited storage capacity. Each external request, after passing a validity check, results in the controller generating a unique session ID as part of the response, which is then stored in the local session table. Consequently, sending a large number of REST requests to the controller can quickly fill the session table, causing new external requests to be rejected until previous sessions expire or the controller is restarted. This significantly impacts performance during testing.

To address this limitation, we extract two constraints that external requests must meet during interaction with industrial robots through the REST API: legitimacy constraints and resource constraints. Firstly, legitimacy constraints require that the functionality and authentication fields of external requests pass checks by the RESTful service. Therefore, we use a sniffing message `"GET /rw"` with the correct function field and authentication information to establish a connection with the robot service. Secondly, to tackle resource constraints related to session quantity, we extract the `"http-session"` and `"ABBCX"` fields from the returned information of the sniffed message and insert the relevant fields into the header of each subsequent request. This process is repeated after every controller restart. The forwarder ensures minimal use of computing resources. Therefore, even if other manufacturers' robots do not have session limitations, using the Forwarder module can smoothly conduct testing without redesigning the solution.

3.4 Rob-Monitor

Flaw Identification. First, we introduce the concept of checking the state of industrial robot controllers. We observed a unique anomaly indicator within industrial robots: the controller state. This is because requests sent through the REST API can impact the controller, so when an internal error occurs in the controller, it may intuitively reflect on the controller state and its availability. When a request triggers a system vulnerability in the controller, causing an exceptional crash, the controller state changes from the normal `"started"` to `"closed"`. This state change occurs in the physical world, is applied to real industrial robot equipment, and can be intuitively seen by the operator. In order to automatically identify the occurrence of this physical event, we need to use other programmable events that occur simultaneously with this event as indicators. Since the controller is actually a Windows-based system, this process generates *Windows Socket Error (WSA)*. To capture the controller state in industrial robots, we identify Windows socket error codes related to industrial robot controllers, such as *WSAECONNABORTED* and *WSAECONNRESET*. Secondly, to maintain generality, we use common HTTP status codes in REST API testing as bug indicators. If Rob-Monitor receives a response message from the service with a status code of 50X, it is considered an error. We store the request sequence and response that can trigger errors for later analysis.

Post-processing. Triggering the vulnerability causes the controller to fall into a denial of service state, so we need to restore the controller state. In the test environment, the controller's startup, shutdown, and restart are managed through

GUI buttons in the simulation software. Therefore, when the `"closed"` state is detected, we simulate GUI operations to complete a mechanical restart of the controller device.

4 Evaluation

Our evaluation aims to address the following research questions:

RQ1 (Flaw Detection) How effective is the flaw detection capability of ROBREST?

RQ2 (Coverage) How well does ROBREST explore API operations?

RQ3 (Ablation Study) How does the parameter candidate value extraction strategy affect the ability to discover flaws of ROBREST?

4.1 Experimental Setup

Implementation. We implemented the prototype ROBREST. In the *Specification Parser* and *Generator* modules, we first parsed the OpenAPI specification and then used GPT-4 [28] to generate the parameter dictionary and conduct request construction. Our approach was based on the state-of-the-art method MOREST [27] for dependency extraction, which was then used to generate request sequences. Furthermore, we implemented the *Forwarder* module based on mitmproxy [22]. For the *Rob-Monitor* module, we wrote scripts that utilize HTTP status codes and Windows socket error codes to identify flaws. Then, we developed guiAutolits scripts for the post-processing stage based on AutoIt [6], which are used to recover the controller state after triggering controller vulnerabilities.

Compared Tools. To assess the capability of Rob-Monitor in detecting vulnerabilities in Industrial Robots, we constructed a prototype without the Rob-Monitor, named ROBREST_NR. We compared ROBREST and ROBREST_NR with state-of-the-art tools on RESTful API testing and a popular vulnerability scanner. RESTful API testing tools are designed to identify bugs within a service but are not capable of revealing deeper vulnerabilities. Moreover, current testing tools [5,27] cannot be directly applied to test RESTful services of industrial robot controllers because they do not consider the limitations of embedded devices, leading to a large number of requests being rejected by the service. Vulnerability scanners are designed to assess web applications and APIs. Given the request for an API endpoint, they send modified requests to these endpoints and report potential vulnerabilities directly. For effective evaluation, we selected the latest scalable RESTful API testing approach and a vulnerability scanner, both of which support tampering with request headers. Finally, we chose the following two tools and modified their request sending logic for high-throughput testing.

(1) MOREST [27] is a state-of-the-art black-box testing tool that fully utilizes dependencies in the specification to build and dynamically maintain an RPG for bug detection in RESTful services.

(2) ZAP [31] is the most popular black-box vulnerability scanning tool. We use its OpenAPI add-on to parse specifications and conduct security tests on RESTful services.

Evaluation Object. We evaluated the above tools on ABB RWS [7] with all 19 REST APIs as shown in Table 2. The reasons for choosing RWS are as follows. Firstly, ABB is one of the world's largest manufacturers of industrial robots, and ABB robots are widely used. Assessing their security is meaningful for both suppliers and users. Secondly, RWS contains multiple complex APIs, each interacting with different types of resources in the controller. Testing these APIs allows for a comprehensive evaluation of the performance of our approach. Thirdly, ABB provides a virtual simulation environment [1] that enables the quick reset or creation of new industrial robot controllers to eliminate any side effects produced by previous tests. Furthermore, we omitted the OPTIONS operations in the RWS specification because these operations are REST-unrelated and are used solely to query the server for supported methods without transferring actual data.

Evaluation Criteria. We use two criteria to evaluate the flaw detection capabilities of RobRest and other tools.

(1) **Flaws:** Identifying flaws within Industrial Robot RESTful service is the aim of this paper, so the number of detected flaws is a necessary criterion. The flaws we focus on include bugs that can cause "Internal Server Error" and vulnerabilities that can lead to the loss of controller availability.
(2) **Operation coverage:** Successful API operations can directly demonstrate the exploration ability of the RESTful service. We use successfully requested operations (SROs) as the criterion to evaluate operation coverage, because this indicates whether the tool can generate effective requests to reach and test the deeper states of the service.

Experiment Settings. All experiments were conducted on a Windows 11 machine equipped with an Intel Core i9-11950H, 64GB RAM, and Python 3.8.2. Each experiment lasted for 12 h and was repeated five times. Additionally, to avoid the side effects of the industrial robot controller, we used a new controller for each test round.

4.2   Flaw Detection (RQ1)

To evaluate the real performance of each tool in discovering flaws, we manually verified all the flaws. We also filtered out errors that could not be replicated due to changes in controller states, as discussed in Sect. 5. The details of the flaws discovered by the four methods in RWS are shown in Table 1.

RobRest identified 19 flaws in RWS, including 15 unknown bugs, 2 unknown vulnerabilities, and 2 vulnerabilities that had been discovered and fixed by the manufacturer but had not been publicly disclosed. The results in Table 1 indicated that RobRest completely covers and surpasses the flaws discovered by

Table 1. Flaws in ABB robot web service identified by RobRest, RobRest_Nr, Morest and ZAP. ✔ indicates that this tool can identify the flaw, while ✗ indicates it cannot.

API	No.	Operation	Endpoint	Flaw Type	Status	RobRest	RobRest_Nr	Morest	ZAP
Devices	1	POST	/rw/devices/search	Bug	Confirmed	✔	✔	✔	✗
	2	POST	/ctrl/restart	Bug	Confirmed	✔	✔	✔	✔
Controller	3	POST	/ctrl/compress	Vulnerability	Fixed	✔	✗	✗	✗
	4	POST	/ctrl/decompress	Vulnerability	Fixed	✔	✗	✗	✗
Motion System	5	POST	/rw/motionsystem/mechunits/{mechunit}/lead-through/load	Bug	Confirmed	✔	✔	✔	✔
	6	POST	/rw/iosystem/devices/{network}/add	Bug	Confirmed	✔	✔	✗	✗
IO	7	GET	/rw/iosystem/devices/{network}/{device}/upgradeinfo	Bug	Confirmed	✔	✔	✗	✗
	8	POST	/rw/iosystem/devices/{network}/recover	Bug	Confirmed	✔	✔	✗	✗
File	9	POST	/fileservice/{device|env_variable|directory}/rename	Bug	Confirmed	✔	✔	✗	✗
	10	GET	/rw/vision/{camera-identifier}/info	Bug	Confirmed	✔	✔	✔	✔
	11	GET	/rw/vision/{camera-name}/status	Bug	Confirmed	✔	✔	✔	✔
	12	GET	/rw/vision/{camera-name}/jobname	Bug	Confirmed	✔	✔	✔	✔
Vision	13	POST	/rw/vision/{camera-name}/flash	Bug	Confirmed	✔	✔	✔	✔
	14	POST	/rw/vision/{camera index}/user	Bug	Confirmed	✔	✔	✔	✗
	15	POST	/rw/vision/{camera-name}/hostname	Bug	Confirmed	✔	✔	✔	✗
	16	POST	/rw/vision/{camera-index}/name	Bug	Confirmed	✔	✔	✔	✗
	17	POST	/rw/rapid/execution/start	Vulnerability	CVE-2024-1914	✔	✗	✗	✗
RAPID	18	GET	/rw/rapid/tasks/{task}/modules/{module}/rulesinstr	Bug	Confirmed	✔	✔	✗	✗
	19	GET	/rw/rapid/tasks/{task}/motion/robtarget	Vulnerability	CVE-2024-1913	✔	✗	✗	✗

other tools. Compared to RobRest_Nr, Morest, and ZAP, RobRest was able to identify 4, 9, and 13 additional flaws, respectively. Since the actions and task completions of any industrial robot model are controlled by controllers, there are currently only two mainstream types of ABB controllers: IRC5 and Omnicore, both of which have deployed RESTful services. Therefore, any vulnerabilities we discover can affect almost all industrial robot models, a fact confirmed during our communication with the manufacturer. Meanwhile, among the four vulnerabilities discovered by RobRest, two were 0-day vulnerabilities caused by 1) out-of-bounds write and 2) null pointer dereference. Exploiting these vulnerabilities can affect the physical world, causing most ABB industrial robots to experience denial of service, arbitrary code execution, and severe disruption of industrial production.

After analysis, we identified that the failure of other tools to detect certain flaws can be attributed to the poor quality of input parameters and the lack of appropriate flags for capturing flaws. For Morest although it generated operation request sequences that could access API endpoints with flaws, the absence of knowledge on how to assign values to parameters, or incorrect examples in the specification, made it difficult to generate correct input parameter values. This limitation made it challenging to trigger the defect swiftly with high-quality requests. ZAP cannot be used to generate meaningful request sequences and is limited to testing single API operations. For APIs that impose special character restrictions, ZAP cannot gather any further information or accurately modify the service to accept valid fields, thus it performs the worst. Moreover, Morest and ZAP use HTTP status codes as indicators for bug detection, thereby failing to capture vulnerabilities within the controller. The experimental results demonstrate that (1) RobRest is capable of discovering more flaws within the same amount of time. (2) Utilizing Windows socket error as feedback, the uniquely designed Rob-Monitor for industrial robot controllers helps us better detect and capture controller crash phenomena, thereby uncovering denial of service vulnerabilities caused by certain operations.

Case Study. We use two flaws detected in RWS to illustrate why ROBREST can find more flaws. To trigger bug No.6 `/rw/iosystem/devices/{network}/add` in Table 1, it is necessary to ensure that `{network}` is a fixed value such as `SC_Feedback_Net`. However, the value `SC_Feedback_Net` of parameter `{network}` cannot be obtained from the dependent response and is difficult to generate through random mutation. In contrast, ROBREST can obtain the description information of the parameter when parsing the document, thereby obtaining the parameter's value. Additionally, we discovered that the API specification might be inconsistent with the implementation. A parameter that is stated to exist in the specification for a certain operation is not implemented, causing all dependencies that have the parameter to be interrupted, which has an impact on dependency-based REST testing tools. Because they rely on the response of the previous request to construct their own request parameter values, this situation will cause them to be unable to obtain the producer parameters, causing the request to fail. In contrast, ROBREST can extract valid parameter candidate values from the specification description and construct the correct request directly. For instance, to trigger No.19 in Table 1, three constraints must be met: (1) `{task}=T_ROB1`, (2) the URL contains parameters `tool` or `wobj`, and (3) the parameter length in the URL is greater than 90. When this vulnerability is triggered, the industrial robot controller will fall into a denial of service state, at which point the service will not return a 500 status code. Although the documentation states that executing the `GET /rw/rapid/modules` operation successfully will return a response containing `task-name`, which will be used in the dependency relationship to obtain the value of the parameter `task`, the actual test process revealed that this parameter is not included in the response. Therefore, baseline tools struggle to obtain the correct value of the parameter `task` and fail to discover this flaw.

4.3   Coverage (RQ2)

Table 2 describes the number of successfully requested operations and the number of flaws detected under different methods. Compared to the state-of-the-art RESTful API testing tool MOREST, ROBREST and ROBREST_NR achieved an average increase of 30.94% in operation coverage, and were able to discover 90% and 50% more flaws than MOREST. The above results show that ROBREST significantly outperforms the baseline in terms of the number of successful operation requests and the number of flaws detected. We analyzed the operations sent by ROBREST and MOREST and drew the following conclusion.

The parameter candidate value extraction strategy we proposed has significantly improved the quality of requests. Firstly, ROBREST extracts available parameter values from the OpenAPI specification, enhancing the effectiveness of requests, reducing the number of parameter mutations, and increasing the frequency of access to different sequences. Secondly, the use of a parameter dictionary reduced the dependency on responses, especially when there is no dependency between parameters. The correct parameter candidate value can make the request directly pass the parameter constraints. As described in the case study of § 4.2, due to the lack of dependency on the critical parameter {task}, all end-points in the RAPID API containing /rw/rapid/tasks/{task}/ were challenging to generate valid values for the parameter task. In contrast, RoBREST can extract the correct candidate value of the parameter from the description of an operation, which led to RoBREST and RoBREST_NR achieving a 95.83% higher operation coverage rate in the RAPID API compared to the MOREST method. Since the only difference between RoBREST and RoBREST_NR is that RoBREST includes a monitor for industrial robot controllers, they were able to construct the same access request sequences and request parameters, resulting in the same operation coverage rate.

Table 2. The coverage of MOREST, RoBREST_NR and RoBREST.

REST API	# Total Operation	MOREST			RobREST_NR			RobREST		
		Pass Rate	# Covered Operation	Flaws	Pass Rate	# Covered Operation	Flaws	Pass Rate	# Covered Operation	Flaws
Root Resource	2	100.0%	2	0	100%	2	0	100%	2	0
User	11	45.45%	5	0	54.55%	6	0	54.55%	6	0
UAS	26	42.31%	11	0	42.31%	11	0	42.31%	11	0
Subscription	4	00.00%	0	0	00.00%	0	0	00.00%	0	0
Controller	83	43.37%	36	1	49.40%	41	1	49.40%	41	3
File	12	8.33%	1	0	25.00%	3	1	25.00%	3	1
RobotWare	1	100.0%	1	0	100.0%	1	0	100.0%	1	0
CFG	15	13.33%	2	0	13.33%	2	0	13.33%	2	0
DIPC	6	50.00%	3	0	83.33%	5	0	83.33%	5	0
Devices	2	50.00%	1	1	50.00%	1	1	50.00%	1	1
Elog	7	57.14%	4	0	57.14%	4	0	57.14%	4	0
IO	32	21.88%	7	0	25.00%	8	3	25.00%	8	3
Mastership	9	77.78%	7	0	77.78%	7	0	77.78%	7	0
MotionSystem	55	20.00%	11	1	21.82%	12	1	21.82%	12	1
Panel	18	38.89%	7	0	77.78%	14	0	77.78%	14	0
RAPID	92	26.09%	24	0	51.09%	47	1	51.09%	47	3
System	7	71.43%	5	0	85.71%	6	0	85.71%	6	0
Retcode	1	100.0%	1	0	100.0%	1	0	100.0%	1	0
Integrated Vision	13	84.62%	11	7	84.62%	11	7	84.62%	11	7
Total	396	35.10%	139	10	45.96%	182	15	45.96%	182	19

4.4 Ablation Study (RQ3)

To explore how the parameter candidate value extraction strategy enhances the performance of RoBREST by improving the quality of generated parameters, we conducted ablation experiments on the two main components of the strategy: operation-based candidate dictionary and parameter-based candidate dictionary. To examine the distinct roles of each component, we implemented three variants: (1) RoBREST_NoCAND, which removes all candidate dictionaries, (2) RoBREST_OPERATION, which retains only the operation-based dictionary; and (3) RoBREST_PARAMETER, which retains only the parameter-based dictionary. Figure 4 shows the operation coverage on RWS, and Table 3 presents the number of detected flaws.

Overall, RoBREST outperforms the other three variants in both operation coverage and flaw detection. Both RoBREST_PARAMETER and RoBREST_OPERATION have certain limitations in parameter generation. Taking the IO API as an

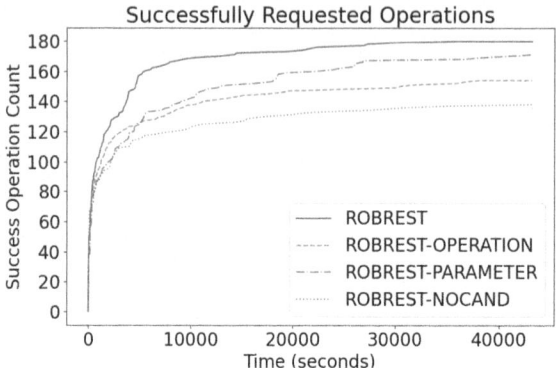

Fig. 4. The successfully requested operations growth of RobRest and its three variants. Growth is calculated from the average time of operations that appear in the same order after removing the maximum and minimum time values.

Table 3. The number of detected flaws in RobRest and its three variants.

REST API	Average Detected Flaws (#)			
	RobRest	RobRest_Parameter	RobRest_Operation	RobRest_NoCand
Controller	3.00	3.00	3.00	3.00
File	1.00	0.00	1.00	0.00
Devices	1.00	1.00	1.00	1.00
MotionSystem	1.00	1.00	1.00	1.00
IO	3.00	2.00	0.00	0.00
RAPID	3.00	1.00	3.00	1.00
Vision	7.00	7.00	7.00	7.00
Average Increased (%)	0.00	26.67	18.75	46.15

example, although RobRest_Operation can generate the correct values for the URL parameter {device}, it is limited by the values of the URL parameter {network}, while RobRest_Parameter does the opposite. This restricts their exploration of complex resources and deep states. Furthermore, RobRest_Parameter covers more operations than RobRest_Operation. This is because, compared to API operations, there are more description fields for parameters, which contain a greater possibility of parameter values. Moreover, due to the acquisition of different parameter values, they possess varying capabilities in detecting flaws in certain APIs (File, IO, RAPID). Thus, we can infer that both operation description and parameter description in API documentation can improve the performance of RESTful API testing, and their combination can bring greater performance improvement.

5   Discussion and Limitation

Input Parameter Generation. RobRest employs the knowledge extraction capabilities of large predictive models to generate potential candidate values for parameters from the OpenAPI specification in advance, enhancing the quality of request generation. Since our method is highly dependent on the description field in the OpenAPI specification, it will not be able to improve the request quality when the description field does not exist in the specification. However, other types of documentation may also contain natural language descriptions relevant to RESTful service request parameters, such as technical specifications and operating manuals. Therefore, extracting actionable knowledge for REST testing from these information-rich documents could be an interesting topic for future work.

Error Identification Across Sequences. Due to changes in the state of robot controllers, some unique errors can only be triggered under specific states of industrial robots and cannot be replicated in subsequent analyses. This is a common issue of error reproduction in REST API testing. For instance, executing an operation request to access a resource triggers an error when the resource is affected by a request sequence that is either temporally distant or occurred a long time ago. Moreover, subsequent testing processes may modify the resource in various ways, making it impossible to reproduce the error. Therefore, determining whether there is a correlation between distant sequences or different sequences, and how to reproduce such errors, are potential research directions in testing RESTful services.

6   Related Work

Security of Industrial Robot. Security issues in the industrial robot ecosystem have received attention from researchers. Quarta et al. [34] systematically analyzed the security issues of ABB IRB140 and proposed the corresponding attack chain. Moreover, researchers applied program analysis techniques to conduct security vulnerability analysis of industrial robot programming languages [32,33]. RoBREST is the first to conduct research on the security issues of RESTful communication services for industrial robots, which directly affects the security of robot controllers.

RESTful Service Testing. Several approaches have been proposed for the automated testing of RESTful services. Some black-box testing tools [3,5,9,13, 20,27] construct and expand request sequences by extracting dependency information from the OpenAPI specification through different strategies. EvoMaster [4] is a white-box testing technique that uses instrumentation technology to collect feedback information during testing to guide the generation of operation sequences. Although it can reach deeper states in RESTful services, it can only instrument and test services written in languages like Java and Scala. Moreover, since industrial robot controllers are completely black-boxed, white-box testing is not applicable for testing the RESTful services of industrial robots. In addition, some approaches generate data payloads embedded in REST API requests. Godefroid et al. [16] extracts usable data from the parameter examples in the

API specification, but this is not suitable for industrial robots, as their example values are all "string". Grammarinator [23] optimizes the mutation process, which is orthogonal to our work. These testing techniques rely on the HTTP status codes returned by the service as the criteria for evaluating the reliability of the service, and they are not designed to test for security vulnerabilities. In general, these methods cannot effectively address the challenges mentioned in Sect. 1.

Penetration Testing Techniques. Due to the close similarity in implementation between the backend of RESTful APIs and traditional web services, researchers aim to expand web penetration techniques to conduct vulnerability detection in RESTful APIs. Zed Attack Proxy (ZAP) [31] is primarily used for web vulnerability assessment and penetration testing. Its current OpenAPI extension supports parsing API specifications, enabling web vulnerability testing for API services. However, such penetration tools are limited to testing individual API endpoints, thus requiring significant manual effort, with restricted depth and scope of testing. Moreover, they are unable to identify vulnerabilities that require interaction with multiple APIs. NAUTILUS [11] incorporates annotation strategies into API specifications to mark operation relationships and parameters, generating meaningful sequences of logical operations and uncovering injection vulnerabilities in REST APIs. VoAPI^2 [12] identifies and detects security vulnerabilities in RESTful services by extracting common string keywords from API endpoint specifications to determine functionality and pinpointing vulnerability types through differentiating functionalities, thereby reducing the potential input space. However, due to the specificity of industrial robot controllers, such functional vulnerabilities fall outside the scope of this study.

7   Conclusion

In this paper, we present ROBREST, a flaw detection methodology designed for industrial robot RESTful service. First, ROBREST generates high-quality requests by parsing the description fields in the API specification. ROBREST then modifies the requests based on the sniffed message to circumvent restrictions in industrial robots for effective testing. Finally, ROBREST leverages the physical status and service status of the industrial robot controller to realize flaw automated monitoring. We conducted an exhaustive evaluation of ROBREST. ROBREST identified a total of 4 vulnerabilities and 15 bugs, and 2 of them have been assigned CVE IDs. This result outperforms state-of-the-art RESTful API testing and penetration testing tools.

Acknowledgements. We thank the anonymous reviewers for their insightful comments on our work, and we also thank Roger Dahlgren and the cybersecurity team at ABB for their assistance. This work was supported by the National Natural Science Foundation of China under Grant No. 62472302, and Beijing Natural Science Foundation under Grant No. L234033.

References

1. ABB: Abb robotstudio (2023). https://new.abb.com/products/robotics/zh/robotstudio
2. AG, K.: Kuka (2024). https://www.kuka.com
3. APIFuzzer: Apifuzzer (2022). https://github.com/KissPeter/APIFuzzer
4. Arcuri, A.: Restful api automated test case generation with evomaster. ACM Trans. Softw. Eng. Methodol. (TOSEM) **28**(1), 1–37 (2019). http://dx.doi.org/10.1145/3293455
5. Atlidakis, V., Godefroid, P., Polishchuk, M.: Restler: stateful rest api fuzzing. In: 2019 IEEE/ACM 41st International Conference on Software Engineering (ICSE), pp. 748–758 (2019). http://dx.doi.org/10.1109/icse.2019.00083
6. Bennett, J.: Autoit scripting language (2024). https://www.autoitscript.com/site/autoit/
7. Center, A.D.: Robot web service. https://developercenter.robotstudio.com/api/RWS (2020)
8. Commission, I.E.: Iec 61508-1:2010, functional safety of electrical/electronic/programmable electronic safety-related systems. Tech. rep, IEC (2010)
9. Corradini, D., Zampieri, A., Pasqua, M., Viglianisi, E., Dallago, M., Ceccato, M.: Automated black-box testing of nominal and error scenarios in restful apis. Softw. Testing Verification Reliability **32**(5), e1808 (2022). http://dx.doi.org/10.1002/stvr.1808
10. Dailymail: "tesla robot attacks an engineer at company's texas factory during violent malfunction" (2023). https://www.dailymail.co.uk/sciencetech/article-12869629
11. Deng, G., et al.: Nautilus: automated restful api vulnerability detection. In: 32nd USENIX Security Symposium (USENIX Security 23), pp. 5593–5609 (2023). https://dlnext.acm.org/doi/10.5555/3620237.3620550
12. Du, W., et al.: Vulnerability-oriented testing for restful apis. In: 33rd USENIX Security Symposium (USENIX Security 24), pp. 739–755 (2024)
13. Ed-Douibi, H., Izquierdo, J.L.C., Cabot, J.: Automatic generation of test cases for rest apis: A specification-based approach. In: 2018 IEEE 22nd international enterprise distributed object computing conference (EDOC). pp. 181–190. IEEE (2018), http://dx.doi.org/10.1109/edoc.2018.00031
14. Fielding, R.T., Taylor, R.N.: Architectural Styles and the Design of Network-based Software Architectures. Ph.D. thesis, University of California, Irvine (2000)
15. Gamez-Diaz, A., Fernandez, P., Ruiz-Cortes, A.: An analysis of restful apis offerings in the industry. In: International Conference on Service-Oriented Computing, pp. 589–604. Springer (2017). https://doi.org/10.1007/978-3-319-69035-3_43
16. Godefroid, P., Huang, B.Y., Polishchuk, M.: Intelligent rest api data fuzzing. In: Proceedings of the 28th ACM Joint Meeting on European Software Engineering Conference and Symposium on the Foundations of Software Engineering, pp. 725–736 (2020). http://dx.doi.org/10.1145/3368089.3409719
17. Gosewehr, F., Wermann, J., Borsych, W., Colombo, A.W.: Specification and design of an industrial manufacturing middleware. In: 2017 IEEE 15th International Conference on Industrial Informatics (INDIN), pp. 1160–1166. IEEE (2017). http://dx.doi.org/10.1109/indin.2017.8104937
18. Group, A.: Abb (2024). https://global.abb/group/en

19. Hägele, M., Nilsson, K., Pires, J.N., Bischoff, R.: Industrial robotics. Springer handbook of robotics, pp. 1385–1422 (2016). https://doi.org/10.1007/978-3-319-32552-1_54

20. Hatfield-Dodds, Z., Dygalo, D.: Deriving semantics-aware fuzzers from web api schemas. In: Proceedings of the ACM/IEEE 44th International Conference on Software Engineering: Companion Proceedings, pp. 345–346 (2022), http://dx.doi.org/10.1109/icse-companion55297.2022.9793781

21. Heyer, C.: Human-robot interaction and future industrial robotics applications. In: 2010 IEEE/RSJ International Conference on Intelligent Robots and Systems, pp. 4749–4754. IEEE (2010). http://dx.doi.org/10.1109/iros.2010.5651294

22. Hils, M.: An interactive https proxy (2024). https://mitmproxy.org

23. Hodován, R., Kiss, Á., Gyimóthy, T.: Grammarinator: a grammar-based open source fuzzer. In: Proceedings of the 9th ACM SIGSOFT International Workshop on Automating TEST Case Design, Selection, and Evaluation, pp. 45–48 (2018). http://dx.doi.org/10.1145/3278186.3278193

24. ISO: Robots and robotic devices - Safety requirements for industrial robots - Part 2: Robot systems and integration. Standard ISO 10218-2:2011(E) (2011)

25. ISO: Robotics - Vocabulary. Standard ISO 8373:2021(E) (2021)

26. Li, C., Park, J., Kim, H., Chrysostomou, D.: How can i help you? an intelligent virtual assistant for industrial robots. In: Companion of the 2021 ACM/IEEE International Conference on Human-Robot Interaction, pp. 220–224. ACM (2021). http://dx.doi.org/10.1145/3434074.3447163

27. Liu, Y., et al.: Morest: model-based restful api testing with execution feedback. In: 2022 IEEE/ACM 44th International Conference on Software Engineering (ICSE), pp. 1406–1417 (2022). http://dx.doi.org/10.1145/3510003.3510133

28. OpenAI: Gpt-4 (2024). https://openai.com/gpt-4

29. OpenAPI: Openapi initiative (2024). https://www.openapis.org

30. OpenAPI: Swagger (2024). https://swagger.io

31. OWASP: The owasp zed attack proxy (zap) (2024). https://www.zaproxy.org

32. Pogliani, M., Maggi, F., Balduzzi, M., Quarta, D., Zanero, S.: Detecting insecure code patterns in industrial robot programs. In: Proceedings of the 15th ACM Asia Conference on Computer and Communications Security (2020). http://dx.doi.org/10.1145/3320269.3384735

33. Pogliani, M., Quarta, D., Polino, M., Vittone, M., Maggi, F., Zanero, S.: Security of controlled manufacturing systems in the connected factory: the case of industrial robots. J. Comput. Virology Hacking Tech. **15**, 161 – 175 (2019). http://dx.doi.org/10.1007/s11416-019-00329-8

34. Quarta, D., Pogliani, M., Polino, M., Maggi, F., Zanchettin, A.M., Zanero, S.: An experimental security analysis of an industrial robot controller. In: 2017 IEEE Symposium on Security and Privacy (SP), pp. 268–286 (2017). http://dx.doi.org/10.1109/sp.2017.20

35. of Robotics, I.F.: World robotics 2023 report (2023). https://ifr.org/ifr-press-releases/news/world-robotics-2023-report-asia-ahead-of-europe-and-the-americas

36. Sandiland, D.: Stop spending millions on robot downtime now (2022). https://www.robotics247.com/article/stop_spending_millions_on_robot_downtime_now/supply_chain

37. Souza, R., Pinho, F., Olivi, L., Cardozo, E.: A restful platform for networked robotics. In: 2013 10th International Conference on Ubiquitous Robots and Ambient Intelligence (URAI), pp. 423–428. IEEE (2013). http://dx.doi.org/10.1109/urai.2013.6677301

Formally Verifiable Generated ASN.1/ ACN Encoders and Decoders: A Case Study

Mario Bucev[1], Samuel Chassot[1], Simon Felix[2]([✉]), Filip Schramka[2], and Viktor Kunčak[1]

[1] EPFL IC LARA, Lausanne, Switzerland
{mario.bucev,samuel.chassot,viktor.kuncak}@epfl.ch
[2] Ateleris GmbH, Brugg, Switzerland
{simon.felix,filip.schramka}@ateleris.ch

Abstract. We propose a verified executable Scala backend for ASN1SCC, a compiler for ASN.1/ACN. ASN.1 is a language for describing data structures widely employed in ground and space telecommunications. ACN can be used along ASN.1 to describe complex binary formats and legacy protocols. To avoid error-prone and time-consuming manual writing of serializers, we show how to port an ASN.1/ACN code generator to generate Scala code. We then enhance the generator to emit not only the executable code but also strong enough preconditions, postconditions, and lemmas for inductive proofs. This allowed us to verify the resulting generated annotated code using Stainless, a program verifier for Scala. The properties we prove include the absence of runtime errors, such as out-of-bound accesses or divisions by zero. For the base library, we also prove the invertibility of the decoding and encoding functions, showing that decoding yields the encoded value back. Furthermore, our system automatically inserts invertibility proofs for arbitrary records in the generated code, proving over 300'000 verification conditions. We establish key steps towards such proofs for sums and arrays as well.

Keywords: formal verification · serialization · Scala · Stainless

1 Introduction

Values of structured data types are a key for programming above the level of assembly. To transmit structured data across a communication channel, programs need to serialize data into bits on one side and deserialize it on the other side. Writing serialization and deserialization of protocols requires great expertise and is error-prone. Buffer overflows during deserialization may result in security exploits [9,18]. Data corruption and loss may also arise due to hard-to-test implementation errors in serialization or deserialization. Automated code generation mitigates these issues by letting protocol designers define the messages rather than writing the encoders and decoders by hand. However, bugs may still arise

© The Author(s), under exclusive license to Springer Nature Switzerland AG 2025
K. Shankaranarayanan et al. (Eds.): VMCAI 2025, LNCS 15530, pp. 185–207, 2025.
https://doi.org/10.1007/978-3-031-82703-7_9

in such an approach due to errors in the code generator or bugs in the used base libraries. We aim to prevent such errors using formal verification.

We use ASN.1 [17], an interface description language (IDL) that specifies protocol messages, as well as their serialization and deserialization. ASN.1 describes messages using a declarative specification similar in purpose to algebraic data types, JSON, or XML schemas. ASN.1 is widely used in many applications, for example, to define the format of HTTPS certificates [12] or 5G network packets.

In contrast to other general message formats, such as Protocol Buffers [11] or Apache Thrift [26], ASN.1 supports the serialization of messages in multiple concrete representations. While this design introduces additional complexity, it also means that users can optimize the wire format for each use case. Out of all ASN.1 wire formats, ACN [19] is the most flexible one because it gives users granular control of most aspects of serialization. This flexibility means that most existing and legacy protocols can be described with ASN.1/ACN.

In this paper, we present the development of an automated generator for verifiable serializers and deserializers for messages conforming to a given ASN.1/ACN specification. The generator emits code in Scala along with inductive specifications in such a way that the generated code automatically verifies using the Stainless verifier [15]. Stainless is non-interactive: if it times out during verification, the programmer needs to update the source code with additional assertions, preconditions, and postconditions and re-run the verification.

We can consider several levels of verification, depending on the level of properties we expect from code given to the Stainless verifier:

Level 1. The generated subset of Scala corresponds to what Stainless accepts in terms of syntactic constructs and types. This implies the absence of null dereferences, which are ruled out by construction, thanks to use of case class constructors and required declaration-time initialization.

Level 2. Stainless proves all automatically generated verification conditions for generated Scala programs. These conditions guarantee function termination and the absence of run-time errors such as pattern matching failures, array out-of-bounds accesses, division by zero, or unsafe casts. As the verification processes each function individually, in practice, this requires adding appropriate preconditions and postconditions as part of generated functions.

Level 3. Verifying additional partial specification properties. A particularly interesting property verified at this level is the exact number of bits encoded and decoded by each function.

Level 4. Verification of key functional correctness characterization properties, such as the fact that decoding an encoded value (with an appropriate position of the decoder) recovers the original value.

In our work, we first developed an executable Scala backend and a suite of unit tests. The test suite checks whether the generated Scala code produces bit-wise identical results as other ASN.1 serializers. To achieve the first verification level, we needed to modify the code generator to ensure that the initialization of data structures is compositional. Due to limitations regarding nested mutable types in Stainless, we moved to immutable vectors instead of arrays to make verification more feasible. For level 2, we enhanced the code generator to emit sufficient preconditions and postconditions that ensure the absence of run-time errors. To ensure verification succeeds in a reasonable amount of time, we modified the code generator to be more compositional in the generated code. We also automatically generate the declarations and uses of lemmas for inductive properties needed to establish the absence of errors. Moving to level 3 required stronger postconditions and additional lemmas. These lemmas are parametric in the data types that are being generated, so they also need to be automatically generated, as opposed to being developed once and for all.

We completed level 3 for generated code, ensuring no run-time errors and that the expected amount of data is written. We also made significant progress towards level 4: our base library has full specifications of invertibility that are proven by Stainless. Furthermore, we generate verifiable code with specifications that preserve the invertibility properties for records that are encoded serially. This development required additional lemmas about encoding in array buffers. We believe that extending the invertibility proof for the remaining recursive cases is feasible. During the verification effort, we also identified and fixed errors in the existing code generator.

1.1   Contributions

This paper makes the following contributions:

– We extend the ASN1SCC compiler with a Scala backend alongside a runtime library for encoding and decoding primitive ASN.1/ACN constructs and provide a testing framework to test the interoperability of the newly developed Scala backend with the existing C backend.
– We discover and address bugs in existing and new parts of the code generator and library, in part thanks to formal verification (Sect. 8.2).
– We prove the runtime safety of the serialization for primitives and the safety of the generated code. Furthermore, we prove strong properties such as invertibility for all library functions except for floating point and string-related operations. We also prove invertibility for records in the generated code. For all of the generated code, Stainless obtains our proofs automatically because the generated code contains sufficient assertions and other proof hints for Stainless to succeed. We evaluate the automation of our approach by generating encoding and decoding functions for the PUS-C format that is of practical interest in aerospace applications.

Our code generator that produces verifiable Stainless code is available as part of the ASN1SCC generator written in $F^{\#}$:

https://github.com/maxime-esa/asn1scc

Generated Scala serializers and desearializers for packets defined by the PUS-C standard [10] are available at:

https://github.com/epfl-lara/fovcom

1.2   Related Work

Narcissus project [8] developed flexible and trustworthy combinators for parsing and unparsing within the Coq proof assistant, showing that the approach works well enough to replace packet processors for a full Internet protocol stack in the Mirage operating system. Our approach is based on a code generator and starts from ASN.1 and ACN as existing definition languages. We consider it a promising yet challenging future work to develop verified invertible combinators that could express ACN constructs.

The authors of EverParse [28] present a formally verified library and framework for binary formats in F*. This framework supports proofs for non-malleability (unique representation of values), safety and inversion of encoding. EverParse can then emit high-performance C code from the formally proven F*.

Based on EverParse and related to ASN.1 is ASN1* [23], which supports the Distinguished Encoding Rule (DER) format. This encoding ensures non-malleability, a characteristic that is, for instance, particularly desired in cryptography. The ASN1* authors formalized the DER semantics in F* within the EverParse framework and proved its non-malleability. The parser combinators are non-malleable and correct by construction. In contrast to DER, our work supports serialization in the highly flexible ACN [19] format, more intended towards legacy protocols or complex binary format where non-malleability is not required.

Promiwag [20] is a library to generate protocol deserialization code. The authors prove that the generated code will terminate and is free of out-of-bounds memory accesses. Our work proves stronger invariants for the generated code, for example, invertibility, and also covers serialization.

2   Use of ASN.1 in Space Missions

Reliable communication is crucial for spacecrafts due to their long missions, large distances, and high costs. In the past, projects relied on human-readable protocol specifications and manually written protocol code on the spacecraft and in the ground software. This made it challenging to iterate a protocol quickly and reliably.

In addition, commercial or military space applications are becoming increasingly attractive targets for attackers. In recent years, the resilience of satellites and ground stations has gained considerable attention from operators and space agencies. According to the CWE Top 25 list [9], bugs in the deserialization of untrusted data (CWE-502) are among the most common security vulnerabilities.

To reduce these risks, missions increasingly rely on automated code generators such as ASN1SCC. The European Space Agency (ESA) has announced its cybersecurity strategy [24], which also encourages the use of code generators as part of the secure-as-built principle. ESA has produced the PUS-C standard [10], which defines the communication protocol for all ESA missions. The PUS-C standard was later formally specified in ASN.1/ACN [22]. Multiple ESA missions already use automatically generated code, for example, CHEOPS or PROBA-3 [29], produced by the ASN1SCC code generator from the PUS-C ASN.1/ACN specification.

The ASN.1 encoding schemes such as PER, uPER or BER control the binary format of the messages. ASN.1 Control Notation [19] (ACN) is an encoding scheme that allows developers to specify the binary layout of data structures, for instance, the determinant of CHOICE (a sum type), the length of SEQUENCE OF (an ordered collection), integer bit size, or alignment of fields. This detailed control means that legacy formats with non-uniform formatting rules can be described, for example, PUS-C [22].

As an example, consider the TC[2, 7] telecommand, which can be defined with ASN.1 as follows:

```
TC-2-7-DistrPhysicalDevCmds ::= SEQUENCE {
  physicalDevCmds SEQUENCE (SIZE(1 .. 63)) OF PhysicalDevCmd
}
PhysicalDevCmd ::= SEQUENCE {protoData ProtoData, cmdData CmdData}
CmdData        ::= CHOICE {dev1 INTEGER (0 .. 255)}
ProtoData      ::= CHOICE {dev1 INTEGER (0 .. 255)}
```

TC-2-7-DistrPhysicalDevCmds is a record containing an array of Physical-DevCmd, whose size ranges between 1 and 63. CmdData and ProtoData are both a sum type with only one variant, an integer ranging from 0 to 255 (note that these are mission specific and are expected to be tailored).

However, the TC[2, 4] telecommand mandates a certain binary format. In particular, both CmdData and ProtoData have their determinant specified by a "physical device ID" residing outside of their definition (i.e. the determinant is not embedded within the CHOICE). This can be achieved by parameterizing both CHOICEs and inserting an *ACN field* named physicalDev-ID within PhysicalDevCmd:

```
TC-2-7-DistrPhysicalDevCmds [] {n PUSC-UINT32 [], physicalDevCmds [size n]}
PhysicalDevCmd [] {
  physicalDev-ID PhysicalDev-ID [] ,
  protoData⟨physicalDev-ID⟩ [], cmdData⟨physicalDev-ID⟩ []
}
CmdData⟨PhysicalDev-ID: device⟩ [determinant device] {
  dev1 [/*...*/ ]
}
ProtoData⟨PhysicalDev-ID: device⟩ [determinant device] {
  dev1 [/*...*/ ]
}
```

The ACN specification resides in a corresponding .acn file. Each definition, field, or variant may have an encoding property specified within brackets; if none are desired, these must be empty. ACN allows inserting additional fields, for instance, the n in `TC-2-7-DistrPhysicalDevCmds` and `physicalDev-ID` in `PhysicalDevCmd`. The former is encoding the size of `physicalDevCmds` while the latter is encoding the determinant of both `protoData` and `cmdData`. Setting the determinant of a `CHOICE` is done by parameterizing its definition (by adding a parameter `PhysicalDev-ID: device`) and indicating it as an encoding property (done with `[determinant device]`). Finally, each reference to such parameterized `CHOICE` must be "instantiated"; in the above example, this corresponds to `protoData<physicalDev-ID>` and similarly for `cmdData<physicalDev-ID>`.

3   The Stainless Verification Framework

We use the Stainless program verifier[1] to verify generated encoders and decoders. Stainless is an open-source deductive verifier for the Scala programming language. The verifier runs the Scala compiler and interprets certain annotations and function calls (such as `require, ensuring`) as specifications. The foundation of Stainless is System FR, a dependent-type extension of System F with refinement types and indexed recursive data type definitions [15]. Stainless transforms syntax trees obtained from the Scala compiler into a simpler form, eliminating non-aliased imperative state [3,4] and encoding non-disjoint types [30, Chapter 5], then uses a type checker for System FR [15] to generate proof obligations that contain calls to higher-order and recursive functions, which it passes to its subsystem named Inox[2]. Inox unfolds recursive functions [27] and resolves higher-order function calls [31] until it generates quantifier-free first-order queries for Z3 [7], CVC4 [2], cvc5 [1] and Princess [25] solvers. In each unrolling, Inox alternates checks for validity and counterexamples of verification conditions, allowing Stainless to provide useful feedback to users. It also uses formula simplification, normalization and caching to improve verification efficiency [13].

Stainless programs are valid Scala programs and can thus be compiled to run on the JVM, to native code via Scala Native for LLVM, and to JavaScript via Scala.js. Furthermore, Stainless can transpile a subset of programs to simple C code suitable for embedded applications [14]. In recent years, Stainless was used to verify several case studies, including key parts of a flash file system for the X-ray spectrometer of the Solar Orbiter satellite [14], a hash table implementation from Scala standard collection library [6] (revealing implementation errors hidden for many years and now fixed), and correctness of an implementation of an encoder and decoder for a popular recent lossless image compression format [5].

4   A Verified BitStream Implementation

To introduce the nature of our correctness properties, we present the key mutable data structure and functions for serializing and deserializing basic data types,

---

[1] https://github.com/epfl-lara/stainless.

[2] https://github.com/epfl-lara/inox.

such as signed and unsigned machine integers. This data structure, along with the proof of its correctness that we describe in this section, represents around 3700 lines of code. It is invoked by the generated code for serializers of complex data types, whose overall correctness we wish to prove.

We start with a data structure implementation representing a stream of bits. This data structure is used both as an output medium when serializing data and as input when deserializing.

As the structure needs to offer good runtime performance, it is based on a buffer in the form of an array of bytes (i.e., 8-bit integers). The structure then keeps track of the current index, i.e., the head of the stream, by keeping two variables: *currentByte* and *currentBit*. The *currentByte* value gives the index of the byte in the array where the head is located, and the *currentBit* gives the bit index within this byte. We can then state an invariant that these variables have to satisfy for the structure to be valid:

```
def invariant(currentBit: Int, currentByte: Int, buffLength: Int): Boolean =
    currentBit ≥ 0 && currentBit < 8 && currentByte ≥ 0 &&
    ((currentByte < buffLength) || (currentBit == 0 && currentByte == buffLength))
```

This invariant states that the *currentByte* points to a byte within the underlying buffer bounds and that the *currentBit* points to a bit within the range of a byte (i.e., between 0 and 7).

Let us now discuss what operations this *BitStream* data structure offers. We divide them into three categories.

First of all, we have functions that work on the index. Some of these functions are moving the head of the stream back and forth, such as `increaseBitIndex`, which increases the index by one while taking care of the *currentByte* and *currentBit* arithmetic, or `moveBitIndex` which moves the head by a given offset (positive or negative). Let us focus on one of those functions, `resetAt`, which has a different use case. It takes another `BitStream` as an argument and returns a new instance of `BitStream`, with the buffer of the current instance but the head index of the argument `BitStream`. This function is used only in specification, and we will discuss how it is used in the section about proofs.

A second type of function offered by the `BitStream` class is predicates about the index. These predicates test the available space in the `BitStream` after the current head to ensure enough room when encoding or decoding. Several are offered, but they all take size as an argument (in bytes or bits) and return a boolean value indicating whether this space exists between the stream's current head and the buffer's end.

Finally, the `BitStream` class offers functions to encode and decode bits to and from the stream. Table 1 summarises these functions, showing the encoding and corresponding decoding functions. There are two main classes: the functions encoding and decoding one or more bits and those encoding and decoding one or more bytes. Some functions are implemented by calling other functions in a loop. For example, `appendNBits` uses `appendBit`.

Table 1. `BitStream` Encoding and corresponding Decoding Functions

Bit-Level Operations	
Encoding Functions	**Decoding Functions**
`appendBit, appendBitOne, appendBitZero`	`readBit`
`appendNBits, appendNZeroBits, appendNOneBits`	`readBits`
`appendBitFromByte`	`readBit`
`appendBitsLSBFirst`	`readNBitsLSBFirst`
`appendLSBBitsMSBFirst`	`readNLSBBitsMSBFirst`
`appendBitsMSBFirst`	`readBits, peekBit`
Byte-Level Operations	
`appendPartialByte`	`readPartialByte`
`appendByte`	`readByte`
`appendByteArray`	`readByteArray`

4.1 Specification and Verification

We next present our specifications and formal verification of the functions. We split the specifications into two steps, representing two levels of complexity and guarantees: the absence of crashes (runtime safety) and semantic correctness.

Runtime safety corresponds to the program not crashing. It ensures that all array accesses are within bounds, that integer operations do not overflow, and that no division by zero can occur. To be able to prove the in-bound accesses of arrays, we prove for each function which encodes or decodes data to or from the stream that it is moving the current head by the required number of bits, and we add as a precondition that the number of available bits is greater or equal to the size that will be written or consumed from the stream. For example, for the function `appendBitsMSBFirst`, we prove that the head index after the call is equal to the head index before plus `nBits`, which is the number of bits written by the function. To prove the absence of overflows and division by zero, we add preconditions on arguments and some runtime sanitization checks when the property cannot be ensured statically. For example, these properties ensure the number of bits to encode or decode received as arguments is greater than zero and not too large or that indices are greater or equal to zero and smaller than the biggest integer value.

Once the runtime safety is guaranteed, we can prove some semantic properties. In this case study, the property of interest is the *invertibility*. In the context of the `BitStream` class, it can be intuitively summarized by saying that after encoding some value, reading from the stream returns the same value. This second part of the specification is more challenging to prove. More specifically, the postcondition of the encoding function states that the return value of a call to their corresponding decode function (see Table 1) is equal to what has been written. For some functions, what is written is not exactly the input value. For example, `appendBitFromByte` writes only one bit of the received byte, so the postcondition states only that the written bit is the expected one.

We explore in more detail the proof of the function `appendBitsMSBFirst`, which has the most interesting proof and was the most challenging to verify. First, here is the function's implementation and postcondition part about invertibility (note that the precondition, proof lines, and parts of the postcondition are omitted):

```scala
def appendBitsMSBFirst(srcBuffer: Array[UByte], nBits: Long, from: Long = 0): Unit = {
    appendBitsMSBFirstLoop(srcBuffer, from, from + nBits)
}.ensuring: // ...
    val (r1, r2) = reader(w1, w2) // returns a new bitstream with the buffer
                                  // of w1, but the head of w2
    validateOffsetBitsContentIrrelevancyLemma(w1, w2.buf, nBits)
    val vGot = r1.readBits(nBits)
    byteArrayBitContentSame(srcBuffer, vGot, from, 0, nBits) // Compare the bit
    // content of the two arrays between from and from + nBits, respectively 0 and nBits.
```

```scala
def appendBitsMSBFirstLoop(srcBuffer: Array[UByte], i: Long, to: Long): Unit = {
    if i < to then
        appendBitFromByte(srcBuffer((i / 8).toInt).toRaw, (i % 8).toInt)
        appendBitsMSBFirstLoop(srcBuffer, i + 1, to)
}.ensuring:
    val (r1, r2) = reader(old(this), this)
    validateOffsetBitsContentIrrelevancyLemma(old(this), this.buf, to - i)
    val listBits = bitStreamReadBitsIntoList(r1, to - i)
    val srcList = byteArrayBitContentToList(srcBuffer, i, to - i)
    listBits == srcList
```

The function is effectively a loop calling `appendBitFromByte` for each index i between `from` and `from + nBits`, writing the ith bit to the stream. To aid verification, we implemented the loop as a tail-recursive function (Scala compiler transforms such recursion back to a while loop, so no efficiency is lost).

To give a sense of the size of the proof compared to the size of the code, this function comprises less than 10 lines of code. However, this number increases to 150 when accounting for the proof annotations without even considering the implementations of the different lemmas. If we include those, this number increases further to almost 300 lines of code. This function was one of the most challenging to verify and consequently has one of the worst ratio lines of code versus lines of proof annotations.

The corresponding decode function is `readBits`, which reads a given number of bits from the stream and returns them in an array of bytes. This function is also implemented as a tail-recursive function calling `readBit`.

Proof of invertibility is by induction. The induction hypothesis at iteration i states that the recursive call will correctly encode the bits from $i+1$ to to, meaning that the postcondition is correct. It is then enough to prove that encoding the bit i is correct, i.e., the `readBits` function i^{th} iteration would correctly read the bit i. However, the data are represented by arrays of bytes, not lists of bits, which makes the proof more complicated. Indeed, decoding the bits from $i + 1$ to *to* gives an entirely different array of bytes from decoding i to *to*, as all bits are shifted. Figure 1 illustrates this mismatch.

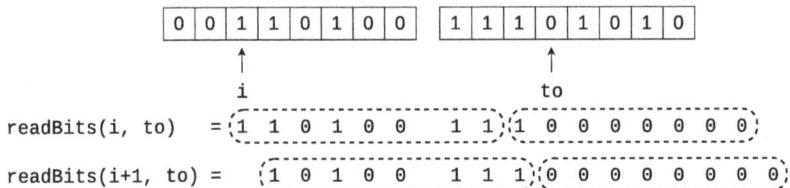

Fig. 1. Applying `readBits` at i and i+1 yields different arrays.

Applying the induction hypothesis in this context is not automatically possible. We, therefore, wrote a proof using a detour through the list of bits. Concretely, we implemented two functions: `bitStreamReadBitsIntoList` which reads `nBits` from a given `BitStream` instance and returns a list of boolean values, and `byteArrayBitContentToList` which transforms a array of bytes into a list of booleans. Proceeding with the previous example, Fig. 2 shows the result of decoding using the function returning a list of booleans.

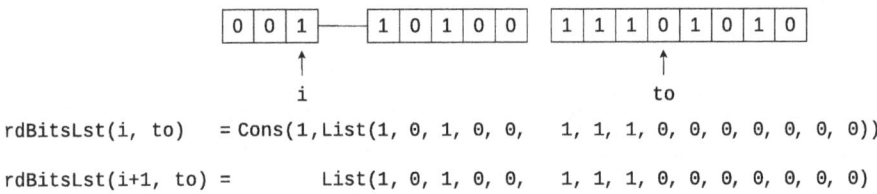

Fig. 2. The result of `bitStreamReadBitsIntoList` (abbreviated `rdBitsLst`) at i+1 is a suffix of the resulting list at i.

We can see that the result as a list has a form that makes applying the induction hypothesis possible. Indeed, the list at iteration i equals the list of the result of the recursive call, with the ith bit prepended. At that stage, the specification of `appendBitsMSBFirstLoop` ensures that the content of the bitstream after the call, read as a list by `bitStreamReadBitsIntoList`, is equal to the content of the `srcBuffer` also read as a list. We then write a lemma to prove that two arrays of bytes for which `bitStreamReadBitsIntoList` returns the same list have the same content when compared with `byteArrayBitContentSame`. This lemma can be applied when calling `appendBitsMSBFirstLoop` in `appendBitsMSBFirst` to finalize the proof.

5   Codecs

ASN1SCC relies on static sets of functions that encode and decode some basic datatypes on and from a `BitStream` instance based on some format. These sets of functions are called *codecs*, and there are three different codecs in this project, corresponding to three formats: UPER, PER, and ACN. We focus on the ACN codec as this is the one used by the generated code we verify in Sect. 7.

Codecs are implemented as decorators of the `BitStream` class, i.e., their interfaces are a superset of the `BitStream` interface. This means that each instance of a codec contains a `BitStream` instance and forwards function calls to it for functions of the `BitStream` interface. The common functionality between the three abovementioned codecs is extracted in a `Codec` class. Therefore, the `Codec_ACN` class contains an instance of `Codec` itself, which, in turn, contains an instance of `BitStream`. The following is an example of a pair of functions implemented in `Codec`:

```
def encodeConstrainedPosWholeNumber(v: ULong, min: ULong, max: ULong): Unit
def decodeConstrainedPosWholeNumber(min: ULong, max: ULong): ULong
```

These functions encode (respectively decode) an unsigned 64-bit integer in the interval $[min, max]$.

The complete list of functions implemented in `Codec_ACN` used in the code generated by the ASN1SCC compiler is shown in Table 2. The common class `Codec` and the specific class `Codec_ACN`, along with the proof annotations, comprises around 4000 lines of code.

We performed similar verification work as for the `BitStream` class described in Sect. 4. We proved runtime safety for all of the functions listed in Table 2 and proved invertibility for most of them. We did not prove the invertibility of the functions working with `Real` numbers (following the IEEE754 norm) because Stainless does not support floating point numbers (unlike a related Daisy tool [16]). We also did not yet prove the invertibility of String encoding and decoding functions.

The approach used to write the proof is the same as for the `BitStream` class (explained in Sect. 4) and relies a lot on the properties proved on `BitStream`'s operations.

While writing the proof of invertibility, we performed an interesting refactoring to one auxiliary function: `uint2int`. This function converts an unsigned integer of a given number of bytes to a 32-bit signed integer. The unsigned integer is a 64-bit integer in which only the given number of bytes are considered. This function was implemented using a `while` loop to iterate over the number of bytes. Stainless supports verification of `while` loops, but verifying functions that use them requires writing invariants that can be complex and, therefore, hard to come up with. A basic invariant was insufficient to prove the needed properties in this case. Instead of trying to find a sufficient invariant, given that the number of iterations the loop could perform was bound by 0 and 7, we decided to unroll the loop by hand in the source code. While preserving the code's conciseness, this manual unrolling allows automatic verification of properties that would have otherwise required a non-trivial invariant. This illustrates that refactoring can simplify verification while preserving the semantics.

Table 2. Encoding and Corresponding Decoding Functions

ACN Codec	
Encoding Function	**Decoding Function**
enc_Int_PositiveInteger_ConstSize_8	dec_Int_PositiveInteger_ConstSize_8
enc_Int_PositiveInteger_ConstSize_big_endian_{16,32,64}	dec_Int_PositiveInteger_ConstSize_big_endian_{16,32,64}
enc_Int_PositiveInteger_ConstSize	dec_Int_PositiveInteger_ConstSize
enc_Int_TwosComplement_ConstSize_8	dec_Int_TwosComplement_ConstSize_8
enc_Int_TwosComplement_ConstSize_big_endian_{16,32,64}	dec_Int_TwosComplement_ConstSize_big_endian_{16,32,64}
enc_Real_IEEE754_{32,64}_big_endian	dec_Real_IEEE754_{16,32,64}_big_endian
enc_Real_IEEE754_{16,32,64}_little_endian	dec_Real_IEEE754_{16,32,64}_little_endian
enc_String_Ascii_Null_Terminated_multVec	dec_String_Ascii_Null_Terminated_multVec
enc_String_CharIndex_private	dec_String_CharIndex_private
enc_IA5String_CharIndex_External_Field_DeterminantVec	dec_IA5String_CharIndex_External_Field_DeterminantVec
enc_IA5String_CharIndex_Internal_Field_DeterminantVec	dec_IA5String_CharIndex_Internal_Field_DeterminantVec

6   Tailoring ASN1SCC to Verification

We next present our design of a generator that produces verifiable code along with inductive specifications, while preserving the spirit of existing C and Ada generators. Our approach illustrates that verification is best done in a situation when it is possible to adjust the code, balancing efficiency and verifiability. We first discuss the decoders. Whereas C and Ada decode the result in-place, we redesigned the Scala backend to construct fresh values. We review below the replacement of mutable arrays by immutable sequences, the alternatives we considered, and the use of recursion instead of imperative loops. We then consider the generation of alignment-aware size methods for SEQUENCE, SEQUENCE OF and CHOICE. We finally examine the case of ACN-specialized type assignments, the result in the generated code, their negative impact on verification, and the countermeasures.

6.1   Alternatives to In-Place Mutation

ASN1SCC generated code for existing targets (C and Ada) leverages in-place mutation for decoding. The caller of any decoding function is expected to provide a structure. Arrays within structures (e.g. strings or SEQUENCE OF) are expected to be allocated with the maximum size. As the decoding progresses, the structure is updated with the decoded values. Of particular interest is decoding a CHOICE. The C template leverages union. Since these are unchecked, they can be written without additional constraints. Assuming we are decoding a SEQUENCE named S in an instance s with a field choice whose variant is decided by kind, the generated C would be:

```
union { C1 c1; /* ... */ Cn cn;} C_union; struct { C_selection kind; C_union u; } C;
struct { /* ... */ C choice } S;
switch (kind) {
    case C1_tag: s→choice.kind = C1_tag; decodeC1(&s→choice.u.c1, bitStream);
    // ...
    case Cn_tag: s→choice.kind = Cn_tag; decodeCn(&s→choice.u.cn, bitStream);}
```

In the above snippet, the field `kind` stores the tag of the `CHOICE` variant, and
the field `u` is a union of all `CHOICE` variants.

The Ada template uses variant records. It deviates from the C backend
because it must first default-initialize the variant. The rest, however, remains
similar. In particular, the variant is selected and decoded in place.

This template where the choice variant is selected and mutated is not possible
in Scala, as it results in a typing error:

```
enum C { case C1(...); /* ... */ case Cn(...) }
kind match
    case C1_tag ⇒ s.choice=C1_Init(); decodeC1(s.choice, bitStream)//expect C1, got C
    case Cn_tag ⇒ s.choice=Cn_Init(); decodeCn(s.choice, bitStream)//expect Cn, got C
```

A solution is to declare and initialize a local variable, assign it to `s.choice`
and pass it to the decoding function. Unfortunately, Stainless does not allow
this kind of mutation as it is not part of the aliasing fragment it supports. This
aspect motivated the change of the Scala decoding functions to return decoded
values instead of relying on mutation. For instance, the above example becomes:

```
val choice = kind match
    case C1_tag ⇒ decodeC1(bitStream); case Cn_tag ⇒ decodeCn(bitStream)
S(..., choice, ...) // Return the decoded S with the decoded choice and other fields
```

When decoded, individual elements of `SEQUENCE OF` are also returned as
values. There are, however, two possible ways to decode the overall `SEQUENCE
OF`. One way is to create a local array of the appropriate size, update it in place,
and return it. Another one is to append the decoded elements to a collection.
This possibility can be split depending on whether we choose the collection to
be immutable or not. We detail the approach we retain in the following section.

6.2   Data Structures Representing SEQUENCE OF

ASN1SCC employs mutable arrays to store the elements of a `SEQUENCE OF`. For
decoding, the array is pre-allocated to the maximum size stated in the ASN.1
specification. Individual elements are decoded in place by dereferencing the element for both the C and Ada backends.

Due to our choice of decoding elements (relying on constructing values instead
of mutating them in place, as discussed earlier), the template for decoding individual `SEQUENCE OF` elements also needs to be adapted. One solution would be
to create a local array and update it with the decoded elements:

```
val arr = Array.fill(len)(elemInit()) // Default−initialized elements
for (i ← 0 until len) { val elem = decode(bitStream); arr(i) = elem }
S(..., arr, ...) // Return the structure wrapping the array with other decoded fields
```

Stainless will accept this snippet as long as the type of the element does not contain any other `Array`. If, however, the type transitively contains an `Array`, we will stumble upon the same issue of unsupported aliasing, as mentioned earlier. It is possible to create an alias analysis escape hatch. We need to uphold the unchecked assumption that the decoded elements are not mutated for the rest of the program. While suboptimal, this could be considered an acceptable solution because the code we generate will maintain this invariant. Such a solution still fails, however, when it comes to actual verification conditions. Namely, Stainless encodes the JVM `Array` leveraging the SMT theory of generalized arrays with map combinators [21], a `SEQUENCE OF` is therefore represented in part as an SMT array. The encoding becomes more complex in the presence of nested `SEQUENCE OF` and uses a non-standard array combinator. While Z3 supports this combinator, it sometimes times out for intermediate queries. On the other hand, cvc5 does not support it.

Our solution is to wrap a Scala `Vector` in a class that exposes commonly used operations. ASN1SCC relies on random access on some procedures, which is an acceptable $O(\log n)$ for `Vector`. We specify vector operations using a `List`. This specification has the advantage of relying on the SMT theory of ADT. The Stainless library provides many properties and lemmas regarding operations on `List`. Our solution thus achieves a usable combination of efficiency and verifiability.

We also lift the recursive function for encoding and decoding a `SEQUENCE OF` to the top-level. This enables the encoding function to refer to its decoding counterpart in the postcondition, to specify the invertibility property.

6.3   Size Computation

ASN1SCC computes a lower and an upper bound of their size in bits for each type assignment. These statically known bounds are sufficient to prove runtime safety. For invertibility, however, the exact position of the bitstream must be known. This, in turn, requires computing the exact size of structures, some of which may be dynamic (e.g. variable-sized strings, `SEQUENCE OF` or `CHOICE`).

We generate size methods for all class definitions. ACN allows fields to have an alignment restriction, impacting the size of the overall structure. Therefore, the size of such a structure depends on the bitstream position. The generated size methods take an offset corresponding to the bitstream position as an argument to account for this. The size of structures with no alignment restriction in any of their component is invariant in the bitstream position. Structures having a component with a byte alignment restriction and word (16 bits) or double word alignment (32 bits) have a size invariant in the offset modulo 8, 16 and 32. A similar observation can be made for word (invariant modulo 16 and 32) and double word alignment (invariant modulo 32). We generate lemmas stating these properties and apply them whenever the bitstream must align its cursor. The size methods are implemented by recursively computing the size of each element and threading the offset with the accumulated size.

6.4 The Case of ACN-Specialization

For type assignments whose binary format is described with ACN, ASN1SCC will specialize the encoding and decoding functions. It does so by inlining calls to these type assignments' encoder and decoder functions and by inserting the necessary logic for the ACN-specific part. This specialization by inlining does not hinder verification much for small structures and simple properties. However, for invertibility, this becomes an issue because the verification conditions (VCs) are already complex and large enough. We solve this problem by "restoring" some modularity and hoisting (or outlining) the inlined code into top-level functions. We also parameterize the functions by their ACN dependencies and the ACN fields they return.

7 Proof Generation for SEQUENCE Invertibility

When invoked with the -invertibility flag, our code generator produces additional postconditions, assertions and lemmas to prove SEQUENCE invertibility. The extra postconditions essentially state that decoding the encoded message yields the original result. These are also generated for CHOICE and SEQUENCE OF, even though currently, no proofs are produced for the invertibility of these recursive cases.

For a structure s (whether a SEQUENCE, CHOICE or SEQUENCE OF), the postcondition is as follows:

```
@opaque def S_Encode(s: S, codec: ACN) = { /* ...implementation... */ }.ensuring:
  case Left(_) ⇒ true // 1.
  case Right(_) ⇒ // 2.
    old(codec).buf.length == codec.buf.length && // 3.
    codec.bitIndex == old(codec).bitIndex + s.size(old(codec).bitIndex) && // 4.
    old(codec).isPrefixOf(codec) && locally: // 5.
      val r1 = codec.resetAt(old(codec)) // 6.
      val (r2Got, decRes) = S_Decode_pure(r1) // 7.
      decRes match
        case Left(_) ⇒ false // 8.
        case Right(resGot) ⇒ r2Got == codec && resGot == s // 9.
```

At line 1, no properties are stated if the encoding fails. If it succeeds (line 2), line 3 states that the codec before entering this function (denoted by old(codec)) has its buffer length unchanged. Line 4 states that the cursor of the codec is precisely advanced by the size of s. As discussed previously, a structure may have some alignment restriction in one or multiple of its fields. Its overall size, therefore, depends on the starting position of the bitstream, which we pass to s.size. These stated properties are always generated, even if -invertibility is not passed. Line 5 essentially indicates that the function did only append in the buffer (hence, the old codec is a prefix of the new one). The condition wrapped in a local block starting at line 5 is the inversion property. In line 6, we rewind the bitstream back to the original position. We then decode what we encoded using a pure version of the corresponding decoding function at line 7. It conceptually makes a copy of the given bitstream and returns the mutated version in c2Res and the decoded result in decRes. Line 8 states that the decoding cannot fail, and line 9 specifies that the decoded value is the same as we started with.

The same template applies to ACN-specialized hoisted functions, discussed earlier. The difference lies in the additional parameters, and the returned decoded ACN fields (if any). The invertibility accounts for these differences. In particular, it ensures that the decoded ACN fields are equal to what was encoded.

We furthermore annotate the encoding function with @opaque, effectively hiding the body to the solver, as the postcondition exactly states the behavior of the function. This improves the verification performance since the solver does not need to unfold the body (which, in turn, would lead to unfolding other encoding calls). To improve performance further, we also annotate the decoding functions as opaque. The postcondition of the decoding functions is incomplete (they only state properties 1 through 4); therefore, we need to explicitly unfold them when proving their invertibility (inside the encoding function) or the lemmas related to them. Failing to do so would lead to "false" counterexamples.

Proving SEQUENCE invertibility relies on the composition of the invertibility of each field but is insufficient on its own. The postcondition for a field f_i only states that decoding from the bitstream at $i + 1$ (rewound at i) yields the same result, while we need it to hold for the bitstream at $n + 1$ rewound at i. We intuitively need a "prefix lemma" stating that if one bitstream is a prefix of another one up to the size of the structure to decode, then the results are equivalent. The size of the structure to decode is unknown in the context of this lemma, we need to correctly "guess" it (which we detail next). When applying the lemma, the size simply corresponds to the size of the field we encoded, which we know.

The template of a prefix lemma follows. It applies to SEQUENCE, CHOICE, and SEQUENCE OF. Only SEQUENCE OF does not have a proof. ACN-specialized decoding functions may have additional parameters and are accounted for.

```
def T_prefixLemma(c1: ACN, c2: ACN, sz: Long): Unit =
  require(c1.buf.length == c2.buf.length && c1.validate_offset_bits(T_MAX_SIZE) &&
    0L ≤ sz && sz ≤ T_MAX_SIZE) // 1.
  require(arrayBitRangesEq(c1.buf, c2.buf, 0L, c1.bitIndex + sz)) // 2.
  val c2Rst = c2.resetAt(c1) // 3.
  val (c1Res, decRes1) = T_Decode_pure(c1) // 4.
  val (c2Res, decRes2) = T_Decode_pure(c2Rst) // 5.
  val v1Size = decRes1 match // 6.
    case Right(v1) ⇒ v1.size(c1.bitIndex); case Left(_) ⇒ 0L
  { /* proof */ }.ensuring:
    decRes1 match
      case Right(v1) if v1Size == sz ⇒ // 7.
        decRes2 match
          case Right(v2) ⇒ c1Res.bitIndex == c2Res.bitIndex && v1 == v2 // 8.
          case Left(_) ⇒ false // 9.
      case Left(_) ⇒ true // 10.
```

The lemma assumes the following. The two codecs must have the same buffer length. c_1 needs sufficient space to decode any T, with T_MAX_SIZE being the statically known maximum size for any instance of T. The size parameter must be positive and no greater than T_MAX_SIZE. c_1 and c_2 must be equal in content up the position of c_1 plus the size to be "guessed". Then, the lemma states the following. We rewind c_2 at the position of c_1 at line 3. Line 4 and 5 decode a T with c_1, and c_2 rewound at c_1. The resulting codecs are stored in c1Res and c2Res respectively. The decoding results are bound to decRes1 and decRes2. We compute the size of the decoded message from c_1 at line 6. In case of failure, it is arbitrarily set to 0. The lemma result is stated in the ensuring clause. If decoding from c_1 fails or if the guessed size is different from the resulting size, there are no claims (line 10). Otherwise (line 7), the lemma states that decoding from c_2 cannot fail (line 9). Furthermore, the decoded value is the same as the one from c_1, and the codecs end up in the same position (line 8).

Proving this property is a matter of applying the corresponding prefix lemma for each field along with other lemmas. Without further proof engineering, performance unfortunately greatly suffers even with the outlining of ACN-specialized function. The main culprit is in the need to unfold both the body of T_Decode(c1) and T_Decode(c2Rst) at the beginning of the proof. This is necessary to prove that decoding f_i cannot fail thanks to the end-to-end decoding success. Additionally, many intermediate assertions cause the VCs for subsequent fields to grow.

As a first step to make verification feasible, we wrap the proof for each field in a local opaque function and state the desired property in the postcondition. We apply these sublemmas sequentially, then unfold the body of T_Decode applied with c1 and c2Rst to "glue" everything together. The proof becomes:

```
decRes1 match
  case Right(v1) if v1Size == sz ⇒
    @opaque def proof_f_i(c1_i: ACN, c2_i: ACN): Unit = {
      require(c1_i.buf == c1.buf && c2_i.buf == c2.buf) // 1.
      val offset = size_1 + ... + size_{i-1} // 2.
      require(c1_i.bitIndex == c1.bitIndex+offset && c1_i.bitIndex == c2_i.bitIndex)// 3.
      arrayBitRangesEqSlicedLemma(c1_i.buf, c2_i.buf,
        0L, c1_i.bitIndex + v1Size - offset, 0L, c1_i.bitIndex + size_i) // 4.
      // ... assertions to prove requirements ...
      T_i_prefixLemma(c1_i, c2_i.withMovedBitIndex(size_i), size_i) // 5.
      // ... other assertions
    }.ensuring:
      val (c1_{i}+1, dec1) = T_i_Decode_pure(c1_i) // 6.
      val (c2_{i}+1, dec2) = T_i_Decode_pure(c2_i) // 7.
      val f_i_1 = dec1 match case Right(res) ⇒ res; case Left(_) ⇒ ??? // 8.
      val f_i_2 = dec2 match case Right(res) ⇒ res; case Left(_) ⇒ ??? // 9.
      f_i_1.size(c1_i.bitIndex) == size_i && f_i_1 == f_i_2 && c1_{i+1}.buf == c1.buf &&
      c2_{i+1}.buf == c2.buf && c1_{i+1}.bitIndex == c1.bitIndex + offset &&
      c1_{i+1}.bitIndex == c2_{i+1}.bitIndex // 10.
    unfold(T_Decode(snapshot(c1))) // 11.
    proof_f_i(c1_i, c2_i)
```

```
    val(c1ᵢ₊₁, dec1ᵢ₊₁)=Tᵢ_Decode_pure(c1ᵢ);val(c2ᵢ₊₁, dec2ᵢ₊₁)=Tᵢ_Decode_pure(c2ᵢ)
    unfold(T_Decode(snapshot(c2Rst))) // 12.
    decRes2 match
      case Right(v2) ⇒ check(c1Res.bitIndex == c2Res.bitIndex && v1 == v2)
      case LeftMut(_) ⇒ check(false)
  case _ ⇒ () // vacuous
```

For each field f_i, we generate a corresponding proof_f_i. It takes two codecs $c1_i$ and $c2_i$ corresponding to the outer c1 and c2 rewound at the offset of f_i, conditions stated in 1 and 3. The offset computed at 2 refers to the sizes of each previous field from the decoded v1 and is computed outside of this function. For $i = 1$, proof_f_i is parameterless and directly uses c1 and c2Rst. Application of the lemma at 4 allows to deduce that $c1_i$ and $c2_i$ are equal in $[0,$ $c1_i$.bitIndex $+$ size$_i)$ where size$_i$ is the size of f_i. This property is needed to be able to apply the prefix lemma at 5. proof_f_i states that decoding from $c1_i$ and $c2_i$ (lines 6 and 7) cannot fail (lines 8 and 9) and that the values are equal, alongside other properties (line 10). Intuitively, proving that decoding from $c1_i$ succeeds requires unfolding the body of T_Decode(c1) in order to unveil that this step is part of T_Decode. It is, however, insufficient since the $c1_i$ of proof_f_i has no correspondence with the state of the codec in T_Decode at i (having the same position and buffer is insufficient). We discuss one solution afterward. On the other hand, T_i_prefixLemma guarantees that decoding from $c2_i$ cannot fail provided decoding from $c1_i$ succeeds.

The actual proof of T_prefixLemma starts at 11. We unfold T_Decode applied with c1. Note that unfolding T_Decode_pure is futile because it calls T_Decode, which will not be unfolded. That said, T_Decode mutates its parameter. We therefore need to make a copy of c1 (done with snapshot). For each field f_i, we then apply the corresponding proof and call the decoding function. We thread the resulting codecs to the following field. For ACN-specialized outlined functions, functions returning ACN fields have their value extracted and forwarded where needed. At line 12, we unfold T_Decode applied with c2Rst to prove that decoding from c2 (rewound at c1) cannot fail and yield the same values.

Going back to proving the infallibility in proof_f_i, a solution consists in first creating an "origin function" which threads the various codecs:

```
@opaque def fᵢ_codec_origin(c1ᵢ: ACN): Boolean =
    val (c1₂_got, dec₂_got) = T₁_Decode_pure(c1)
    dec₂_got match
      case Right(_) ⇒ // ... decode T₂ and so on
        val (c1ᵢ_got, decᵢ_got) = Tᵢ_Decode_pure(cᵢ₋₁)
        decᵢ_got match case Right(_) ⇒ c1ᵢ_got == c1ᵢ; case Left(_) ⇒ false
      case Left(_) ⇒ false
```

Then, we add f_i_codec_origin($c1_i$) as a precondition to proof_f_i and unfold it along with T_Decode to prove infallibility.

8   Experience with Case Study

8.1   Verified Properties and Statistics

Codec Classes. We present the verification of the `BitStream` and codecs classes in Table 3, column *Library*. Runtime safety is proven for all functions. Invertibility is proved for all functions used by the PUS-C services, minus the floating point and string-related functions. We conducted the experiment with a timeout of 6 min, with Z3 v4.12.2, cvc5 1.1.2 and CVC4 1.8.

We report for each VC category the number of Verified, Undecided, and Invalid VCs. *Measures* VCs ensure that the measure annotations for recursive functions and `while` loops are positive and strictly decreasing. *Class invariants* check that instances of classes with invariant uphold the stated properties (whether on construction or on mutation). It mostly relates to updates of codec instances. *Pos(itive) array size* refers to VCs checking array allocation size to be positive. *Miscellaneous* contains all VCs introduced by some Stainless phases.

Generated Scala Code. We present two sets of results reported in Table 3. The first one, column *PUS-C services*, is a run over all but one PUS-C services without the `-invertibility` option. The verified properties include runtime safety, exact structure size computation, and precise bitstream position. We excluded PUS-C service number 4 because it uses floating point arithmetic, which Stainless does not support. Additionally, all services share the ASN.1/ACN definitions within `ccsds` and `common`. Consequently, these VCs are counted multiple times. We did not, however, deduplicate them because the different services had different numbers, which was caused by some light variation in the generated code. When generated and verified in isolation, they have 11,268 VCs, all valid.

The second set of results is the verification of `TC-Packet` with the `-invertibility` flag and is reported in column *TC-Packet*. This `SEQUENCE` is the largest in all services. Note that the individual components are not reported since they contain structures for which no proofs are generated at the moment of writing (such as `CHOICE` or `NullType` with special encoding).

It should be noted that no in-bound array access VCs are present in both columns due to the usage of the `Vector` class: in-bound access is instead verified as a precondition of the indexing method `apply`. Furthermore, the underlying buffer of the codec is never directly accessed by the generated code.

Both experiments were run with Z3 v4.12.2 and cvc5 1.1.2 in the portfolio, with timeouts of 3 min and 15 min, respectively.

Regarding column *PUS-C services*, services 8 and 18 have two invalid VCs related to preconditions. They both concern the encoding of `IA5String`: the recursive function in charge of encoding the string requires the size of the string (represented as a `Vector[Byte]`) to have a certain size dictated by the ASN.1 definition. The caller may not satisfy this condition even in the presence of a constraint check automatically inserted by ASN1SCC. The latter only validates the position of a null-terminator without knowing the pointer or array's underlying capacity. Integrating this check is surprisingly non-trivial since it would impact

Table 3. Statistics of Verification Conditions.

VCs	Library			PUS-C services			TC-Packet		
	# V	# U	# I	# V	# U	# I	# V	# U	# I
Preconditions	4,252	0	0	152,201	1	2	529	0	0
Overflows/casts	936	0	0	82,037	0	0	230	0	0
Assertions	544	0	0	23,284	0	0	167	0	0
Postcondition	443	0	0	22,365	1	0	30	0	0
Arithmetic ops	183	0	0	3,711	0	0	0	0	0
Array access	181	0	0	0	0	0	0	0	0
Measures	132	0	0	2,796	0	0	0	0	0
Class invariant	54	0	0	1,722	0	0	0	0	0
Match exh.	39	0	0	38,283	0	0	101	0	0
Pos. array size	5	0	0	0	0	0	0	0	0
Miscellaneous	2	0	0	918	0	0	0	0	0
Total	**6,771**	**0**	**0**	**327,317**	**2**	**2**	**1,057**	**0**	**0**

the C and Ada backends. In particular, the generated C code uses raw pointers without embedding any capacity or size information. Finally, the two timeouts, both appearing in service 15, are of the same nature, with the difference being that Stainless cannot find a counterexample.

Though it does not stand out from *TC-Packet*, we note that this SEQUENCE has many ACN-inserted fields, nested and direct. A prior attempt at proving invertibility for a subset shows that the outlining discussed in 6.4 is necessary.

8.2 Identified Bugs

We give an overview of some issues we have found thanks to verification.

Incorrect Treatment of NaN. While translating the floating point encoding and decoding functions from C to Scala, we have discovered that an assertion did not hold when the bit pattern represented a NaN. The original C and Ada code did not handle this case, and we have opened an issue that was swiftly addressed[3]. Note that we represent the floating point number as an uninterpreted Long since Stainless does not support Double or Float.

Improper Alignment for Padding. SEQUENCE whose fields have an alignment requirement were not correctly aligned with padding bits. Additionally, the buffer was not appropriately checked for remaining space in such cases[4].

---

[3] https://github.com/maxime-esa/asn1scc/issues/287.

[4] https://github.com/maxime-esa/asn1scc/issues/283.

Erroneous Decoding for Optional CHOICE. When prototyping a proof for invertibility, we discovered that optional CHOICEs were not correctly decoded when the determinant and presence bits were specified with ACN[5].

Missing Validation Check for 7-bit Strings. For all backends, 7-bit strings are represented as byte arrays. However, the generated encoder does not validate the bytes' values and assumes they are within the range $[0, 127]$, which leads to incorrectly written data.

9 Conclusion

We developed a Scala backend for the ASN1SCC compiler with an accompanying library for ASN.1/ACN primitives. We proved the absence of runtime errors for the library and the generated code. We also proved that the deserialization functions from the library used by the PUS-C services are the inverse of the corresponding serialization functions. For the generated code, we also established the precise amount of data written or read. Furthermore, we proved the invertibility property for records. Our verification and specification process led us to discover several bugs in the existing code generator. The errors have been addressed, leading to more reliable communication infrastructure.

Acknowledgements. This research was supported by the European Space Agency Open Space Innovation Platform, 4000140196/22/NL/GLC/ov, New Concepts for Onboard Software Development. We thank Maxime Perrotin for overseeing the project.

References

1. Barbosa, H., et al.: cvc5: a versatile and industrialstrength SMT solver. In: TACAS (1). Lecture Notes in Computer Science, vol. 13243, pp. 415–442. Springer (2022)
2. Barrett, C., Conway, C.L., Deters, M., Hadarean, L., Jovanović, D., King, T., Reynolds, A., Tinelli, C.: CVC4. In: Gopalakrishnan, G., Qadeer, S. (eds.) CAV 2011. LNCS, vol. 6806, pp. 171–177. Springer, Heidelberg (2011). https://doi.org/10.1007/978-3-642-22110-1_14
3. Blanc, R.W., Kneuss, E., Kuncak, V., Suter, P.: An overview of the Leon verification system: verification by translation to recursive functions. In: Scala Workshop (2013)
4. Blanc, R.W.: Verification by Reduction to Functional Programs. Ph.D. thesis, EPFL, Lausanne (2017). https://doi.org/10.5075/epfl-thesis-7636, http://infoscience.epfl.ch/record/230242
5. Bucev, M., Kunčak, V.: Formally verified quite OK image format. In: Formal Methods in Computer-Aided Design (FMCAD) (2022)
6. Chassot, S., Kunčak, V.: Verifying a realistic mutable hash table - case study (short paper). In: International Joint Conference on Automated Reasoning (IJCAR) (2024)

---

[5] https://github.com/maxime-esa/asn1scc/issues/289.

7. De Moura, L., Bjørner, N.: Z3: An efficient SMT solver. In: International Conference on Tools and Algorithms for the Construction and Analysis of Systems, pp. 337–340. Springer (2008)
8. Delaware, B., Suriyakarn, S., Pit-Claudel, C., Ye, Q., Chlipala, A.: Narcissus: correct-by-construction derivation of decoders and encoders from binary formats. Proc. ACM Program. Lang. **3**(ICFP) (2019). https://doi.org/10.1145/3341686
9. Enumeration, M.C.W.: CWE top 25 most dangerous software weaknesses. https://cwe.mitre.org/top25/ (2023). Accessed 4 Sep 2024
10. ESA-ESTEC, E.S.: Telemetry and telecommand packet utilization. Standard, European Cooperation for Space Standardization (April 2016)
11. Google: Protocol buffers. https://protobuf.dev/
12. Group, N.W.: Internet x.509 public key infrastructure certificate and certificate revocation list (crl) profile. https://www.rfc-editor.org/rfc/rfc5280
13. Guilloud, S., Bucev, M., Milovančević, D., Kunčak, V.: Formula normalizations in verification. In: Computer-Aided Verification (CAV) (2023)
14. Hamza, J., Felix, S., Kunčak, V., Nussbaumer, I., Schramka, F.: From verified Scala to STIX file system embedded code using Stainless. In: NASA Formal Methods (NFM), pp. 18 (2022). http://infoscience.epfl.ch/record/292424
15. Hamza, J., Voirol, N., Kunčak, V.: System FR: formalized foundations for the Stainless verifier. Proc. ACM Program. Lang. **3**(OOPSLA) (Oct 2019). https://doi.org/10.1145/3360592
16. Isychev, A., Darulova, E.: Scaling up roundoff analysis of functional data structure programs. In: SAS. Lecture Notes in Computer Science, vol. 14284, pp. 371–402. Springer (2023)
17. ITU-T Study Group 17: Abstract syntax notation one (ASN.1) recommendations. Standard ITU-T X.680, International Telecommunication Union (ITU), Geneva, CH (2008). https://www.itu.int/ITU-T/studygroups/com17/languages/
18. Laboratory, N.I.T.: National vulnerability database CVE-2024-37305 detail. https://nvd.nist.gov/vuln/detail/CVE-2024-37305
19. Mamais, G., Tsiodras, T., Lesens, D., Perrotin, M.: An ASN.1 compiler for embedded/space systems. In: Embedded Real Time Software and Systems (ERTS2012). Toulouse, France (Feb 2012), https://hal.science/hal-02263447
20. Mondet, S., Alberdi, I., Plagemann, T.: Generating optimised and formally checked packet parsing code. In: IFIP International Information Security Conference, pp. 173–184. Springer (2011)
21. de Moura, L.M., Bjørner, N.S.: Generalized, efficient array decision procedures. In: FMCAD, pp. 45–52. IEEE (2009)
22. N7 Space: ASN.1 PUS-C types library. https://n7space.github.io/asn1-pusc-lib/. Accessed 13 Sep 2024
23. Ni, H., Delignat-Lavaud, A., Fournet, C., Ramananandro, T., Swamy, N.: ASN1*: provably correct, non-malleable parsing for ASN.1 DER. In: Proceedings of the 12th ACM SIGPLAN International Conference on Certified Programs and Proofs, pp. 275–289. ACM, Boston MA USA (Jan 2023). https://doi.org/10.1145/3573105.3575684
24. Office, E.S.: ESA cyber security resilience achievement. Tech. rep., European Space Agency (10 2023)
25. Rümmer, P.: A constraint sequent calculus for first-order logic with linear integer arithmetic. In: Cervesato, I., Veith, H., Voronkov, A. (eds.) LPAR 2008. LNCS (LNAI), vol. 5330, pp. 274–289. Springer, Heidelberg (2008). https://doi.org/10.1007/978-3-540-89439-1_20

26. Slee, M., Agarwal, A., Kwiatkowski, M.: Thrift: scalable cross-language services implementation. Facebook white pap. **5**(8), 127 (2007)
27. Suter, P., Köksal, A.S., Kuncak, V.: Satisfiability modulo recursive programs. In: Static Analysis Symposium (SAS) (2011)
28. Swamy, N., et al.: Hardening attack surfaces with formally proven binary format parsers. In: Proceedings of the 43rd ACM SIGPLAN International Conference on Programming Language Design and Implementation, pp. 31–45. PLDI 2022, Association for Computing Machinery, New York, NY, USA (2022). https://doi. org/10.1145/3519939.3523708
29. Tsiodras, T.: TASTE - an ESA-led toolchain that uses model-driven code generation to create correct-by-construction sw for safety-critical targets. In: MeTRiD 2018: First International Workshop on Methods and Tools for Rigorous System Design (2018)
30. Voirol, N.: Verified Functional Programming. Ph.D. thesis, EPFL, Switzerland (2019)
31. Voirol, N., Kneuss, E., Kuncak, V.: Counter-example complete verification for higher-order functions. In: Scala Symposium (2015)

ExpectAll: A BDD Based Approach for Link Failure Resilience in Elastic Optical Networks

Gustav S. Bruhns, Martin P. Hansen, Rasmus Hebsgaard, Frederik M. W. Hyldgaard, and Jiří Srba[⊠]

Department of Computer Science, Aalborg University, Aalborg, Denmark
srba@cs.aau.dk

Abstract. Constantly growing demands on higher bandwidth and quality of service in modern communication networks motivate the introduction of fully optical network technologies that can eliminate the bottlenecks of optical to digital signal conversions. Recent advances in elastic optical networks enable fine-grained resource allocation technologies for traffic demands, which introduces the Routing and Spectrum Allocation (RSA) problem. In order to improve network resilience for multiple link failures while avoiding double light-spectrum allocation, we present *ExpectAll*—a novel approach and a tool for resilience and path/spectrum allocation based on binary decision diagrams (BDDs). Our method efficiently computes and stores all solutions to the RSA problem in the BDD data structure, facilitating optimal and fast failover protection for failure scenarios even with multiple failing links. *ExpectAll* surpasses the state-of-the-art methods in both the speed of finding a single optimal solution for a currently occurring failure scenario as well as in the preparation time required to precompute all optimal route and spectrum assignments.

1   Introduction

With more than two-thirds of the global population having access to the internet and the increasing amount of data that originates from more and more devices being connected, modern data networks are put under pressure, heightening the need for increased bandwidth and network resilience [16]. Elastic, all-optical networks [10] are a possible solution to deal with these challenges. Traditionally, optical networks use wavelength-division multiplexing (WDM) [33] in order to split the frequency spectrum into slots of 50 GHz. As the amount of traffic has increased and is only predicted to increase more in the future, a new flexible paradigm has been proposed using elastic *Flexgrid* technology to enable more fine-grained splitting of the bandwidth down to 6.25 GHz slots [30]. In elastic optical networks, data is transported along *lightpaths*, which are all-optical connections between two access points in the network using one or more spectrum slots. Given a set of traffic demands, the routing and spectrum allocation (RSA) problem [31] involves finding a lightpath in the form of a route through the

K. Shankaranarayanan et al. (Eds.): VMCAI 2025, LNCS 15530, pp. 208–230, 2025.
https://doi.org/10.1007/978-3-031-82703-7_10

network and a set of spectrum slots for each demand. A solution to the RSA problem must comply with the constraints of:

- *Continuity*: A lightpath must use the same spectrum slots throughout its entire flow through the network.
- *Contiguity*: The spectrum slots used on a lightpath must be consecutive (follow each other).
- *Non-overlapping*: For each link in the network, a spectrum slot can be used by at most one lightpath.

With the increasing scale of networks and the amount of data that can be transported through an elastic optical network, the consequences of link outages become more severe [27,28]. Examples include loss of business revenue and disruption of safety-critical networks [12,17,24]. It is therefore important that networks quickly recover from link failures to reduce the consequences by finding an alternative routing for the demands affected by the link failures [19]. Current approaches in all-optical networks achieve quick recovery times for link failures by reserving a backup path for each demand [3,9] through over-allocation, thereby wasting network resources. Furthermore, this approach can only handle one-link failures, but multiple links are likely to fail [25]; hence, preparing for only one link failures can be inadequate. Ensuring resilience to multiple link failures with quick recovery times in elastic optical networks therefore presents a relevant challenge in the foreseeable future.

Our Contributions. We design and implement *ExpectAll*, a novel approach for solving the RSA problem using Binary Decision Diagrams (BDDs) to ensure failure resilience in elastic optical networks. The tool efficiently finds and compactly stores *all* solutions to the RSA problem, which can be leveraged to quickly provide real-time solutions to multiple link failure scenarios. To this end, our contributions are as follows.

First, we investigate the possible application of existing approaches relying on integer linear programming (ILP) for ensuring optimal failure resilience in optical networks. We experiment on two real network topologies in multiple-link failure scenarios and show that ILP is impractical for ensuring failure resilience, both in terms of the time for synthesizing new network configurations as well as in the excessive memory requirements.

As our second contribution, we leverage the BDD technology to solve the RSA problem for the purpose of failure resilience, culminating in an open-source tool *ExpectAll*. We prove that our approach finds all optimal solutions and present improvements to increase its scalability w.r.t. the number of traffic demands.

Finally, we showcase two applications of the BDD technology for ensuring resilience under multiple-link failures. The first application can handle an arbitrary number of link failures, whereas the second application provides quicker recovery times at the cost of being able to handle only an a priori given maximum number of failing links. The two applications are compared to the ILP approach, where it is clear that they both outperform the ILP solver when comparing how quickly they are able to recover from link failures, as well as how they scale when increasing the number of link failures.

Related Work. The Routing and Wavelength Assignment (RWA) problem is a well-studied [43,46,61] NP-complete problem [14] in optical networks that use Wavelength Division Multiplexing (WDM) technology [33] to partition the spectrum into a fixed number of wavelengths. Later on, the notion of fine-grained spectrum allocation was introduced for elastic optical networks as an improvement on the WDM optical networks, generalizing the problem to the routing and spectrum allocation (RSA). This more recent RSA problem is, as expected, also NP-hard [15].

Integer Linear Programming (ILP) has originally been used to formalize and solve the RSA lightpath constraints with the goal of minimizing the maximum slot index used in the light spectrum [15,42,58]. The ILP formulations were later improved to be more concise and efficient; example encodings are by Zhang et al. [62] improving the work done in [58], and Esteban et al. [56] who introduce even more efficient ILP encoding by employing the notion of channels to handle the spectrum contiguity constraint. Wang et al. [57] contribute with a relaxed ILP problem that establishes a lower bound of the optimal solution. In contrast to our approach, the ILP implementations are relatively effective in finding a single optimal solution to a given RSA problem. However, finding all optimal solutions (as needed for the failover protection) is not feasible using the ILP method as we demonstrate in our paper. While ILP formulations can find optimal solutions, they are generally not applicable for time-critical purposes, such as reacting to link failures in the order of miliseconds [19]. For more scalable approaches to solve the RSA problem, heuristics have been proposed [15,34,41], as well as genetic algorithms [29,30,39] and reinforcement learning models [54,60]. These approaches trade off optimality for computation speed, where the genetic algorithms tend to be closer to achieving more optimal solutions than the heuristics and reinforcement learning approaches at the cost of being computationally slower. In contrast to these approaches, we are able to find and efficiently store all optimal solutions for path and spectrum allocations, which can prove useful for handling changes during the network operation, such as quickly recovering from link failures.

While the study of fast failover protection [11] in IP networks [5,45], MPLS networks [44,55], Segment Routing networks [21,22] and software-defined networks [13,53] has been well-studied, these approaches cannot be easily transferred to elastic optical networks due to the orthogonally different way of packet forwarding in traditional networks (allowing us to e.g. analyze/modify packet headers) and fully optical ones. The current approaches for ensuring link failure resilience in optical networks generally only prepare for one-link failures by preallocating backup paths [3,9,23,50,51]. For example, Castro et al. [9] propose a MILP formulation that maximizes the total bitrate recovered in case of a single-link failure scenario by allocating a backup path for each demand such that when a demand is affected by a link failure, it can quickly switch over to its backup path, and Singhal et al. [51] and Gao et al. [23] expand upon this idea with the notion of a more resource-friendly cross-sharing, where groups of demands with link-disjoint primary paths are allowed to share a backup path. These approaches based on preallocated backup paths assume only one link failure, but some research (see e.g. Athe and Singh [4] and Li et al. [40]) has been carried out to extend the failure resilience to two link and network-bound link failures. Common for the current approaches to ensuring failure resilience is that they require additional spectrum resources to always be allocated for both the primary and backup paths in the network to ensure quick failover, which entails that they do not provide optimal solutions. Our approach finds optimal solutions while being able to handle multiple link failures without having to resort to resource over-allocation.

Organization. The rest of the paper is organized as follows. First, we formally define the RSA problem and the problem of handling link failures in Sect. 2. Then, in Sect. 3, we present and evaluate how to use an ILP formulation to handle link failures. In Sect. 4, we encode the RSA problem into BDDs which we then use in Sect. 5 to handle link failures. Finally, in Sect. 6, we compare our BDD-based approach with the ILP approach, and conclude on our work in Sect. 7.

2   Problem Definition

Let us first introduce formally the routing and spectrum allocation problem (RSA) and the problem of failover protection. We start with the definition of a network and a routing path.

A *network topology* is a tuple $G = (V, E, src, tgt)$ where V is a finite set of nodes, E is a finite set of edges and $src, tgt : E \rightarrow V$ denote the source and target of an edge, excluding self-loops by imposing $src(e) \neq tgt(e)$ for every $e \in E$. A *path* in G is a sequence of connected edges $\pi = e_1 e_2 e_3 ... e_n \in E^*$ such that $tgt(e_i) = src(e_{i+1})$ for all i, $1 \leq i < n$. Let $e \in \pi$ denote that an edge e is present in the path π and let $\pi \cap \pi'$ denote the set of all edges that the two paths share. Let $first(\pi) = e_1$ and $last(\pi) = e_n$ be mappings that return the first and the last edge on the path π, respectively. A *simple path* is a path where $src(e') \neq src(e)$ and $tgt(e) \neq tgt(e')$ for all pairs of distinct edges $e, e' \in \pi$; let **Paths** be the set of all simple paths in the topology G.

We assume a finite set of *demands* D with source and target nodes represented by the mappings *ingress, egress* : $D \rightarrow V$ respectively, and *size* : $D \rightarrow \mathbb{N}$ representing the amount of data of a given demand. Furthermore, we assume a finite set of *spectrum slots* $F = \{1, 2, .., f_{max}\}$. Finally, a *channel* is any subset $C \subseteq F$ of consecutive slots, and let $\mathbb{C}$ be the set of all channels. All possible channels that can be used to transfer a given demand are represented by the mapping *channels* : $D \rightarrow 2^{\mathbb{C}}$.

Definition 1 (Routing and Spectrum Allocation Problem). *Given an input consisting of*

- a network topology $G = (V, E, src, tgt)$ with simple paths **Paths**,
- a set of demands $D = \{d_1, d_2, ..., d_m\}$,
- a mapping $DPaths : D \rightarrow 2^{Paths}$ of available paths for each demand $d \in D$, where for every $\pi \in DPaths(d)$ it holds that $src(first(\pi)) = ingress(d)$ and $tgt(last(\pi)) = egress(d)$,
- a finite set of available slots $F = \{1, 2, .., f_{max}\}$, and
- a modulation mapping $\Delta : \textbf{Paths} \rightarrow \mathbb{N}$ which returns the number of slots required per sent unit of data on a given path,

our task is to find a solution (P, ω) where

- $P : D \rightarrow \textbf{Paths}$ is the route assignment function such that $P(d) \in DPaths(d)$ for every demand $d \in D$, and
- $\omega : D \rightarrow 2^F$ is the spectrum allocation function such that $\omega(d) \in channels(d)$ and $|\omega(d)| = \Delta(P(d)) \cdot size(d)$ meaning that the allocated channel has enough spectrum slots depending on the chosen modulation and the size of the demand,

such that for all pairs of distinct demands $d, d' \in D$ either $P(d) \cap P(d') = \emptyset$ or $\omega(d) \cap \omega(d') = \emptyset$, i.e. either the chosen paths do not intersect or, if they do, then the allocated frequency slots may not overlap.

Let us remark that the input to the RSA problem contains also the set of allowed paths for each demand (typically the shortest paths from the ingress to egress nodes). This allows us to restrict the routing to an (often small) subset of simple paths and we assume that these can be enumerated to become a part of the input.

Different metrics have been used to define optimal solutions to the RSA problem, such as minimizing unserved bandwidth of demands [56], the number of frequency slots used [15], and the highest frequency slot index used by any demand [42]. We are aiming to minimize the highest used frequency slot as this supports our intended application discusssed later on. Hence for a given spectrum allocation function ω, let $usage(\omega) = \max\{f \in \omega(d) \mid d \in D\}$ define the highest used frequency slot in ω. A network optimal solution is then a solution (P, ω) to the RSA problem where $usage(\omega) \leq usage(\omega')$ for any other solution to the RSA problem (P', ω').

Example 1. Figure 1a illustrates a simple example of the RSA problem with a small network topology consisting of six nodes and seven edges and a spectrum width of two frequency slots. There are two demands, d_1 from v_1 to v_5 and d_2 from v_2 to v_6, both of size 1. Each demand has two possible paths. If the demand d_1 uses the longer path π_1, then it must be allocated two frequency slots due to the modulation. The same holds true for the demand d_2 if it uses the path π_3. Hence, there are three different possible channels for both demands, as seen in Fig. 1a. An example of an optimal solution to this RSA problem is Solution 1 where the demand d_1 is assigned the path π_0 and channel $\omega(d_1) = \{1\}$ while the demand d_2 is assigned the path π_2 and channel $\omega(d_2) = \{1\}$. Solution 2 is not optimal as its highest used frequency slot is 2.

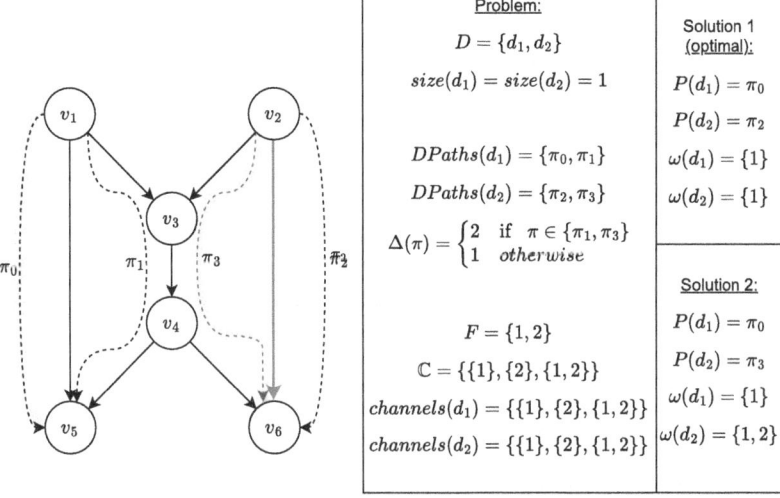

Problem:		Solution 1 (optimal):

$$D = \{d_1, d_2\}$$

$$size(d_1) = size(d_2) = 1$$

$$DPaths(d_1) = \{\pi_0, \pi_1\}$$

$$DPaths(d_2) = \{\pi_2, \pi_3\}$$

$$\Delta(\pi) = \begin{cases} 2 & if \ \pi \in \{\pi_1, \pi_3\} \\ 1 & otherwise \end{cases}$$

$$F = \{1, 2\}$$

$$\mathbb{C} = \{\{1\}, \{2\}, \{1, 2\}\}$$

$$channels(d_1) = \{\{1\}, \{2\}, \{1, 2\}\}$$

$$channels(d_2) = \{\{1\}, \{2\}, \{1, 2\}\}$$

Solution 1 (optimal):

$$P(d_1) = \pi_0$$
$$P(d_2) = \pi_2$$
$$\omega(d_1) = \{1\}$$
$$\omega(d_2) = \{1\}$$

Solution 2:

$$P(d_1) = \pi_0$$
$$P(d_2) = \pi_3$$
$$\omega(d_1) = \{1\}$$
$$\omega(d_2) = \{1, 2\}$$

(a) Example of RSA problem and two corresponding solutions. Solution 2 is highlighted in yellow, and uses 2 slots. Solution 1 uses only 1 slot. In the case where the thick red edge fails, only solution 2 is valid.

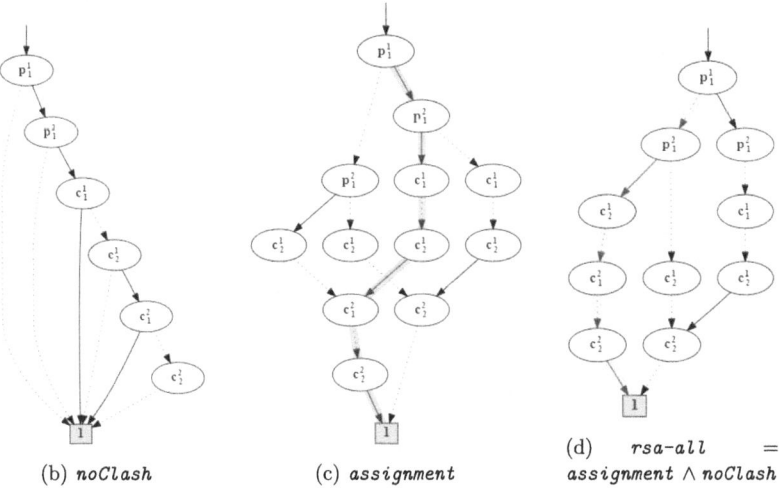

(b) *noClash*   (c) *assignment*   (d)   *rsa-all*   = *assignment* ∧ *noClash*

Fig. 1. BDD for *noClash*, *assignment*, and *rsa-all* for the example shown in Fig. 1a. The red highlighted path in 1c shows the path assignments $P(d_1) = \pi_1$ and $P(d_2) = \pi_3$, and the channel assignments $\omega(d_1) = \omega(d_2) = \{1, 2\}$ in which the two demands clash. Clashing assignments are filtered in the BDD *rsa-all*. The yellow highlighted path in Fig. 1d corresponds to two solutions: Solution 2 from Fig. 1a and a similar one where we instead have $\omega(d_1) = \{2\}$.

Having introduced the RSA problem, we now formally define the problem of protecting a network for up to k link failures.

Definition 2 (k-Link Failover Problem). *Given a set of link failures $E_{fail} \subseteq E$ where $|E_{fail}| \leq k$, we want to find a solution (P, ω) to the RSA problem where $P(d) \cap E_{fail} = \emptyset$ for all $d \in D$.*

Example 2. Let us assume that the link between v_2 and v_6 has failed in our running example from Fig. 1a. Our goal is to find a solution where no demand is assigned a path that uses the failed link. Clearly, Solution 1 from Fig. 1a is now not a valid option anymore. However, Solution 2 is still a valid solution as the demand d_2 is assigned the path π_3 which does not use the failed link. Now Solution 2 becomes an optimal solution in this failure scenario.

The goal is now to be able to quickly react to all possible k-link failure scenarios and suggest an alternative solution that does not use any of the failed links. In the next section, we shall discuss the possible applicability of integer linear programming to adress the failover problem.

3   Failure Resilience via ILP

The goal of fast failover resiliece is to provide an optimal RSA solution for any failure scenario. According to the Metro Ethernet Forum [19], we aim for an average reaction time of less than 50 ms with an upper limit of 200 ms when link failures occur. In order to achieve this goal, we first examine whether the classical approach via integer linear programming (ILP) can be employed. We implement a slight modification (relaxation of the requirement of biderectional demands) of the state-of-the-art ILP formulation of the RSA problem from [42]. The ILP program presented in [42] minimizes for the highest used spectrum slot. For completeness, we now present the implemented encoding that uses the integer variables $x_{d\pi f} \in \{0, 1\}$ where $x_{d\pi f} = 1$ signifies that the demand d uses the path $\pi \in DPaths(d)$ with the start frequency slot $f \in F$. Additionally, it uses the parameters $n_{d\pi}$ to denote the number of frequency slots that the demand d needs to be transmitted along the path $\pi \in DPaths(d)$, i.e. $n_{d\pi} = size(d) \cdot \Delta(\pi)$. The ILP formulation is then as follows:

$$minimize \; f_{max} \tag{1}$$

$$\sum_{f \in F} \sum_{\pi \in DPaths(d)} x_{d\pi f} = 1, \qquad\qquad \forall d \in D \tag{2}$$

$$\sum_{\substack{d \in D}} \sum_{\substack{\pi \in DPaths(d) \\ e \in \pi}} \sum_{\substack{f' \in F \\ f - n_{d\pi} + 1 \leq f' \leq f}} x_{d\pi f'} \leq 1, \qquad\qquad \forall e \in E, \forall f \in F \tag{3}$$

$$\sum_{f \in F} (f + n_{d\pi} - 1) \cdot x_{d\pi f} \leq f_{max} \qquad\qquad \forall d \in D, \forall \pi \in DPaths(d) \tag{4}$$

Constraint 2 assigns a single path and a start slot to each demand. Constraint 3 ensures that there is no spectrum clash between demands on any edge of the paths. Finally, Constraint 4 checks that the slots used by the demands are not exceeding the variable f_{max} (highest used frequency slot) that we are minimizing.

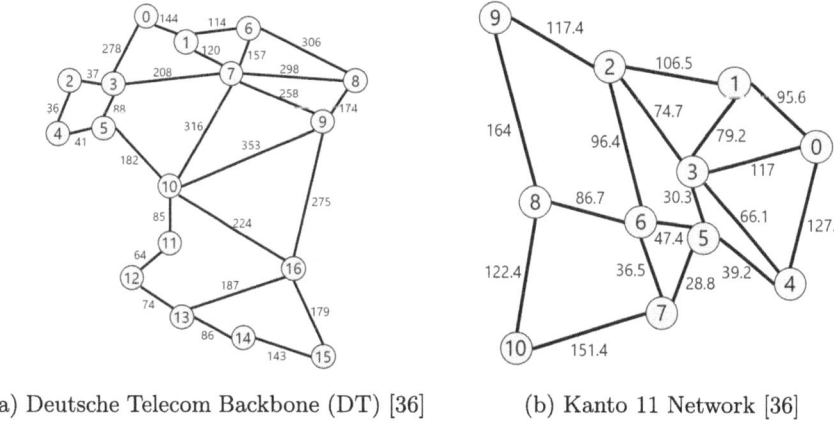

(a) Deutsche Telecom Backbone (DT) [36]   (b) Kanto 11 Network [36]

Fig. 2. Evaluation networks; numbers on edges indicate the distance in kilometers

When link failures occur in the network, the ILP formulation can be used to compute an optimal route and spectrum assignment that avoids the failed links. We evaluate this approach through an experiment, using the Gurobi ILP solver in Python [26], and run the experiment on a Ubuntu 18.04.5 cluster with 2.3 GHz AMD Opteron 6376 processors, with a memory limit of 30GB. For the experiment, we use the Deutsche Telecom Backbone (DT) and Kanto 11 network topologies (see Fig. 2), two highly connected topologies where nodes/cities have their correct population sizes as referenced in [49,59]. The population sizes are used to generate demands using the standard gravity model [48]. Two shortest paths are generated for each demand using the semi-disjoint path generation algorithm from [32], and we assume a standard 320 slot spectrum capacity [36] with modulation 1 for all paths.

We generate random k-link failure scenarios with uniform probability of link failures and Figs. 3a and 3b show box plots of the time it takes to compute a new solution for 1–5 link failures for 3, 6, and 9 demands in the DT network and Kanto network respectively. We see that the response time for finding a solution to the problem is in seconds, even for a smaller number of demands, clearly exceeding the expected average of 50 ms response time. As a conclusion, the ILP approach is too slow (by several orders of magnitude) in order to be applied for fast failover recovery.

As an alternative approach, one can consider to precompute and store all possible optimal solutions for any k-or-fewer link failure scenarios such that a

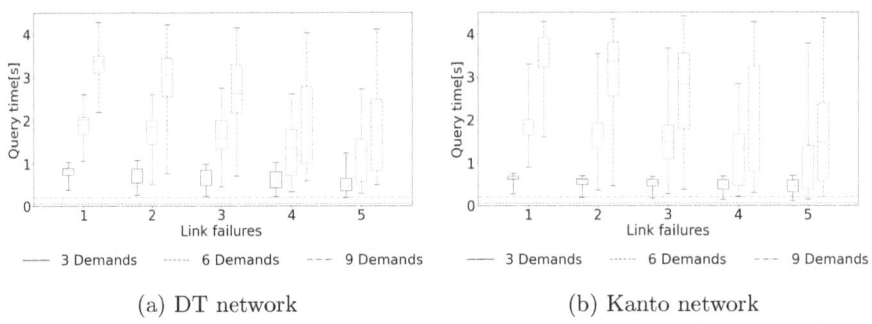

(a) DT network   (b) Kanto network

Fig. 3. Box plots of ILP query times based on 1000 random link-failure scenarios; dotted and dashed vertical lines show the 50 ms and 200 ms threasholds

solution can be provided almost instantaneously once we encounter link failures. This implies that for each possible combination of failed edges $E_{fail} \subseteq E$ where $|E_{fail}| \leq k$, we need to precompute a solution using the ILP formulation. This approach clearly does not scale (neither in CPU time nor the required memory) due to its high combinatorial complexity. Table 1 shows the estimated precomputation times, based on the average of 1000 randomly generated failure scenarios, for up to 9 demands and 5 link failures (the first number is for DT and the second for Kanto). We can observe that already for three failing links, the precomputation takes hours, for four link failures days and for five several weeks, even for as little as three demands. Additionally, the memory required to store all these solutions grows exponentially too. We hence conclude that precomputing solutions to link failures using ILP is not feasible either.

In the next sections, we present a different method based on binary decision diagrams to efficiently compute and store all solutions to a given RSA problem; our method scales orders of magnitude better than the ILP approach, both in the reaction time as well as the precomputation time/memory.

Table 1. Estimated precomputation time for failure resilience using ILP

DT/Kanto		Demands		
		3	6	9
Failures	1	42 s/23 s	2 m/2 m	3 m/3 m
	2	17 m/7 m	40 m/19 m	2 h/35 m
	3	5 h/2 h	11 h/4 h	17 h/6 h
	4	2 d/10 h	5 d/23 h	8 d/2 d
	5	19d/3d	39d/5d	60d/8d

4   BDD Encoding of RSA

We shall now discuss our method and tool *ExpectAll* that relies on binary decision diagrams for computing and compactly storing all solutions to the RSA problem. Binary Decision Diagrams (BDDs) were introduced in [1,38] as a data structure to efficiently represent and manipulate Boolean functions. The concept was later refined in [7] with more efficient Boolean operations. As such, BDDs have been revolutionary at efficiently representing finite state spaces [8] and have by now been applied in various problem domains [18,27,28,37,47].

A BDD is a rooted, directed and acyclic graph [2]. Every non-leaf node in a BDD is labeled with a Boolean variable, while there are exactly two leaf nodes labelled with the value 0 (*False*) and 1 (*True*). Each non-leaf node u has two outgoing edges denoted as $low(u)$ and $high(u)$; following an edge corresponds to setting the variable at node u to either *False* (low edge) or *True* (high edge). The edges are commonly visualized by using a solid line for $high(u)$ and a dotted line for $low(u)$. In the graphical notation, the leaf node 0 and all its incomming arcs are usually omitted. An example of a BDD can be seen in Fig. 1b. Following the right-most branch from the root to the terminal node 1 represents a single assignment where all variables are true except for c_1^1, c_1^2 and c_2^2. The left-most branch, on the other hand, represents a whole set of assignments where p_1^1 is false and all other variables can be set arbitrarily.

A BDD is ordered (OBDD) if the variables in the BDD come in the same order $x_1 < x_2 < ... < x_n$ on all paths from the root to a leaf. An OBDD can be reduced by merging nodes with identical subgraphs, and by deleting nodes where the subgraphs for $low(u)$ and $high(u)$ are equivalent. Such a BDD is called a reduced OBDD (ROBDD) [2]. In the rest of the paper, we write simply BDD instead of ROBDD. In our encoding, we shall use a BDD extension that supports first-order quantifiers and is closed under all Boolean operations and quantifiers; for details see e.g. [38].

4.1   Representing a Set as a Boolean Function

Given a finite set $S = \{s_0, s_1, ..., s_{|S|-1}\}$, let $\overline{x} = [x_k, ..., x_1]$ be a vector of Boolean variables where $k = \lceil log_2(|S|) \rceil$. Now any truth assignment α to $\overline{x}$ can be interpreted as a natural number $n(\alpha) \in \mathbb{N}$ written in binary notation. Thus $\overline{x}$ encodes the $n(\alpha)$'th element of S. Let $\overline{x}(s)$ denote the Boolean expression over $\overline{x}$ with just the single truth assignment corresponding to the element s. Given a Boolean expression $b(\overline{x})$, let $[\![b(\overline{x})]\!]$ denote the encoded subset $\{s_{n(\alpha)} \mid \alpha$ satisfies $b(\overline{x})\} \subseteq S$ such that $[\![b(\overline{x})]\!]$ is a set consisting of all elements that are encoded by all possible Boolean assignments satisfying the expression b.

Example 3. Let us consider the set of paths $\{\pi_0, \pi_1, \pi_2, \pi_3\}$ from Fig. 1a. We need two boolean variables $\overline{p} = [p_2, p_1]$ to encode any of the given paths. For instance, we encode the path π_0 by $\overline{p}(\pi_0) = \neg p_2 \wedge \neg p_1$. In a Boolean expression such as $b = p_1$, where the variable p_2 is free, the Boolean function $b(\overline{p})$ is satisfied both when $[p_2 \mapsto 0, p_1 \mapsto 1]$ and $[p_2 \mapsto 1, p_1 \mapsto 1]$, which means $[\![b(\overline{p})]\!] = \{\pi_1, \pi_3\}$.

4.2   Construction of BDDs with All Valid RSA Assignments

Table 2. Variables used in the BDD encodings of RSA problem

Variables	Description
$\overline{\mathbf{c}^d} = [\mathbf{c}_n^d, \mathbf{c}_{n-1}^d, ..., \mathbf{c}_1^d]$ where $n = \lceil \log_2(|channels(d)|) \rceil$	Encoding of $channels(d)$
$\overline{\mathbf{p}^d} = [\mathbf{p}_n^d, \mathbf{p}_{n-1}^d, ..., \mathbf{p}_1^d]$ where $n = \lceil \log_2(|DPaths(d)|) \rceil$	Encoding of paths $DPaths(d)$

We shall now describe how, for a given instance of RSA problem, we construct a BDD representation of all its solutions. As shown in Table 2, for each demand d, we use a vector of Boolean variables $\overline{\mathbf{p}^d}$ to encode the path assigned to demand d, and another vector of Boolean variables $\overline{\mathbf{c}^d}$ to encode the channel assigned to demand d. To encode a solution (P, ω) to the RSA problem (see Definition 1), we use the vector of vectors $\overline{\mathbf{p}^D} = [\overline{\mathbf{p}^{d_1}}, \overline{\mathbf{p}^{d_2}}, ..., \overline{\mathbf{p}^{d_{|D|}}}]$ to encode P and the vector of vectors $\overline{\mathbf{c}^D} = [\overline{\mathbf{c}^{d_1}}, \overline{\mathbf{c}^{d_2}}, ..., \overline{\mathbf{c}^{d_{|D|}}}]$ to encode ω.

Example 4. We can see that the BDD `assignment` shown in Fig. 1c represents all $3^2 = 9$ possible combinations of P and ω for the running example. For example the assignment where $P(d_1) = \pi_1$, $P(d_2) = \pi_3$ and $\omega(d_1) = \omega(d_2) = \{1, 2\}$ is highlighted in red.

We now define three BDDs which enforce that $\overline{\mathbf{p}^D}$ and $\overline{\mathbf{c}^D}$ encode only valid solutions of the RSA problem. Specifically, the BDDs must ensure that each demand d is assigned a path $\pi \in DPaths(d)$ and a channel $C \in channels(d)$ such that $|C| = \Delta(\pi) \cdot size(d)$. Moreover, whenever a demand shares an edge with another demand, they cannot be assigned overlapping channels.

To enforce that no two demands are allowed to clash, we define the BDD `noClash` (Fig. 1b) by

$$noClash\left(\overline{\mathbf{p}^D}, \overline{\mathbf{c}^D}\right) = \tag{5}$$

$$\bigwedge_{\substack{d,d' \in D, \\ d \neq d' \\}} \bigwedge_{\substack{\pi \in DPath(d), \\ \pi' \in DPaths(d'), \\ \pi \cap \pi' \neq \emptyset}} \left(\neg\left(\overline{\mathbf{p}^d}(\pi) \wedge \overline{\mathbf{p}^{d'}}(\pi')\right) \vee \bigwedge_{\substack{C \in channels(d), \\ C' \in channels(d'), \\ |C| = \Delta(\pi) \cdot size(d), \\ |C'| = \Delta(\pi') \cdot size(d'), \\ C \cap C' \neq \emptyset}} \neg\left(\overline{\mathbf{c}^d}(C) \wedge \overline{\mathbf{c}^{d'}}(C')\right) \right)$$

satisfying that $(P, \omega) \in \llbracket noClash\left(\overline{\mathbf{p}^D}, \overline{\mathbf{c}^D}\right) \rrbracket$ iff for all $d, d' \in D$ where $d \neq d'$ either $P(d) \cap P(d') = \emptyset$ or $\omega(d) \cap \omega(d') = \emptyset$.

Example 5. Consider again the network from Fig. 1a. There is only one way the two demands can clash. The BDD `noClash` therefore has to encode that it is satisfied by all path and channel assignments, except when d_1 and d_2 are assigned the paths π_1 and π_3 respectively, and the channels $\omega(d_1) = \omega(d_2) = \{1, 2\}$. As shown in Fig. 1b, the BDD *noClash* encodes exactly this, as the only way not to satisfy this BDD is taking the right-most path down to the node labeled $\mathbf{c}_2^2$, and then set the value of the Boolean variable $\mathbf{c}_2^2$ to *True*, corresponding to this single clashing assignment.

Next, we must enforce that each demand d is assigned a correct path and channel assignment pair. To this end, we define the BDD *assignment* as

$$assignment(\overline{\mathbf{p}^D}, \overline{\mathbf{c}^D}) = \bigwedge_{d \in D} \bigvee_{\pi \in DPath(d)} \left(\overline{\mathbf{p}^d}(\pi) \wedge \left(\bigvee_{\substack{C \in channels(d), \\ |C| = \Delta(\pi) \cdot size(d)}} \overline{\mathbf{c}^d}(C) \right) \right) \quad (6)$$

and clearly $(P, \omega) \in [\![assignment(\overline{\mathbf{p}^D}, \overline{\mathbf{c}^D})]\!]$ iff for all $d \in D$, it holds that $P(d) \in DPaths(d)$, $\omega(d) \in channels(d)$ and $|\omega(d)| = \Delta(P(d)) \cdot size(d)$.

Finally, we use *assignment* and *noClash* to define the BDD *rsa-all* as

$$rsa\text{-}all(\overline{\mathbf{p}^D}, \overline{\mathbf{c}^D}) = assignment(\overline{\mathbf{p}^D}, \overline{\mathbf{c}^D}) \wedge noClash(\overline{\mathbf{p}^D}, \overline{\mathbf{c}^D}). \quad (7)$$

Example 6. In our running example, we can see in Fig. 1d that *rsa-all* contains all valid solutions for the given RSA problem. All the assignments from the BDD *assignments* are valid solutions, except the assignment that is marked red in Fig. 1c, since it is the only assignment that does not satisfy the BDD *no-clash*, as noted in Example 5. Hence, this is the only assignment that does not appear in *rsa-all*. The highlighted path in *rsa-all* corresponds to the two valid solutions, one solution being Solution 2 from Fig. 1a, the other being the same as Solution 2 except for that the demand d_1 is assigned the channel $\{2\}$ instead of $\{1\}$.

We can now state the correctness theorem of our BDD construction.

Theorem 1. *The pair (P, ω) is a solution to the RSA problem iff $(P, \omega) \in [\![rsa\text{-}all(\overline{\mathbf{p}^D}, \overline{\mathbf{c}^D})]\!]$.*

Proof (Proof sketch). " $\Longrightarrow$ " Assume (P, ω) is a solution to the RSA problem. By Definition 1, the encodings $\overline{\mathbf{p}^D}, \overline{\mathbf{c}^D}$ of (P, ω) satisfy both the constraints enforced by *assignment* and *noClash*, and it thus follows that $(P, \omega) \in [\![rsa\text{-}all(\overline{\mathbf{p}^D}, \overline{\mathbf{c}^D})]\!]$.

" $\Longleftarrow$ " Let $(P, \omega) \in [\![rsa\text{-}all(\overline{\mathbf{p}^D}, \overline{\mathbf{c}^D})]\!]$. By the constraints enforced by *assignment* we know that each demand has been assigned a valid path and channel combination given the used modulation, and due to the constraints enforced by *noClash* we know that the path and channel assignments given to each demand result in a solution, with no clashing between any of the demands. Hence (P, ω) is a solution to the RSA problem. $\square$

As the BDD $rsa\text{-}all\,(\overline{\mathbf{p}^D}, \overline{\mathbf{c}^D})$ now contains all solutions to the RSA problem, it is possible for a network operator to query it to find for example an optimal solution or a solution matching any particular property desired by the operator. For the k-link failover problem, there is no need to store all possible solutions, as we need only to represent all optimal path and channel assignments for each possible scenario of up to k link failures. To limit the set of solutions, we can disallow a fragmenation of the spectrum space i.e. restrict possible gaps of slots between the assigned channels. Concretely, we require that a channel assignment ω must assign all slots smaller than or equal to $usage(\omega)$ to at least one demand. This is enforced by specifying that the channel assigned to a demand either uses the first slot, or follows directly after a channel assigned to another potentially path-overlapping demand.

Definition 3 (Gap-Free Solution). *A solution to the RSA problem* (P, ω) *is gap-free if for all* $d \in D$, *either* $min(\omega(d)) = 1$ *or there exists a* $d' \in D, \pi' \in DPaths(d')$ *and* $\pi \in DPaths(d)$ *s.t.* $\pi \cap \pi' \neq \emptyset$ *and* $min(\omega(d)) = max(\omega(d'))+1$.

Theorem 2. *If* (P, ω) *is a solution to an RSA problem, then there exists a gap-free solution* (P, ω') *such that* $usage(\omega') \leq usage(\omega)$.

Proof. Assume a channel allocation ω for the demands $d_1, d_2, ..., d_m$ such that $d_i \leq d_j$ iff $min(\omega(d_i)) \leq min(\omega(d_j))$. A new ω' can now be constructed, such that ω' is *gap-free*. For d_i, let D_i be the set of demands $d_k < d_i$ where $\omega(d_k) \cap \omega(d_i) = \emptyset$ and there exists $\pi_k \in DPaths(d_k)$ and $\pi_i \in DPaths(d_i)$ s.t. $\pi_k \cap \pi_i \neq \emptyset$. Let ω' initially be undefined for all demands. We now add each demand to ω' one by one in the specified order, such that if $D_i = \emptyset$ then $\omega'(d_i) = \{1, ..., |\omega(d_i)|\}$, otherwise let the highest slot assigned by ω' to a demand in D_i be $\Omega = \max_{d \in D_i}(max(\omega'(d)))$ and then the new channel of d_i is specified as $\omega'(d_i) = \{\Omega + 1, \Omega + 2, ..., \Omega + |\omega(d_i)|\}$. It follows that $usage(\omega') \leq usage(\omega)$ since at any step $min(\omega'(d_i)) \leq 1 + |\bigcup_{d \in D_i} \omega(d)|$. $\qquad\square$

The gap-free property can be imposed on the solutions encoded by the BDD $rsa\text{-}all$ using the following *gapfree* BDD defined as

$$gapfree\,(\overline{\mathbf{c}^D}) = \tag{8}$$

$$\bigwedge_{d \in D} \left(\left(\bigvee_{\substack{C \in channels(d), \\ min(C)=1}} \overline{\mathbf{c}^d}(C) \right) \vee \left(\bigvee_{\substack{d' \in D, \\ \exists \pi \in DPaths(d), \\ \exists \pi' \in DPaths(d'), \\ \pi \cap \pi' \neq \emptyset}} \bigvee_{\substack{C \in channels(d), \\ C' \in channels(d'), \\ min(C)=max(C')+1}} \overline{\mathbf{c}^d}(C) \wedge \overline{\mathbf{c}^{d'}}(C') \right) \right)$$

where we have $\omega \in [\![gapfree\,(\overline{\mathbf{c}^D})]\!]$ iff ω satisfies the gap-free property as described in Definition 3. Additionally, since the gap-free property preserves an optimal solution for each routing assignment, it can handle the same link failure scenarios as $rsa\text{-}all$.

We now define a new BDD $rsa\text{-}gapfree$ using $gapfree$ as follows

$$rsa\text{-}gapfree\,(\overline{\mathbf{p}^D},\overline{\mathbf{c}^D}) =$$

$$assignment\,(\overline{\mathbf{p}^D},\overline{\mathbf{c}^D}) \wedge gapfree\,(\overline{\mathbf{c}^D}) \wedge noClash\,(\overline{\mathbf{p}^D},\overline{\mathbf{c}^D}) \quad (9)$$

and clearly $(P,\omega) \in [\![rsa\text{-}gapfree\,(\overline{\mathbf{p}^D},\overline{\mathbf{c}^D})]\!]$ iff (P,ω) is a valid solution to the RSA problem and ω satisfies the gap-free property. Since $assignment$, $gapfree$ and $noClash$ are just Boolean expressions, their order in the conjunction is semantically irrelevant; in the implementation, the conjunction is evaluated in a left-to-right order, resulting in an improved performance.

In addition to removing spectrum gaps, any optimal solution will never occupy any spectrum slot higher than $max_f = \sum_{d \in D} \max_{\pi \in DPaths(d)} \big(\Delta(\pi) \cdot size(d)\big)$ and thus all channels where $max(C) > max_f$ can be disregarded. By enforcing this upper limit, we effectively prune the candidate channels for each demand (and hence improve the performance) without losing optimality. A more greedy approach to approximate the candidate channels is to assign channels based on an established ordering of the demands. The idea is to remove symmetry in the channel assignments by imposing the following property.

Definition 4 (Limited). *Let $D = \{d_1, d_2, ..., d_m\}$ be a list of demands in a fixed ordering. A solution to the RSA problem (P,ω) is limited if $min(\omega(d_i)) \le c_{max} + \sum_{j<i} |\omega(d_j)|$ for all $d_i \in D$, where $c_{max} = \max_{d \in D}|\omega(d)|$.*

This means that the demand d_i should start at a frequency slots that is no larger than the sum of cardinalities of the already scheduled frequency slots plus the size of the largest possible demand. Both the maximum spectrum slot limit as well as the limited property can be applied by restricting the mapping $channels$ in the definition of $gapfree$ in Eq. 8. We note that the greedy approach can in some rare instances make it impossible to find a channel assignment for a particular path assignment, however, we never experienced this issue on any of the topologies from the Topology Zoo benchmark [35] (with more than 250 ISP topologies) as long as the demands are ordered from the largest one to the smallest. From now on, we shall use $rsa\,(\overline{\mathbf{p}^D},\overline{\mathbf{c}^D})$ to refer to the BDD from Eq. 9 where both the limited property as well as the highest spectrum slot bound are applied.

4.3 Extraction of Optimal Solutions from a BDD

Once we built a BDD representing valid solutions to an RSA problem, we are interested in finding an optimal solution that minimizes the highest used spectrum slot. To this end, we introduce a new vector of Boolean variables $\overline{\mathbf{s}^F} = [\mathbf{s}_1, \mathbf{s}_2, \ldots, \mathbf{s}_{f_{max}}]$ with the meaning that $\overline{\mathbf{s}^F}$ encodes a subset $F' \subseteq F$ of spectrum slots such that $f \in F'$ iff $\mathbf{s}_f$ is true. The intuition behind the use of these variables is that if a variable $\mathbf{s}_f$ is false, then we know that no demand has been assigned a channel with a slot greater than or equal to f, meaning that the usage is at most $f - 1$. We can enforce this using the BDD $rsa\text{-}slotBound$ defined as

$$rsa\text{-}slotBound\left(\overline{\mathbf{p}^D}, \overline{\mathbf{c}^D}, \overline{\mathbf{s}^F}\right) \qquad (10)$$

$$= rsa\left(\overline{\mathbf{p}^D}, \overline{\mathbf{c}^D}\right) \wedge \left(\bigwedge_{d \in D} \bigvee_{C \in channels(d)} \left(\overline{\mathbf{c}^d}(C) \wedge \bigwedge_{f \leq max(C)} \mathbf{s}_f \right) \right)$$

and we observe that $(P, \omega, F') \in [\![rsa\text{-}slotBound\left(\overline{\mathbf{p}^D}, \overline{\mathbf{c}^D}, \overline{\mathbf{s}^F}\right)]\!]$ iff $(P, \omega) \in [\![rsa\left(\overline{\mathbf{p}^D}, \overline{\mathbf{c}^D}\right)]\!]$ and F' contains all $f \leq usage(\omega)$.

An optimal solution can now be found by iteratively (e.g. using the binary search) identifying the smallest value of $f \in F$ for which we can still find a solution

$$(P, \omega, F') \in [\![rsa\text{-}slotBound\left(\overline{\mathbf{p}^D}, \overline{\mathbf{c}^D}, \overline{\mathbf{s}^F}\right)]\!]$$

where the corresponding variable $\mathbf{s}_f$ can take the value false, i.e. $f \notin F'$, since it can then be inferred that the $usage(\omega) = f - 1$.

5   Failure Resilience via BDD Encoding

As shown in the previous section, we can build a BDD that stores all optimal solutions of the RSA problem. We now describe how to quickly extract optimal solutions from such a BDD for the purpose of providing time-critical responses to multiple links failing. In particular, the BDD rsa as defined in Sect. 4.2 contains at least one solution for all failure scenarios, if such a solution exists. Furthermore, we know that at least one of these solutions is optimal under the corresponding failure scenario. To extract these solutions, we present two approaches. The first method supports an arbitrary number of link failures based on the idea of deleting solutions that use invalid paths given the link failures. The second method uses a precomputation in which link failures are directly encoded into the BDD.

5.1   Pruning by Deletion

Solutions encoded in a given BDD rsa that use invalid paths based on the set of link failures E_{fail} are no longer valid solutions. The invalid solutions can be deleted from rsa as follows

$$path\text{-}pruned\text{-}rsa\left(\overline{\mathbf{p}^D}, \overline{\mathbf{c}^D}\right) = rsa\left(\overline{\mathbf{p}^D}, \overline{\mathbf{c}^D}\right) \wedge \bigwedge_{d \in D} \bigwedge_{\substack{\pi \in DPaths(d), \\ \exists e \in E_{fail}, \\ e \in \pi}} \neg \overline{\mathbf{p}^d}(\pi) \quad (11)$$

and clearly $(P, \omega) \in [\![path\text{-}pruned\text{-}rsa\left(\overline{\mathbf{p}^D}, \overline{\mathbf{c}^D}\right)]\!]$ iff $(P, \omega) \in [\![rsa\left(\overline{\mathbf{p}^D}, \overline{\mathbf{c}^D}\right)]\!]$ and it holds for all $d \in D$ that $P(d) \cap E_{fail} = \emptyset$. Since there is no limitation on the size E_{fail}, this pruning method can be used to prune the BDD for an arbitrary number of failed links.

5.2   Pruning by Precomputation

As failure scenarios with a high number of concurrently failing links are less likely, we can focus on preparing a failover protection for an a priori given maximum number k of failing links. We hence construct a parameterized BDD with additional variables representing the failed edges; these can be specified to retrieve the valid solutions for any given link failure scenario up to some sufficiently large k.

Let $E_u = E \cup \{e_{unused}\}$ where the auxiliary edge e_{unused} signifies a non-failing link. For a given k-link failover problem, we introduce a vector of k variable encodings $\overline{\mathbf{c}^K} = [\overline{\mathbf{e}^1}, ..., \overline{\mathbf{e}^k}]$ such that $\overline{\mathbf{e}^i} = [\mathbf{e}^i_n, \mathbf{e}^i_{n-1}, ..., \mathbf{e}^i_1]$ for $i \in K = \{1, 2, ..., k\}$, where $n = \lceil \log_2(|E_u|) \rceil$. The variable $\overline{\mathbf{e}^i}$ thus either encodes a specific link-failed edge or the auxiliary edge e_{unused} signifying that it does not encode any link failing. This makes it possible to encode up to k link failures, rather than being limited to exactly k links failing.

We shall first define a BDD $path\text{-}edge\text{-}overlap$ to associate each path with its edges

$$path\text{-}edge\text{-}overlap\left(\mathbf{p}^D, \overline{\mathbf{e}}\right) = \bigvee_{d \in D} \quad \bigvee_{\pi \in DPaths(d)} \quad \bigvee_{e \in \pi} \overline{\mathbf{p}^d}(\pi) \wedge \overline{\mathbf{e}}(e) \qquad (12)$$

where $(P, e) \in [\![path\text{-}edge\text{-}overlap\left(\mathbf{p}^D, \overline{\mathbf{e}}\right)]\!]$ iff there exists a $\pi \in P(D)$ such that $e \in \pi$.

We can now define the BDD $failover^k$ that encodes all valid path assignments for every combination of k or fewer link failures:

$$failover^k\left(\overline{\mathbf{p}^D}, \overline{\mathbf{c}^D}, \overline{\mathbf{e}^K}\right) = rsa\left(\overline{\mathbf{p}^D}, \overline{\mathbf{c}^D}\right) \wedge \bigwedge_{m < j \leq k} \overline{\mathbf{e}^j}(e_{unused})$$

$$\bigvee_{\substack{E' = \{e_1, e_2, ..., e_m\} \subseteq E \\ |E'| \leq k}} \quad \bigwedge_{1 \leq i \leq m} \left(\overline{\mathbf{e}^i}(e_i) \wedge \neg path\text{-}edge\text{-}overlap\left(\overline{\mathbf{p}^D}, \overline{\mathbf{e}^i}\right)\right) . \qquad (13)$$

Theorem 3. *Let $E_{fail} \subseteq E$ be a subset of failed edges where $|E_{fail}| \leq k$, and let the vector of variable encodings $\overline{\mathbf{e}^K}$ encode the set E_{fail}. Then $(P, \omega, E_{fail}) \in [\![failover^k\left(\overline{\mathbf{p}^D}, \overline{\mathbf{c}^D}, \overline{\mathbf{e}^K}\right)]\!]$ iff $(P, \omega) \in [\![rsa\left(\overline{\mathbf{p}^D}, \overline{\mathbf{c}^D}\right)]\!]$ is a solution to the k-link failover problem with link failures E_{fail}.*

Proof (Proof sketch). "$\Longrightarrow$" Let $(P, \omega, E_{fail}) \in [\![failover^k\left(\overline{\mathbf{p}^D}, \overline{\mathbf{c}^D}, \overline{\mathbf{e}^K}\right)]\!]$ where $|E_{fail}| \leq k$. From Condition 13, we know that $(P, \omega) \in [\![rsa\left(\overline{\mathbf{p}^D}, \overline{\mathbf{c}^D}\right)]\!]$ and thereby is a valid solution to the RSA problem. Furthermore, as $\overline{\mathbf{e}^K}$ encodes E_{fail}, we know that for every edge e in E_{fail} that there exists a variable encoding $\overline{\mathbf{e}^i}$ from $\overline{\mathbf{e}^K}$ that encodes e. Additionally, we know that for all e in E_{fail} that $(P, e) \notin [\![path\text{-}edge\text{-}overlap\left(\overline{\mathbf{p}^D}, \overline{\mathbf{e}^i}\right)]\!]$ and thus none of the paths assigned to the demands contain any of the failed edges, which means that (P, ω, E_{fail}) is a solution.

"$\Longleftarrow$" Let $(P, \omega) \in [\![rsa(\overline{\mathbf{p}^D}, \overline{\mathbf{c}^D})]\!]$ be a solution to the k-link failover problem for a set of failed edges $E_{fail} \subseteq E$ where $|E_{fail}| \leq k$, and demands D. By Definition 2, we know that for all $\pi \in P(D)$ that $\pi \cap E_{fail} = \emptyset$. Hence, the BDD $\neg path\text{-}edge\text{-}overlap(\overline{\mathbf{p}^D}, \overline{\mathbf{e}^i})$ is satisfied for all e. As such, Condition 13 is satisfied and $(P, \omega, E_{fail}) \in [\![failover^k(\overline{\mathbf{p}^D}, \overline{\mathbf{c}^D}, \overline{\mathbf{e}^K})]\!]$. $\square$

Using the BDD $failover^k$, we can finally define the BDD $path\text{-}pruned\text{-}rsa$ which encodes all valid solutions given a set of specific link failures E_{fail} where $|E_{fail}| \leq k$

$$path\text{-}pruned\text{-}rsa(\overline{\mathbf{p}^D}, \overline{\mathbf{c}^D}) = \tag{14}$$

$$\exists \overline{\mathbf{e}^K}.\left(\bigwedge_{e_i \in E_{fail} = \{e_1, e_2, \ldots, e_m\}} \overline{\mathbf{e}^i}(e_i) \right) \wedge \left(\bigwedge_{m < j \leq k} \overline{\mathbf{e}^j}(e_{unused}) \right) \wedge failover^k(\overline{\mathbf{p}^D}, \overline{\mathbf{c}^D}, \overline{\mathbf{e}^K})$$

satisfying that $(P, \omega) \in [\![path\text{-}pruned\text{-}rsa(\overline{\mathbf{p}^D}, \overline{\mathbf{c}^D})]\!]$ iff $(P, \omega) \in [\![rsa(\overline{\mathbf{p}^D}, \overline{\mathbf{c}^D})]\!]$ and $P(d) \cap E_{fail} = \emptyset$ for all $d \in D$. Note that we reuse the name $path\text{-}pruned\text{-}rsa$ from Eq. 11 since the BDDs in Eqs. 11 and 14 are identical for any given failure scenario of size at most k.

5.3   Lightpath-Preserving Solutions

A property that is often desirable in fast failover protection in a given a failure scenario is that we update the routes and spectrum allocations only for the demands that are affected by some of the failing edges while preserving the remaining lightpaths. This will minimize the number of configuration updates and result in fewer traffic interruptions.

Let (P, ω) be an existing routing and spectrum assignment and let $E_{fail} \subseteq E$ be a given failure scenario. A solution (P', ω') to the failover problem for the set E_{fail} is *lightpath-preserving* if $P(d) \cap E_{fail} = \emptyset$ implies that $P(d) = P'(d)$ and $\omega(d) = \omega'(d)$ for all $d \in D$. We can identify all lightpath-preserving solutions for a given failure scenario E_{fail} as follows:

$$rsa\text{-}lightpath\text{-}preserving(\overline{\mathbf{p}^D}, \overline{\mathbf{c}^D}) = \tag{15}$$

$$path\text{-}pruned\text{-}rsa(\overline{\mathbf{p}^D}, \overline{\mathbf{c}^D}) \wedge \left(\bigwedge_{\substack{d \in D, \\ P(d) \cap E_{fail} = \emptyset}} \overline{\mathbf{p}^d}(P(d)) \wedge \overline{\mathbf{c}^d}(\omega(d)) \right)$$

where $(P', \omega') \in [\![rsa\text{-}lightpath\text{-}preserving(\overline{\mathbf{p}^D}, \overline{\mathbf{c}^D})]\!]$ iff (P', ω') is a *lightpath-preserving* solution to the given RSA solution (P, ω) for the link failures E_{fail}.

6   Implementation and Evaluation

We implemented the BDD encoding of the RSA problem presented in Sect. 4.2 and the two approaches for solving the k-link failover problem from Sect. 5 in our open-source tool *ExpectAll*, using the improvements from Sect. 4.2. The tool is implemented in Python using a Cython wrapper [20] of the library CUDD [52] to perform BDD operations. The source code of *ExpectAll*, including the reproducibility package, is publicly available at [6].

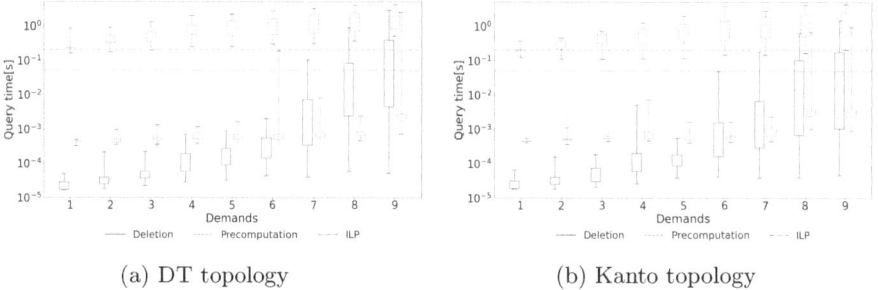

(a) DT topology   (b) Kanto topology

Fig. 4. Query time to find an optimal solution. Lines are shown for 50 ms and 200 ms.

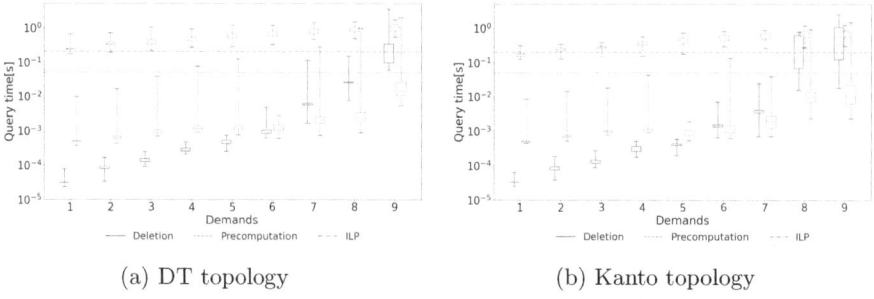

(a) DT topology   (b) Kanto topology

Fig. 5. Query time to find an optimal lightpath-preserving solution. Lines are shown for 50 ms and 200 ms.

We now evaluate our two BDD approaches to the k-link failover problem against the ILP approach presented in Sect. 3. The BDD approach using path-deletion pruning is called *deletion*, and the one using precomputation pruning is called *precomputation*. We compare the query times for finding an optimal solution and an optimal lightpath-preserving solution, by simulating 1000 random k-link failure scenarios using the same experimental setup as described in Sect. 3. The results are depicted in Figs. 4 and 5 for five link failures. We observe that both BDD approaches significantly outperform the ILP model (note the logarithmic scale on the y-axis). The query time of the *deletion* method however increases with the number of demands, exceeding the 200 ms threshold at 8–9 demands. On the other hand, the *precomputation* method has lower variance and maintains average query times consistently below 50 ms, demonstrating its suitability for a reliable, optimal and fast failover recovery. Furthermore, the results in Fig. 5 show that while the ILP approach is slightly faster at finding an optimal lightpath-preserving solution compared to a general optimal solution, both our BDD approaches are still significantly faster.

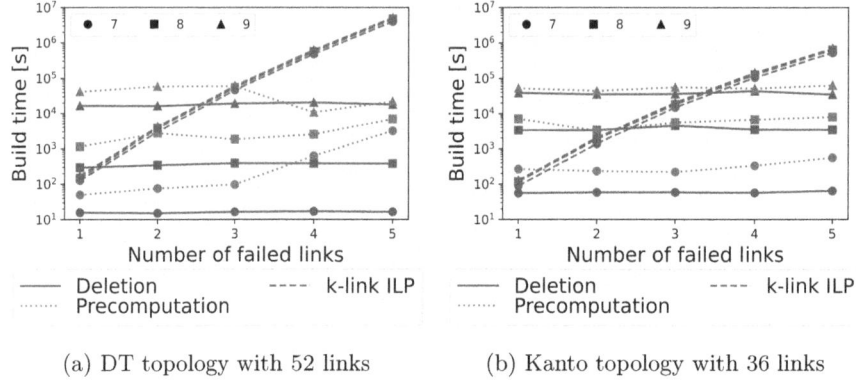

(a) DT topology with 52 links   (b) Kanto topology with 36 links

Fig. 6. Build time plots; the three line shapes denote 7,8 and 9 demands

We shall now focus on the time required to compute the BDD representation of all solutions. While a fast (< 50 ms) reaction time to the link failure is high priority, link failures are in general infrequent and we can afford to spend more time (hours) to precompute the BDDs for a given topology and set of demands. The experiments, presented in Fig. 6 for 7, 8 and 9 demands, indicate that as expected the ILP computation time grows exponentially with the number of failed links. In contrast, the build time for the *deletion* method remains nearly constant because the BDD always represents the same number of solutions, rendering the method independent of the number of link failures. The build time of the *precomputation* method increases with the number of link failures, though still remains within acceptable bounds and clearly outperforms the ILP approach (again the y-axis is logarithmic).

The plots also depict a limitation of our method as the precomputation does not yet scale for instances with 10 or more demands. This indicates that our current method is suitable for fast and effective failover protection of a smaller number of critical, high-bandwidth lightpaths that are stable over time. For the remaining, less critical and short-lived demands, we suggest to use e.g. existing heuristic approaches for failover protection at higher spectrum frequency slots, even though optimality is not guaranteed and this approach can lead to possible issues with lack of spectrum slots.

7   Conclusion

We introduced and implemented *ExpectAll*, a novel approach and a tool to efficiently compute and represent *all* valid route and spectrum allocations for a given all-optical network topology. Our approach guarantees optimality and fast reaction time in case of multiple link failures. It outperforms the state-of-the-art approaches based on integer linear programming by orders of magnitude, both in the response time to link failures as well as the precomputation time for three and more concurrently failing links. If we relax the optimality criterium, heuristic approaches for the route and spectrum allocations can be also considered for fast failover protection and we suggest a combination of the heuristic path allocation with *ExpectAll* so that the critical and long-lived lightpath requests can be dealt with in an optimal way using our method, while the remaining demands can be protected in a less optimal way using the heuristic approaches.

Our experimental evaluation indicates that we can handle up to 9 critical demands in an optimal manner. Even though this is currently the best-scaling technique that guarantees optimality, in the future work we plan to focus on further improving the scalability of the method, e.g. by exploring ideas that can eliminate symmetries in spectrum allocation.

Acknowledgement. This work was supported by the Independent Resarch Fund Denmark (DFF) under the project QASNET.

References

1. Akers, Binary decision diagrams. IEEE Trans. Comput. **100**(6), 509–516 (1978)
2. Andersen, H.R.: An introduction to binary decision diagrams. Lecture Notes, vol. 5. IT University of Copenhagen (1997)
3. Assis, K.D.R., et al.: Protection by diversity in elastic optical networks subject to single link failure. Opt. Fiber Technol. **75**, 103208 (2023)
4. Athe, P., Singh, Y.: Improved double cycle and link pair methods for two-link failure protection. Telecommun. Syst. **74**, 05 (2020)
5. Atlas, A., Zinin, A.: Basic specification for IP fast reroute: loop-free alternates. RFC **5286**, 1–31 (2008)
6. Bruhns, G.S., Hansen, M.P., Hebsgaard, R., Hyldgaard, F.M.W., Srba, J.: Reproducibility package for "ExpectAll: A BDD based approach for link failure resilience in elastic optical networks" (2024). Zenodo: https://doi.org/10.5281/zenodo.14179191
7. Bryant, R.: Graph-based algorithms for boolean function manipulation. IEEE Trans. Comput. **100**(8), 677–691 (1986)
8. Bryant, R.: Symbolic Boolean manipulation with ordered binary decision diagrams. ACM Comput. Surv. **24**, 03 (2003)
9. Castro, A., Velasco, L., Comellas, J., Junyent, G.: On the benefits of multi-path recovery in flexgrid optical networks. Photon Netw. Commun. **28**(3), 251–263 (2014). https://doi.org/10.1007/s11107-014-0443-5
10. Chatterjee, S., Pawlowski, S.: All-optical networks. Commun. ACM **42**(6), 74–83 (1999)
11. Chiesa, M., Kamisiński, A., Rak, J., Rétvári, G., Schmid, S.: A survey of fast recovery mechanisms in the data plane. TechRxiv (2020). https://doi.org/10.36227/techrxiv.12367508.v2

12. Chiesa, M., Kamisinski, A., Rak, J., Retvari, G., Schmid, S.: A survey of fast-recovery mechanisms in packet-switched networks. IEEE Commun. Surv. Tutorials **23**(2), 1253–1301 (2021)

13. Chiesa, M., et al.: PURR: a primitive for reconfigurable fast reroute: hope for the best and program for the worst. In: CoNEXT 2019, pp. 1–14. ACM (2019)

14. Chlamtac, I., Ganz, A., Karmi, G.: Lightpath communications: an approach to high bandwidth optical WAN's. IEEE Trans. Commun. **40**(7), 1171–1182 (1992)

15. Christodoulopoulos, K., Tomkos, I., Varvarigos, E.A.: Elastic bandwidth allocation in flexible OFDM-based optical networks. J. Lightwave Technol. **29**(9), 1354–1366 (2011)

16. Cisco Systems, Inc. Cisco Annual Internet Report (2018–2023) White Paper (2020). https://www.cisco.com/c/en/us/solutions/collateral/executive-perspectives/annual-internet-report/white-paper-c11-741490.html

17. Corfield, G.: British airways' latest total inability to support upwardness of planes caused by amadeus system outage (2018). https://www.theregister.co.uk/2018/07/19/amadeus_british_airways_outage_load_sheet

18. Eachempati, S., Saripalli, V., Vijaykrishnan, N., Datta, S.: Reconfigurable BDD based quantum circuits. In: 2008 IEEE International Symposium on Nanoscale Architectures, pp. 61–67. IEEE (2008)

19. Metro Ethernet. Technical specification MEF 2 requirements and framework for ethernet service protection in metro ethernet networks. Technical report, Metro Ethernet (2004). https://www.mef.net/resources/mef-2-requirements-and-framework-for-ethernet-service-protection/

20. Filippidis, I., Haesaert, S., Livingston, S.C., Wenzel, M.: Python package tulip-control/dd (2024). https://github.com/tulip-control/dd?tab=readme-ov-file

21. Foerster, K.-T., Parham, M., Chiesa, M., Schmid, S.: TI-MFA: keep calm and reroute segments fast. In: IEEE INFOCOM 2018-IEEE Conference on Computer Communications Workshops (INFOCOM WKSHPS), pp. 415–420. IEEE (2018)

22. François, P., Filsfils, C., Bashandy, A., Decraene, B., Litkowski, S.: Topology independent fast reroute using segment routing. Network Working Group (2014). https://datatracker.ietf.org/doc/html/draft-francois-spring-segment-routing-ti-lfa-02

23. Gao, T., Zou, W., Li, X., Guo, B., Huang, S., Mukherjee, B.: Distributed sub-light-tree based multicast provisioning with shared protection in elastic optical datacenter networks. Opt. Switch. Netw. **31**, 39–51 (2019)

24. Gibbs, C.: ATT's 911 outage result of mistakes made by ATT, FCC's Pai says (2017). https://www.fiercewireless.com/wireless/at-t-s-911-outage-result-mistakes-made-by-at-t-fcc-s-pai-says

25. Gill, Ph., Jain, N., Nagappan, N.: Understanding network failures in data centers: measurement, analysis, and implications. In: Proceedings of the ACM SIGCOMM 2011 Conference, pp. 350–361 (2011)

26. Gurobi Optimization, LLC. Gurobi Optimizer Reference Manual (2024). https://www.gurobi.com

27. Gyorgyi, C., Larsen, K.G., Schmid, S., Srba, J.: SyPer: synthesis of perfectly resilient local fast re-routing rules for highly dependable networks. In: IEEE International Conference on Computer Communications (INFOCOM 2024), pp. 2398–2407. IEEE (2024)

28. Gyorgyi, C., Larsen, K.G., Schmid, S., Srba, J.: SyRep: efficient synthesis and repair of fast re-route forwarding tables for resilient networks. In: Proceedings of the 54th Annual IEEE/IFIP International Conference on Dependable Systems and Networks (DSN 2024), pp. 483–494. IEEE (2024)

29. Hai, D.T., Hoang, K.M.: An efficient genetic algorithm approach for solving routing and spectrum assignment problem. In: 2017 International Conference on Recent Advances in Signal Processing, Telecommunications & Computing (SigTelCom), pp. 187–192. IEEE (2017)

30. Hai, D.T., Morvan, M., Gravey, P.: Combining heuristic and exact approaches for solving the routing and spectrum assignment problem. IET Optoelectron. **12**(2), 65–72 (2018)

31. Jinno, M., Takara, H., Kozicki, B.: Dynamic optical mesh networks: drivers, challenges and solutions for the future. In: 35th European Conference on Optical Communication, pp. 1–4. IEEE (2009)

32. Johansen, N.S., et al.: FBR: Dynamic memory-aware fast rerouting. In: IEEE Global Internet (GI) Symposium 2022, pp. 55–60. IEEE (2022)

33. Keiser, G.E.: A review of WDM technology and applications. Opt. Fiber Technol. **5**(1), 3–39 (1999)

34. Klinkowski, M., Walkowiak, K.: Routing and spectrum assignment in spectrum sliced elastic optical path network. IEEE Commun. Lett. **15**, 884–886 (2011)

35. Knight, S., Nguyen, H.X., Falkner, N., Bowden, R., Roughan, M.: The internet topology zoo. IEEE J. Sel. Areas Commun. **29**(9), 1765–1775 (2011)

36. Kubota, K., Tanigawa, Y., Hirota, Y., Tode, H.: Crosstalk-aware resource allocation based on optical path adjacency and crosstalk budget for space division multiplexing elastic optical networks. IEICE Trans. Commun. **107**(1), 27–38 (2024)

37. Larsen, K.G., Mariegaard, A., Schmid, S., Srba, J.: AllSynth: transiently correct network update synthesis accounting for operator preferences. In: International Symposium on Theoretical Aspects of Software Engineering, pp. 344–362. Springer (2022). https://doi.org/10.1007/978-3-031-10363-6_23

38. Lee, C.Y.: Representation of switching circuits by binary-decision programs. Bell Syst. Tech. J. **38**(4), 985–999 (1959)

39. Lezama, F., Martínez-Herrera, A., Castanon, G., Del Valle Soto, C., Sarmiento, A., Munoz de Cote, E.: Solving routing and spectrum allocation problems in flexgrid optical networks using pre-computing strategies. Photonic Netw. Commun. **41**, 1–19 (2021)

40. Li, X., Huang, S., Zhang, J., Zhao, Y., Gu, W., Wang, Y.: Analysis and modeling of k-regular and k-connected protection structure in ultra-high capacity optical networks. China Commun. **12**, 106–119 (2015)

41. Marković, G.Z.: Routing and spectrum allocation in elastic optical networks using bee colony optimization. Photon Netw. Commun. **34**(3), 356–374 (2017). https://doi.org/10.1007/s11107-017-0706-z

42. Miyagawa, Y., Watanabe, Y., Shigeno, M., Ishii, K., Takefusa, A., Yoshise, A.: Bounds for two static optimization problems on routing and spectrum allocation of anycasting. Opt. Switch. Netw. **31**, 144–161 (2019)

43. Mukherjee, B.: Optical communication networks. McGraw-Hill (1997)

44. Pan, P., Swallow, G., Atlas, A.: Fast reroute extensions to RSVP-TE for LSP tunnels. RFC **4090**, 1–38 (2005)

45. Papan, J., Segec, P., Moravcik, M., Kontsek, M., Mikus, L., Uramova, J.: Overview of IP fast reroute solutions. In: ICETA, pp. 417–424. IEEE (2018)

46. Ramaswami, R., Sivarajan, K.N.: Routing and wavelength assignment in all-optical networks. IEEE/ACM Trans. Netw. **3**(5), 489–500 (1995)

47. Rauchenecker, A., Wille, R.: An efficient physical design of fully-testable BDD-based circuits. In: 2017 IEEE 20th International Symposium on Design and Diagnostics of Electronic Circuits & Systems (DDECS), pp. 6–11. IEEE (2017)

48. Roughan, M., Greenberg, A., Kalmanek, Ch., Rumsewicz, M., Yates, J., Zhang, Y.: Experience in measuring backbone traffic variability: models, metrics, measurements and meaning. In: Proceedings of the 2nd ACM SIGCOMM Workshop on Internet Measurment (IMW 2002), pp. 91–92. ACM (2002)
49. Sakano, T., et al.: A study on a photonic network model based on the regional characteristics of Japan. PN2013-1 **113**(91), 1–6 (2013)
50. Shen, G., Wei, Y., Bose, S.K.: Optimal design for shared backup path protected elastic optical networks under single-link failure. J. Opt. Commun. Netw. **6**(7), 649–659 (2014)
51. Singhal, N.K., Ou, C., Mukherjee, B.: Cross-sharing vs. self-sharing trees for protecting multicast sessions in mesh networks. Comput. Netw. **50**(2), 200–206 (2006) Optical Network
52. Somenzi, F.: CUDD: CU decision diagram package release 2.4.1 (2024). https://web.mit.edu/sage/export/tmp/y/usr/share/doc/polybori/cudd/cuddIntro.html
53. Switch Specification 1.3.1. OpenFlow. In: Open Networking Foundation (2013). https://bit.ly/2VjOO77
54. Tanaka, T., Shimoda, M.: Pre-and post-processing techniques for reinforcement-learning-based routing and spectrum assignment in elastic optical networks. J. Opt. Commun. Netw. **15**(12), 1019–1029 (2023)
55. van Duijn, I., et al.: Automata-theoretic approach to verification of MPLS networks under link failures. IEEE/ACM Trans. Netw. **30**(2), 766–781 (2022)
56. Velasco Esteban, L., Klinkowski, M., Ruiz, M., Comellas, J.: Modeling the routing and spectrum allocation problem for flexgrid optical networks. Photonic Netw. Commun. **24**, 177–186 (2013)
57. Wang, J., Shigeno, M., Wu, Q.: ILP models and improved methods for the problem of routing and spectrum allocation. Opt. Switch. Netw. **45**, 100675 (2022)
58. Wang, Y., Cao, X., Pan, Y.: A study of the routing and spectrum allocation in spectrum-sliced elastic optical path networks. In: Proceedings IEEE INFOCOM 2011, pp. 1503–1511. IEEE (2011)
59. Wikipedia. List of cities in germany by population. https://en.wikipedia.org/wiki/List_of_cities_in_Germany_by_population
60. Xu, L., Huang, Y.-C., Xue, Y., Hu, X.: Deep reinforcement learning-based routing and spectrum assignment of EONs by exploiting GCN and RNN for feature extraction. J. Lightwave Technol. **40**(15), 4945–4955 (2022)
61. Zang, H., Jue, J.P., Mukherjee, B.: A review of routing and wavelength assignment approaches for wavelength-routed optical WDM networks. Opt. Netw. Mag. **1**(1), 47–60 (2000)
62. Zhang, J., Miao, P., Zhang, F.: On optimal routing and spectrum allocation in elastic optical networks. In: 2nd International Conference on Big Data, Information and Computer Network (BDICN) 2023, pp. 284–287. IEEE (2023)

Constructing Trustworthy Smart Contracts

Devora Chait-Roth[1] and Kedar S. Namjoshi[2]([⊠])

[1] New York University, New York, NY, USA
dc4451@nyu.edu
[2] Nokia Bell Labs, Murray Hill, NJ, USA
kedar.namjoshi@nokia-bell-labs.com

Abstract. Smart contracts form the core of Web3 applications. Contracts mediate the transfer of cryptocurrency, making them irresistible targets for hackers. We introduce Asp, a system aimed at easing the construction of provably secure contracts. The Asp system consists of three closely-linked components: a programming language, a defensive compiler, and a proof checker. The language semantics guarantee that Asp contracts are free of commonly exploited vulnerabilities such as arithmetic overflow and reentrancy. The defensive compiler enforces the semantics and translates Asp to Solidity, the most popular contract language. Deductive proofs establish functional correctness and freedom from critical vulnerabilities such as unauthorized access.

1 Introduction

Decentralized blockchain-based systems such as Ethereum have ushered in a new computation paradigm dubbed "Web3." Smart contracts form the core of Web3. A smart contract ("contract" for short) is a program whose code and execution are recorded on a blockchain. A contract facilitates an exchange of value, typically cryptocurrency, between contract participants.

Trust is a central concern for Web3 applications, as there is no central trusted authority and the participants need not have prior trust relationships. Blockchain properties play a crucial role in building trust in contract execution. Every contract is stored on the blockchain, ensuring that its code is open and immutable. Every contract transaction is executed in a replicated manner, guarding against the possibility of machine failures or compromised execution engines.

Although these mechanisms provide a solid foundation, contracts are but programs, and programs may have errors. Immutability of code and fault-tolerant execution are of no help if contract code is buggy and vulnerable to attack. As contracts handle large amounts of cryptocurrency, they are irresistible targets for hackers, who stole nearly $4 billion in 2022 and $2 billion in 2023 from buggy contracts [6,8]. In response to this threat, developers audit contracts through standard processes: testing, static analysis, and expert code review. While these methods uncover some bugs, they are inadequate at guaranteeing security.

K. Shankaranarayanan et al. (Eds.): VMCAI 2025, LNCS 15530, pp. 231–252, 2025.
https://doi.org/10.1007/978-3-031-82703-7_11

As a result, [4,19,22] and others have developed *formal verification* methods for contracts. However, formally verifying contracts written in standard languages such as Solidity or Rust is difficult, in part due to complex language constructs and the need to reason about reentrancy.[1] Non-standard contract languages have emerged in response, but have drawbacks. FSolidM [13,14] models contracts as finite state machines extended with Solidity actions, but verification is limited to propositional CTL properties. Scilla [20] compiles contracts to Coq for analysis, but proofs in Coq require substantial expertise and effort. Move [3] has impressive automated verification support, but Move contracts cannot execute on common blockchains such as Ethereum.

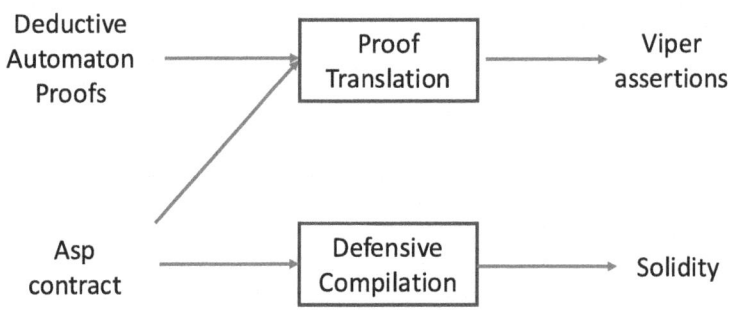

Fig. 1. The Asp pipeline. The Viper system is used to verify proof assertions.

This paper introduces the Asp system, which is aimed at easing the construction of provably secure contracts while addressing these drawbacks. It has three tightly-linked components, as shown in Fig. 1: the Asp contract language, a defensive compiler, and a deductive proof checker. The Asp language provides abstractions that simplify verification; the compiler translates Asp contracts to run on standard blockchains; and the Asp proof checker is used to verify safety and liveness properties that users state directly on the Asp contracts.

Asp combines well-known notions (such as state machines and deductive proofs) but differs from prior work on contract verification in its emphasis on abstraction coupled with defensive compilation. We illustrate the Asp language and particularly its abstraction mechanisms through a variety of examples, show how the use of abstractions simplifies proofs, and describe the implementation of the compiler and proof checker. Our Asp prototype is implemented in about 3000 lines of OCaml and 100 lines of Viper; examples of Asp contracts (with proofs) are available at https://github.com/DebraChait/Asp-example-contracts. We summarize the three components of the Asp system next.

The Asp Language. A programming language typically strikes a balance between ease of expression and ease of analysis. Most contracts today are written in the

---

[1] 'Reentrancy' is an execution pattern where a malicious contract forces an internal contract-invocation loop that drains cryptocurrency from the target contract.

languages Solidity, Vyper, Ink! and Rust. These are full-fledged programming languages, which complicates verification. With Asp, we aim to achieve a good balance through a programming model that employs *abstractions*. Coins, tokens, timers, and addresses – Web3 abstractions ordinarily represented at a low level – are provided as abstract types in Asp. Abstractions ease programming and simplify analysis, as the abstract operations are few in number and have precise semantics. The Asp language semantics inherently forbids arithmetic overflow, out-of-bounds accesses, and reentrancy, eliminating those common vulnerabilities as concerns for a contract programmer.

An Asp contract has a finite-state machine skeleton that is augmented with actions on state variables defined over abstract data types. This structure naturally models real-world contract execution. For instance, consider sending a package through the mail–a contract between the sender and the post office. This contract passes through the stages of preparation, payment, transit, and delivery; these stages are naturally modeled by a finite-state machine. In conventional languages such as Solidity, such structure must be encoded and enforced implicitly, obscuring it and complicating analysis.

Asp contracts interact through synchronized message transfers. That also models the event driven structure of real-world contracts: for instance, the change from the transit to the delivery stage in the prior example is through the event of delivering the package.

The Asp Defensive Compiler. The Asp compiler translates contracts to Solidity, the most popular contract language. Compilation preserves Asp semantics, implements the high-level abstractions, and supports the message-transfer view of communication. The compiler adds auxiliary *defensive* code to enforce the language semantics and check properties dynamically that are difficult to prove statically. Development of a compiler back-end for Ink! is in progress; compilation to other contract languages follows the same design. Compilation assures the portability of verified Asp contracts to a variety of blockchains.

The Asp Proof Checker. Although the Asp semantics eliminates many common vulnerabilities, it cannot rule out all of them. Thus, one must prove that contract behavior does not, for instance, allow unauthorized access to stored cryptocurrency, or that the contract cannot be placed in a 'frozen' state. (These are commonly exploited vulnerabilities.[2]) The first is a safety property, which Asp users can establish through a standard automaton-based proof system. Asp allows auxiliary 'ghost' state to be added to a contract purely for the purposes of proof – this state may be viewed as implementing a deterministic checking automaton. The second is an adversarial liveness property, which requires its own proof system. The Asp proof-checker takes a declarative proof sketch for an Asp contract and turns that sketch into lemmas that encode the requirements of the proof system. The lemmas are checked by translation to an SMT solver, invoked indirectly through the Viper system [15]. This provides Asp users with deductive proofs of contract properties.

---

[2]https://info.merklescience.com/april-2023-hackhub-report.

Deductive proofs are an important mechanism for enhancing trust in the inherently trustless world of Web3, as any contract participant may *independently* validate a claimed proof of a property against the contract code. As contract code is immutable once deployed, a proof remains valid throughout execution.

2   Asp By Example

To illustrate how the design of Asp promotes trustworthy smart contracts, we present an example of an open auction contract. Figure 2 shows the skeleton of the contract written in Asp. (We will fill in the contract gradually.)

```
 1  contract SimpleAuction(beneficiary: address,
 2                          bidding_time: nat)
 3    where beneficiary != Address.none && bidding_time > 0 {
 4
 5    msg start, bid(coin);
 6    var tmr: timer, maxBid: coin,
 7        maxBidder: address := Address.none;
 8
 9    initial StartAuction;
10
11    state StartAuction:
12    | owner??start -> AuctionOpen
13      { Timer.set(tmr, bidding_time); }
14
15    state AuctionOpen:
16    | a??bid(c)
17      when Timer.is_active(tmr)
18        && Coin.value(c) > Coin.value(maxBid)
19      notby beneficiary -> AuctionOpen
20      { /* actions */ }
21
22    | when Timer.has_fired(tmr) -> AuctionClosed
23      { /* actions */ }
24
25    state AuctionClosed: // no transitions
26
27  }
```

Fig. 2. Skeleton of an open auction contract in Asp

2.1   State Machine Structure

An auction is naturally expressed as a state machine. The initial state is denoted by keyword `initial`, the auction moves to a state where it is open and bidders can submit bids, and finally transitions to a state where the auction has closed. Mirroring blockchain execution, transitions can only be triggered by external messages (akin to function calls in function-based languages). The construct `a??` `bid(c)` in line 16 of Fig. 2 denotes a message named `bid`, received from an address dubbed `a`, containing an input parameter `c` of type `coin`. Section 2.2 elaborates on Asp's special abstract types.

An Asp contract starts in its initial state, and execution must follow the contract transition partial function $T : S \times M \rightharpoonup P(S)$, which maps each state and its input messages to the set of its possible next states. Thus, if a contract is in state $s \in S$, a message $m \in M$ cannot be received at that state if $(s, m) \notin T$. (Contrast this with a function-structured language, where functions represent state *changes* but there is no clear delineation of allowed state *sequences*.)

Asp has transition guards to capture the conditions under which a transition is enabled. Transition guards include input guards denoting received messages, such as `a??bid(c)` in line 16; predicates which must evaluate to true, such as `when` `Timer.has_fired(tmr)` in line 22; and access control, such as `notby beneficiary` in line 19. Access control errors represent one of the most highly exploited contract vulnerabilities in 2023.[3] Asp's access guards make access control explicit.

A transition without an input guard is called a τ transition. It is enabled at a state if its boolean guard evaluates to true. Although executable contracts must be deterministic, Asp contracts may be internally non-deterministic – that is, multiple τ transitions may be enabled at a state – to allow for a notion of contract refinement. The compiler enforces determinism by choosing arbitrarily between enabled τ-transitions; correctness proofs are not affected by this choice.

In Fig. 3, we add actions to the transitions from state `OpenAuction` that specify state changes. Transition actions are comprised of a *loop-free* sequence of operations that include message transmissions and state updates. Restricting the action to be loop-free simplifies analysis, in effect by letting the skeleton structure explicitly define any loops.

2.2 Abstractions

We introduce the *abstractions* of common Web3 constructs that are included in Asp. We demonstrate how these abstractions simplify verification by comparing the Asp types and semantics with existing encodings and verification schemes.

Basic Types. Basic types include `int` (integer), `nat` (naturals), `Tuple` (tuples), `Seq` (unbounded sequences), and `Map` (mappings from a key type to a value type).

In a significant departure from a conventional programming language, the number types (`int` and `nat`) are given their *mathematical* definitions. As a consequence, there is no notion of arithmetic overflow in Asp, so a contract programmer need not be concerned with vulnerabilities that arise from such overflow. Of

---

[3] https://blog.merklescience.com/general/hackhub2024.

```
1  contract SimpleAuction(...) {
2    ...
3    state AuctionOpen:
4    | a??bid(c)
5      when Timer.is_active(tmr)
6        && Coin.value(c) > Coin.value(maxBid)
7      notby beneficiary -> AuctionOpen
8        { maxBidder!!bid_lost(maxBid);
9          /* store new maxes */
10         maxBidder = a;
11         Coin.moveall(c,maxBid); }
12
13   | when Timer.has_fired(tmr) -> AuctionClosed
14     { beneficiary!!winner(maxBid, maxBidder); }
15   ...
16 }
17
18 contract Beneficiary() {
19   var auction: Address;
20   ...
21   state AcceptBid:
22   | a??winner(amt, addr) by auction -> FinalState {
23     log!!final_winner(Coin.value(amt), addr);
24   }
25   ...
26 }
```

Fig. 3. Code snippets of open auction contract and receiving contract in Asp

course this is a fiction. It is the task of the defensive compiler to check for violations of this fiction (such as when arithmetic operations overflow) and cancel the
transaction execution when violations occur. Certain operations are inherently
partial: for instance, division by zero is undefined, as is an out-of-bounds access
to a sequence. The defensive compiler also checks that every executed operation
is well-defined. Section 5 elaborates.

Coins. In conventional contract languages, there are two separate ways of
accounting for cryptocurrency. The underlying blockchain maintains an account
balance for each address and ensures that no double-spending can occur. At the
contract level, cryptocurrency is represented as integers. This leaves contract-
level accounting open to arithmetic error. Mistakes at the contract-accounting
level cannot propagate to the blockchain balances, but they can allow malicious
contracts to obscure the amounts of currency sent or received by a contract.

Asp unifies both views in a single Coin datatype. A coin variable is a container
for the native cryptocurrency of the blockchain in its most basic unit, whatever
that may be. This allows Asp contracts to be ported to multiple blockchains,
each with its own cryptocurrency. Coins cannot be created: they may only be

transferred. To enforce this requirement, the Coin datatype provides only three operations: Coin.value(c) is the (non-negative integer) value contained in c; Coin.moveall(c,d) transfers all value from c to d; and the partial operation Coin.move(c,k,d) transfers value k from c to d, if c has value at least k–if not, the operation is undefined.

Example 1. 2vyper [4] provides *resources* as special ghost state that can only be manipulated in certain prescribed ways – somewhat similar to the coin type in Asp. However, 2vyper users must define ghost state operations and set up coupling invariants to prove that the values of uint type used for cryptocurrency accounting remain consistent with the resource values in ghost state.[4] Asp users rely on the guarantees of coin type, without needing extra proofs to verify against malicious accounting manipulation.

All coins that are received in an input message must be transferred to coin state variables–i.e., received coins are not lost. (This is enforced by the defensive compiler.) Sending a message containing a coin variable transfers its entire value to the receiving entity; thus, those variables have zero value after the send action.

Example 2. Line 11 of Fig. 3 moves all coins sent as a bid to the contract's maxBid coin. As line 8 transfers the previous maxBid to the dethroned maxBidder, the final value of maxBid is that of the accepted bid.

The Coin operations in Asp directly modify the value of a coin operand. However, coins stored in Map, Seq, or Tuple types are *copied* when retrieved from their storing structure. To prevent spurious coin creation or destruction, coin operators only accept operands that *reference* coins stored in complex types. Map.ref (and likewise for Seq and Tuple) directly modifies the specified entry, so that total coins are conserved.

Example 3. In Example 5, a map stores Token types (described next; analogous operations to Coin). Note that Token.move expects Map.ref rather than Map.get, as move operations modify the values of their operands.

The coin operations enforce a *coin conservation* law: coins are neither created nor destroyed and all coins in circulation are accounted for. In Solidity contracts, one would have to explicitly prove that the integer-based accounting in contracts matches the blockchain-based accounting; the Asp coin datatype eliminates the need for such proofs.

Tokens. While cryptocurrency conservation is maintained by the blockchain, tokens in conventional languages have no such protection and are exclusively treated as uints. Issuing, burning, and transferring tokens is entirely comprised of arithmetic manipulation, heightening the need to verify against malicious

---

[4]See, for example, Fig. 9 on page 16 of [4].

token contracts. Asp provides an abstract `token` type to prevent honeypot contracts and other exit scams that issue malicious tokens.[5]

A *token* in Asp is essentially a coin, with two important differences: (1) tokens are defined and allocated by an issuer, and (2) tokens can be of different kinds. In Asp, these aspects are handled by designating an Asp contract as a token issuer with the declaration "`issues Token(<limit>)`," where the optional `<limit>` is a natural number indicating the number of available tokens.

A token issuing contract issues just one kind of token. Tokens are issued into a token variable v through `Token.issue(<number>,v)`, which is defined only if a sufficient quantity of tokens are available for issue. The coin operations have analogous token operations (`Token.move(v,k,w)`, `Token.moveall(v,w)`, and `Token.value(v)`) and follow similar token-conservation rules, up to minting (i.e., issuing) and burning (i.e., removal) through a `Token.burn(v,k)` operation.

Example 4. The following code adapted from Certik (See footnote 5) presents a snippet from malicious token contract:

```
function setBalance(address user, uint256 amt) public
    onlyRole(DEFAULTADMINROLE){
_balanceaccs[user] = amt * 10**decimals();}
```

This code allows the contract owner to change a user's token balance to any amount they specify. Such a scam is impossible with Asp's `token` type.

Example 5. The Move [10] language is designed for contract verification. However, Move does not have a built-in notion of tokens: [10] provides an example contract for designing a safe token.[6] To ensure that tokens are not generated or destroyed spuriously in a token transfer, the Move contract defines a transfer function as a withdraw followed by a deposit, both carefully defined to disallow manipulation. Asp condenses this into a few lines of code with the `token` type:

```
1 contract BasicCoin() issues Token() {
2    msg transfer(nat,address);
3    var accounts: map[address,token];
4    ...
5    state Bank:
6    | a??transfer(x,b) when Map.in(a,accounts) && Map.in(b,
       accounts) notby b -> Bank {
7      Token.move(Map.ref(accounts,a),x,Map.ref(accounts,b));}
8 }
```

The Move contract ensures that tokens are not duplicated by defining a unique token storage struct for each address. Asp tokens cannot be duplicated by design. As tokens are explicit types, the Asp contract simply has an `accounts` map from `address` to `token`, through which tokens may be issued, transferred, or burned.

---

[5] https://certik.com/resources/blog/honeypot-scams.
[6] Figures 1,2 on pages 3,4 in [10].

Addresses. An `address` holds the blockchain address of an Asp contract or an external party. A contract contains the special address variables `creator` and `owner`. These are set at the time of instantiation to the address of the creating entity. The `owner` may be modified, but only through an `Address.change_owner` initiated by the current owner. The `creator` cannot be modified. A contract instance has its own unique constant address, referred to as `Address.self`.

Example 6. Line 12 of Fig. 2 specifies that the first transition of the auction contract can only be triggered by the contract owner.

Addresses may be stored, copied, and transferred between contracts. Message receive and send operations refer to addresses. (Interaction between contracts will be addressed in Sect. 2.3.) A special address, `log`, is used to log messages to the blockchain. Log transfers are always enabled and do not change the state of the contract.

Example 7. Line 23 of Fig. 3 logs the amount of the winning bid and the winning bidder after the auction is complete. This is akin to an event in Solidity.

Timers. In blockchains, time is not measured as real time, but rather by the growth of the chain, to ensure that all miners and validators have a common view of time. As a consequence, the progress of time may be uneven and is not guaranteed, which may be a point of fallacy for developers. Time-dependent actions are important for contracts such as auctions, which must terminate, and for contracts such as hashed timed locks, which place a limit on how long cryptocurrency is kept in escrow.

Asp includes a `Timer` data type. A `timer` variable represents a timer that is initially inactive. Timer operations move the (implicit) timer state machine through a sequence of states: `Off`, `Active(k)` (for $k > 0$), and `Fired`. The `Timer.has_fired` predicate determines whether a timer is in its `Fired` state. All timers advance together by the same non-deterministically chosen amount on a transition, which models the way time advances on the underlying blockchain. Timers also simplify proofs of liveness properties; Sect. 4.2 elaborates.

Example 8. VerX [19], a contract verification tool, provides a benchmark of a contract that runs a continuous sale, which divides time into "buckets" of twelve hours (in Solidity) for accounting purposes.[7] Since time varies based on the frequency of blocks added to the chain, the only guarantee that this contract provides is that a bucket will eventually be reset. However, it does not guarantee anything more precise about the time of each bucket, since time is dependent on the length of the blockchain. Asp timers abstract away concrete notions of time so users do not rely on mistaken assumptions. The snippet below replaces the concept of "hours" with a timer that progresses non-deterministically and is reset upon firing.

---

[7] https://github.com/eth-sri/verx-benchmarks/blob/master/Mana/main.sol.

```
1 state TrackSale:
2   | when Timer.is_active(timer_bucket) -> CheckMax
3     { ... }
4
5   | when Timer.has_fired(timer_bucket) -> CheckMax
6     { Timer.reset(timer_bucket);
7       Timer.set(timer_bucket, bucket_size);
8       ...
9     }
```

2.3   Interacting Contracts

In our auction example, we may want to create a separate state machine to describe the actions of the beneficiary, perhaps to model sending the prize or dividing the winning bid among collaborators. Interactions between contracts are specified by synchronized send and receive actions, using notation akin to CCS and CSP. In Fig. 3, the second transition from `AuctionOpen` in the `SimpleAuction` contract sends the message `winner` with parameters of type `coin` and `address` to the beneficiary. [8] The contract `Beneficiary` receives a corresponding transition message `winner` of identical parameter types. When `SimpleAuction` sends the `winner` message to `Beneficiary`, the message will be received by `Beneficiary` only if `Beneficiary` previously set the variable `auction` to the address of `SimpleAuction` and is at state `AcceptBid`.

3   Asp Semantics

At its core, Asp contracts induce reactive state machines which communicate through synchronized message exchanges. We define the semantics of execution and communication. We begin by defining the transition semantics of a single contract instance, then consider the semantics of interactions between multiple contract instances.

For a simpler notation, we assume that every contract transition contains either an input guard and no output actions, or is a τ transition with at most one output action. A contract is easily restructured to meet these requirements by introducing fresh states and edges.

3.1   Single Instance Transition Semantics

A single contract instance defines a labeled transition system with transitions labeled as input, output, or internal. The contract skeleton may be viewed as the tuple (Q, q_0, δ), where Q is the finite set of states, q_0 is the initial state,

---

[8] The lack of withdraw pattern may seem concerning to smart contract developers who are familiar with reentrancy. Asp contracts automatically guard against reentrancy, so refunds can be sent to losing bidders directly.

and $\delta \subseteq Q \times Q$ is the next-state relation. We refer to a pair (q, q') in the next-state relation as an *edge*. The state variables V induce the space of assignments $X = V \to D$. (For simplicity in notation, all variables have domain D.) The initial assignment is denoted x_0.

Every message declaration $m(t_0, \ldots, t_{n-1})$ is viewed semantically as the set of input letters of the form $m(\alpha, d_0, \ldots, d_{n-1})$, where α represents the address of the message sender and each d_i is a value in D. The set of input letters is denoted Σ; this has a corresponding set $\overline{\Sigma}$ of output letters. For a letter e (input or output), $\overline{e}$ represents the matching letter (output or input, resp.), with $\overline{\overline{e}} = e$.

The labeled transition system for a single contract instance is defined as a tuple (S, s_0, Σ, T). The state space S is $Q \times X$. The initial state s_0 is (q_0, x_0). T denotes the set of labeled transitions. A transition from state (q, x) to state (q', x') is defined if (q, q') is an edge labeled with guard g and action a, the predicates in guard g evaluate to true at the state (q, x), the operations in action a are fully defined at state x, and x' is the result of performing the operations in a. This transition is labeled by input letter e if g has an input guard that evaluates to e in the state (q, x), by output letter $\overline{e}$ if the action a is an output action that evaluates to e, and by τ otherwise.

3.2   Multiple Instance Semantics

Consider a collection $M_1, \ldots, M_n$ of contract instances. Intuitively, the instances communicate by synchronizing pairwise on transition labels, i.e., when an output transition of one contract matches with an input transition of another.

As described previously, contract execution on a blockchain is single-threaded and externally triggered. To match this execution model, we define a single-threaded *cascading semantics* for Asp contracts. Intuitively, a cascade is started at a quiescent configuration by invoking an input transition in one of the contracts. (A configuration is a vector of contract states; it is quiescent if no cascade is in progress.) This input transition may trigger a synchronized transition with another contract; the execution thread then moves to that contract. A cascade continues in this manner until no further synchronizations are possible.

The precise formulation of a cascade relies on a pushdown stack of contract indexes and is parameterized by a recurrence limit $R \geq 0$. A *configuration* is a pair (s, γ) where s is a vector with $s(i)$ being the state of M_i, for all i, and γ is a pushdown stack with entries from $\{1, \ldots, n\}$. The stack is represented as a sequence with the left end of the sequence being the top of the stack. It is an invariant of the semantics that in any reachable configuration the stack contains at most $R + 1$ occurrences of each index. A *quiescent* configuration is a pair (s, γ) where the stack γ is empty, denoted by ϵ. We say that an input or output letter e is *directed towards the environment* if its address entry is not one of the M-contract instances. We use the update notation $s' = s[k \leftarrow u]$ to represent the state vector s' which is identical to s except at the k'th entry, which is u.

The transitions from a configuration (s, γ) are as follows.

1. (Local τ-Move) If k is the entry at the top of the stack and $(s(k), \tau, t)$ is a transition of M_k, then $((s, \gamma), \tau, (s', \gamma))$ is a transition, where $s' = s[k \leftarrow t]$.

2. (Synchronized Push) If k is the entry at the top of the stack and there is an output transition $(s(k), \overline{e}, t)$ in M_k and a matching input transition $(s(l), e, u)$ in M_l for some l *that has at most R occurrences in* γ, then $((s, \gamma), \tau, (s', \gamma'))$ is a synchronized transition, where $s' = s[k \leftarrow t, l \leftarrow u]$ is the new state vector, and $\gamma' = l\gamma$ is the new stack.

3. (Environment Output) If k is the entry at the top of the stack and there is an output transition $(s(k), \overline{e}, t)$ in M_k where e is directed towards the environment then $((s, \gamma), \overline{e}, (s', \gamma))$ is a transition, with $s' = s[k \leftarrow t]$.

4. (Pop) If k is at the top of the stack and none of the above types of transitions are enabled at $s(k)$, then $((s, \gamma), \tau, (s, \gamma'))$ is a transition, where $\gamma = k\gamma'$.

5. (Environment Input) Consider a quiescent configuration (s, ϵ). If there is an input transition $(s(k), e, u)$ in M_k for some k where e is directed towards the environment, then $((s, \epsilon), e, (s', \gamma'))$ is a transition, where $s' = s[k \leftarrow t]$ is the new state vector, and $\gamma' = k$ is the new, non-empty stack.

Along a computation, a *cascade* is an execution fragment that starts at a quiescent configuration and ends at the next quiescent configuration. Every infinite computation can be partitioned into either an infinite sequence of cascades, or into a finite sequence of cascades followed by an infinite suffix where the stack is "stuck" and every transition is either a local move or an environment output.

Reentrancy attacks are blocked with $R = 1$ as it is impossible for an attacker contract to trigger an account withdrawal transition twice within a cascade. The extended version [7] contains an illustrative example.

4 Asp Verification Proof Methods

By design and semantics, Asp contracts are inherently free from common vulnerabilities such as reentrancy, out of bounds accesses, and arithmetic overflow. Other security properties, such as proper access control, must be established through verification. In Asp, a user-supplied deductive proof is required for every claimed property of an Asp contract. This has a crucial benefit in the trustless Web3 setting, as any user can *independently* validate claimed proofs before entering into a contract.

The Asp proof-checker reads in a proof sketch for an Asp contract. It translates the sketch into Hoare triples (per contract transition) according to the proof rule specified. The triples are checked for validity through Viper [15], an intermediate verification language based on SMT solving.

Most attacks on smart contracts, including many classified as security violations, target violations of safety properties. Termination is guaranteed in practice by contract execution's reliance on available "gas," cryptocurrency paid to a validator in return for executing a contract transaction. We address other liveness concerns, such as reachability and lockouts. The Asp proof-checker currently supports safety and reachability verification, which we illustrate with examples. In [7] we describe a lockout vulnerability, its resolution, and a formal proof of lockout-freedom; implementation of this proof method is in progress.

4.1 Safety Proofs with Ghost Variables

Informally, a safety property is one that can be falsified in finitely many steps. (A precise formulation is in [1].)

A standard proof method for safety uses a finite-word automaton to detect violations. Instead of defining automata over subsets of atomic state *propositions*, as is typical, we define the automaton directly over the contract state space S. A deterministic finite-word automaton recognizing *violations* of a safety property L is a tuple $A = (Q, q_0, \delta, R)$ where Q is the set of automaton states (not necessarily finite), q_0 is the initial state; $\delta : Q \times S \to Q$ is the transition function, and R is a subset of states, which we refer to as the rejecting states.

A *run* ρ of the automaton on an infinite computation $w = s_0, e_0, s_1, e_1, \ldots$ is a function from Nat to Q, where $\rho(0) = q_0$ and the tuple $(\rho(i), s_i, \rho(i+1))$ is in δ for all i. The run is rejecting if $\rho(k)$ is in R for some k. Computation w violates L if there is a rejecting run of the automaton on w.

Following the automaton-theoretic view of verification, we define a product transition system $M \times A$ from the contract machine M and the safety-violation automaton A. This has state space $S \times Q$, initial state (s_0, q_0), and transitions of the form $((s, q), (s', q'))$ where for some e, (s, e, s') is in T and $\delta(q, s) = q'$. It is straightforward to show that every computation of M is safe if, and only if, a reject state of A is not reachable in $M \times A$; equivalently, if the property "not in R" is invariant over the transition system $M \times A$.

In Asp, the product construction is carried out manually using auxiliary state variables declared with the prefix `ghost`. The type checker ensures that ghost state does not influence normal contract execution: ghost state is never used to modify contract state, ghost state is never communicated to other contracts through messages, and assertions on ghost state cannot be used to control transition enabledness. Ghost variables encode the state of the violation automaton. As the automaton is deterministic, automaton state updates can be added to every contract transition. (Updates to ghost/automaton state can, and do, depend on the contract state variables.) Checking that "not in R" is invariant amounts to showing that an assertion θ on the joint state (contract + ghost) satisfies:

- (Initiality) $\theta(s_0, q_0)$ holds,
- (Inductiveness) If $\theta(s, q)$ holds and $((s, q), (s', q'))$ is a transition of $M \times A$, then $\theta(s', q')$ holds, and
- (Sufficiency) If $\theta(s, q)$ holds, then q is not in R.

A safety proof sketch in Asp partitions θ across the contract skeleton states as a family of assertions $\{\theta_m\}$, where θ_m is associated with the skeleton state m, for all m. Given these assertions, the proof checker carries out the initiality, inductiveness, and sufficiency checks through a translation of Asp constructs to the Viper verifier, as demonstrated in the following example.

Illustrating Asp Safety Proofs. In our auction example, one property we may want to prove is that all bidders receive proper refunds. To prove this property

```
 1 contract SimpleAuction(...) ... {
 2   ...
 3   ghost var bidded: map[address, int] default 0,
 4             refunded: map[address, int] default 0;
 5
 6   state AuctionOpen:
 7     | a??bid(c) ... -> AuctionOpen
 8       {Map.set(bidded,a,Map.get(bidded,a) + Coin.value(c));
 9        Map.set(refunded,maxBidder,
10          Map.get(refunded,maxBidder)+Coin.value(maxBid));
11        maxBidder!!bid_lost(maxBid);
12        ... }
13     | when Timer.has_fired(tmr) -> AuctionClosed
14       {Map.set(refunded,beneficiary,
15          Map.get(refunded,beneficiary)+Coin.value(maxBid));
16        beneficiary!!winner(maxBid, maxBidder);}
17   ...
18 }
```

Fig. 4. Open auction contract with ghost variables

(via invariance), we add ghost variables to the contract (as shown in Fig. 4) to keep track of coins bidded and refunded.

Asp users specify proof sketches in a separate proof file. To prove that all dethroned bidders are refunded their full bids, we write the following assertion using ghost variables:

```
1 always forall b: address :
2 (b != maxBidder && b != beneficiary)
3 ==> Map.get(refunded,b) == Map.get(bidded,b)
```

always denotes an assertion that holds at every skeleton state. To prove that the winning bidder is refunded their previous losing bids, we add:[9]

```
1 always Map.get(refunded,beneficiary) == 0 ==>
2 Map.get(refunded,maxBidder) ==
3 Map.get(bidded,maxBidder) - Coin.value(maxBid)
4 always Map.get(refunded,beneficiary) > 0 ==>
5 Map.get(refunded,maxBidder) ==
6 Map.get(bidded,maxBidder) - Map.get(refunded,beneficiary)
```

We then add that the highest bidder is never the beneficiary, and require the following skeleton-state-specific assertions to support the proof:

```
1 @StartAuction Map.get(refunded,beneficiary) == 0
2 @AuctionOpen Map.get(refunded,beneficiary) == 0
```

The Asp proof-checker checks the initiality of these assertions and their inductiveness over every contract transition. (Sufficiency is not needed, as the assertion

---

[9]This property requires case-specific assertions because Coin.value(maxBid) is 0 after the coin is sent to the beneficiary.

encodes the desired property directly.) The inductiveness checks turn into Hoare triples, which are compiled to Viper and verified automatically.

4.2 Proofs of Timer-Supported Reachability

Timers in an Asp contract can be used to check reachability. However, the state-changes of the corresponding timer state machines are implicit, as is the progress of time which, in a blockchain, is measured by the non-uniform measure of blockchain length. To incorporate these implicit progress measures, we add a self-loop time-progression transition, time, to every skeleton state. The time transition is enabled at state s only if there is some timer in an active state at s. Its effect is to advance time by at least one unit and update the state of all active timers accordingly. (The time transition guard ensures that it is never enabled for a contract without timers.)

With this addition, the reachability proof scheme is formally as follows. A proof consists of a state assertion θ and a partial rank function ρ over a well-founded relation $\prec$ that meet the following conditions (where R represents the set of states for which we wish to prove reachability):

1. θ holds of the initial state,
2. ρ is defined for all states in θ that are not in R, and
3. For every state s in θ but not in R:
   (a) Some transition (either an explicit contract transition or the implicit time transition) is enabled at s, and
   (b) For every transition from s to a state t, it is the case that either R holds at t, or θ holds at t and $\rho(t) \prec \rho(s)$

This proof method is sound for reachability. Consider, to the contrary, that there is a maximal contract computation that does not include a state in R. By the first and third rules, every state on this computation satisfies θ but not R. This computation cannot be finite, as the final state must have an enabled transition, contradicting maximality. Hence it must be infinite: but then it induces an infinite decreasing chain in $\prec$, which contradicts well-foundedness.

Illustrating Liveness Proofs with Timers. We prove that AuctionClosed will eventually be reached, which ensures that the auction will close and the beneficiary will receive the winnings. The following timer-based proof verifies this.

```
reachability auction_closed(2) { // "2" is the rank length
  goal = {
    @AuctionClosed true
    /* other state-specific goals are false by default */
  }
  invariant  =  {
    @StartAuction  Timer.is_off(tmr)
```

```
 8        @AuctionOpen     !Timer.is_off(tmr)
 9    }
10    rank = { /* partial function, defined by cases */
11        @StartAuction
12          | (2,0)
13        @AuctionOpen   /* Order is important */
14          | (1, 0) if Timer.has_fired (tmr)
15          | (1, Timer.value(tmr)) if Timer.is_active (tmr)
16        @AuctionClosed
17          | (0, 0)
18    }
19    witness = {
20        @StartAuction a==owner && a != Address.none
21        @AuctionOpen  a != beneficiary && a != Address.none
                && Coin.value(c) > Coin.value(maxBid)
22    }
23 }
```

The Asp verifier compiles this proof outline to a set of lemmas for the Viper verification tool, which encode all the checks defined for the timer-based reachability proof rule. The enabledness check (3(a)) requires existentially quantifying the free variables of each transition. As existential quantification is not well-supported by Viper[10], we explicitly write the existential witness necessary to show that transitions are enabled at each state. The well-founded set is the set of natural-number tuples (of a fixed length), ordered lexicographically. Verification of the generated lemmas takes less than 2 seconds.

4.3   Proofs of Liveness in Adversarial Environments

A class of attacks on contracts consists not in stealing cryptocurrency, but rather in "freezing up" the contract so that it becomes unusable, so that any funds stored in the contract are inaccessible. Showing that such lockouts are not possible requires reasoning about potential adversarial actions in the multi-agent setting of Web3.

One way to formulate the property is to do so in game-theoretic terms. We consider an external Player (address) x and show that from any reachable contract state, there is a winning strategy for the Player to reach a contract state satisfying property Q (say a state where a message from x must be accepted). The Opponents are the other agents and the contract M itself, as M contains non-deterministic τ-actions which are resolved arbitrarily.

The following deductive proof system establishes this property. It is inspired by similar proof systems (cf. [16]) for μ-calculus properties. A proof consists of a state assertion θ and a partial rank function ρ that meet the following conditions:

1. θ is an invariant of the contract M,
2. ρ is defined for all states in θ, and

---

[10]From https://viper.ethz.ch/tutorial//#quantifiers.

3. For every state s in θ, one of the following holds:
   (a) s satisfies Q, or
   (b) There is a transition for the Player to a state s' that is in θ, and the rank decreases strictly after that transition, or
   (c) Some Opponent transition is enabled and all Opponent transitions lead to states in θ and strictly decrease rank.

The soundness of this proof rule is established as follows. Consider any reachable state, say s. As this state is reachable, it is in the invariant θ. From this state, the choices of the Player and Opponent produce a game subtree where all tree states are in θ. This tree cannot have an infinite branch where Q never holds, as rank decreases strictly on every transition, but the domain of ρ is well-founded. Thus, every branch must end in a state satisfying Q. The proof rule is also relatively complete: in fact, it is deduced from a μ-calculus framing of the property [7]. Implementation of this proof system is in progress.

5   Defensive Compilation

The Asp compiler translates Asp contracts to Solidity, the most popular contract language; a translation to Ink! (and Rust) is in progress. To make the discussion concrete, we focus on Solidity and Ethereum; the translation to other languages and blockchains is similar.

We briefly summarize the aspects of contract execution on Ethereum that are most relevant to compilation. On a blockchain, a contract is passive until one of its interface methods is invoked by an external entity; this invocation is called a *transaction*. The externally-invoked method may recursively invoke other methods, including those of other contracts. A transaction is executed by the EVM (Ethereum Virtual Machine) in a single-threaded manner until completion. On successful completion, changes to the contract state are committed to the blockchain. On failure, which can be due either to an undefined instruction (e.g., divide by zero) or a programmed 'revert' instruction, the state of the blockchain is not changed. Every transaction execution has a cost in cryptocurrency, known as the 'gas' fee.

The compiler must transform the state-machine view of an Asp programmer to the method-invocation view of the underlying contract execution engine. The compiler does so by essentially turning an Asp state machine skeleton inside-out. Every Asp message is transformed into a publicly accessible contract method. Within this method, a case analysis by (skeleton) state determines the transition that is executed and the next skeleton state. If no transition is enabled, the compiled contract reverts. Cascade semantics is implemented through a (private) `tau_closure` method that repeatedly executes τ transitions from the next skeleton state on, until a state without τ transitions is reached. The source Asp contract may be internally non-deterministic; the compiler determinizes execution by making an arbitrary choice between simultaneously enabled τ transitions.

Message sends are converted to a low-level Solidity `call` operation. Ghost variables and ghost operations are not compiled to Solidity; they are used only for proofs.

In addition, the compiler inserts code that, at run time, performs two functions: (a) it ensures that the assumptions underlying verification (such as no arithmetic overflow, no reentrancy, no undefined actions) hold during execution and (b) it checks for properties that may be cumbersome to prove, such as coin conservation. These checks guard against semantics violations and are thus defensive in nature; hence, we refer to the process as *defensive compilation*. The no-reentrancy property is enforced by introducing a "reentrancy counter" that is checked then incremented on entry into every invocable method and decremented on exit. A reentrant call exceeding the reentrancy limit will fail the check and cause the entire transaction to be canceled.

The Asp semantics has no notion of explicit failure. However, a transaction of the compiled Solidity code may be canceled due to the failure of a compiled Asp guard, or due to the failure of an inserted defensive check, or due to a detected arithmetic overflow. The first two cases match the Asp semantics, as the corresponding transition is not defined in Asp. The third case is one where the Asp transition is defined (as arithmetic in Asp is ideal) but the compiled version does not succeed. As a consequence, compilation correctness is expressed not as an equivalence but as a language inclusion. This gives us the following important property for Asp compilation.

Property 1. Let $C_1, \ldots, C_n$ be Asp contracts with corresponding compiled contracts $S_1, \ldots, S_n$. Every successful transaction of the compiled contracts corresponds to a successful cascade of the Asp contracts.

As a corollary, every safety property of an Asp contract is a safety property of the compiled contract. The corollary is important as it ensures that proof effort is required only at the Asp level; assuming correct compilation, there is no need to re-check the code generated for (multiple) target blockchains.

These correspondences also hold for adversarial liveness and reachability properties under the assumption that arithmetic overflow or shortage of gas do not cause a transaction to revert. We then obtain in the other direction that every Asp contract cascade is matched by a successful transaction of the compiled code. This induces a bisimulation between the Asp and compiled Solidity transition systems that preserves adversarial liveness and reachability.

6   Related Work and Discussion

The importance of ensuring that smart contracts are free of bugs was recognized early. In 2016, only a year after Ethereum was created, a hacker stole a large amount of ETH (worth about $3B today) in the "The DAO" exploit. That led to a controversial decision to "hard fork" the Ethereum chain to recover the funds. This recourse is impossible today: there are far too many active (and buggy) contracts to hard-fork the chain on each exploit.

Several companies (e.g., CertiK, Certora) offer smart contract audit services, which look for errors in smart contracts.[11] Tools such as MYTHRIL[12] and the Solidity SMT tools [21] search for errors using bounded model checking and automatic invariant inference. The tools and services are valuable for detecting potential errors, but they do not provide a comprehensive guarantee that a contract is free of vulnerabilities. The VerX [19] tool model-checks safety properties of Solidity contracts, while the SmartPulse [22] tool checks LTL properties. Both tools extensively apply predicate abstraction methods. Fully automated checkers are, however, inherently limited by state explosion.

In Sect. 1, we point to the difficulty of formally proving security for contracts written in full-fledged languages such as Solidity. The solc-verify tool [11] checks user-supplied assertions on Solidity contracts, but is restricted to assertions that are quantifier-free. Hence, one cannot typically write proofs about maps, for instance. 2vyper [4] is a verification system for the Vyper language, a variant of Solidity. As neither the Solidity nor the Vyper language restrict reentrancy, it is necessary to prove either that the contract code blocks reentrancy, or that reentrancy, if it occurs, does not lead to a vulnerability. The 2vyper system includes special proof rules to prove these assertions. While the proof system is technically interesting, such proofs add a substantial burden in practice. A similar reentrancy-sensitive proof system is defined in [5] for contracts written in the Dafny language (and potentially compiled to Solidity, although that is not yet implemented).

Such difficulties have led to the development of several non-standard programming languages for smart contract development (including Asp) which eliminate reentrancy as an issue. Move [3] is an object-oriented language with a non-standard, strict type system, similar to the "borrow" system of Rust. The type system naturally blocks reentrancy. The Move verifier [10,24] checks user-supplied proofs and is impressive in its scope and application. However, the non-standard type system of the Move language must be enforced by the underlying virtual machine execution, which limits portability. Indeed, Move is tightly linked to the Diem blockchain and variants such as Aptos and Sui, and has not been ported to common blockchains such as Ethereum and Solana.

Obsidian [9] is another non-standard object-oriented contract language that uses typestate and linear typing to enable static verification of resource use. However, Obsidian does not in itself support the verification of user-supplied safety or liveness properties.

While the languages discussed so far are conventional in their structure and constructs, FSolidM [13] models a contract as a finite-state machine skeleton that is extended with Solidity state variables and statements. The associated VeriSolid [14] tool checks CTL properties of sequences of transition labels of FSolidM machines, using automated data abstraction and model checking via BIP and the NuSMV tool. While model checking has its advantages, the scope

---

[11]http://www.certik.com and http://www.certora.com.
[12]https://github.com/Consensys/mythril.

of verified properties is restricted to propositional CTL (thus no arithmetic or quantified assertions), and the language is closely tied to Solidity.

Contracts in Scilla [20] are also represented as state machines. However, Scilla is viewed as an intermediate notation, with contracts compiled to Coq for verification. The Tezos blockchain also translates from its bytecode notation to Coq [2]. The KEVM formalization of Solidity semantics [12] is similar in nature. While translations to Coq permit complex properties to be expressed and verified, they also create a substantial proof burden, requiring considerable expertise and manual effort.

With Asp, we explore a different point in the design space. The Asp language inherently blocks several commonly exploited vulnerabilities, eliminating a significant concern for programmers. Like the FSolidM designers, we believe that an explicit state-machine skeleton simplifies contract design. However, Asp contracts do not operate on Solidity data types or expressions. Instead, Asp contracts operate on abstract data types, whose mathematical definitions simplify the writing of contracts, and their analysis. Abstract operations are concretized by the compiler and are therefore correct by construction.

Indeed, its reliance on abstract data types is one of the major distinguishing features of Asp. A further advantage of programming with abstract types is portability: a verified Asp contract can be compiled to multiple languages and blockchains. The Asp type system is otherwise conventional (unlike that of Move), which faciliates compilation to standard languages such as Solidity, Rust, and Ink! and to commonly used blockchains such as Ethereum and Solana.

Asp, like Move, relies on programmer-supplied deductive proofs. We believe that this is crucial for security, as deductive proofs allow greater expressiveness (e.g., quantifiers to express properties of maps) and flexibility (e.g., safety, liveness, and adversarial proofs). A current drawback is that proof sketches for even simple properties must be supplied manually; in the future, we aim to augment programmer-supplied proofs with *proof-generating* model checking [16,18]. As argued previously, we believe that it is important in the context of Web3 to provide explicit, independently-checkable proofs for every security claim. In this sense, the design of Asp follows the proof-carrying-code principle [17] of placing the burden of constructing a proof on the contract creator (who may use automated methods) while making it possible for every user to independently validate a claimed proof.

Our motivation in developing Asp is to explore the language design space, prioritizing ease of reasoning over sophisticated language features. Abstractions compensate for the simpler language structure; the defensive compiler enforces the abstract semantics and strengthens security; and verifiable proofs build trust. We expect Asp to evolve over time.[13] We are considering the introduction of new abstractions, such as commitments, secret inputs, and randomness, and modular

---

[13] We intend to release the Asp system as open source once we have the necessary approvals. At present, we have made several Asp contracts (with proofs) available at https://github.com/DebraChait/Asp-example-contracts.

proof methods. We also plan to explore how verified proof assertions can be used by the compiler to eliminate run-time defensive checks (cf. [23]).

References

1. Alpern, B., Schneider, F.B.: Defining liveness. Inf. Process. Lett. **21**(4), 181–185 (1985)
2. Bernardo, B., Cauderlier, R., Hu, Z., Pesin, B., Tesson, J.: Mi-Cho-Coq, a framework for certifying Tezos smart contracts. In: FM Workshops (1). Lecture Notes in Computer Science, vol. 12232, pp. 368–379. Springer (2019)
3. Blackshear, S., et al.: Move: a language with programmable resources. https://developers.diem.com/papers/diem-move-a-language-with-programmable-resources/2019-06-18.pdf
4. Bräm, C., Eilers, M., Müller, P., Sierra, R., Summers, A.J.: Rich specifications for Ethereum smart contract verification. Proc. ACM Program. Lang. **5**(OOPSLA), 1–30 (2021)
5. Cassez, F., Fuller, J., Quiles, H.M.A.: Deductive verification of smart contracts with dafny. In: FMICS. Lecture Notes in Computer Science, vol. 13487, pp. 50–66. Springer (2022)
6. Hack3d: The web3 security report 2023. At https://www.certik.com/resources/blog/7BokMhPUgffqEvyvXgHNaq-hack3d-the-web3-security-report-2023
7. Chait-Roth, D., Namjoshi, K.S.: Constructing trustworthy smart contracts (2024). https://arxiv.org/abs/2411.14563
8. Crypto investors lost nearly $4 billion to hackers in 2022. At https://www.cnbc.com/2023/02/04/crypto-investors-lost-nearly-4-billion-dollars-to-hackers-in-2022.html
9. Coblenz, M., et al.: Obsidian: typestate and assets for safer blockchain programming. ACM Trans. Program. Lang. Syst. **42**(3) (Nov 2020). https://doi.org/10.1145/3417516
10. Dill, D., Grieskamp, W., Park, J., Qadeer, S., Xu, M., Zhong, E.: Fast and Reliable Formal Verification of Smart Contracts with the Move Prover. Presented at the (2022). https://doi.org/10.1007/978-3-030-99524-9_10
11. Hajdu, Á., Jovanović, D.: SOLC-verify: a modular verifier for Solidity smart contracts. In: Chakraborty, S., Navas, J.A. (eds.) Verified Software. Theories, Tools, and Experiments, Lecture Notes in Computer Science, vol. 12301, pp. 161–179. Springer (2020)
12. Hildenbrandt, E., et al.: KEVM: a complete formal semantics of the ethereum virtual machine. In: CSF, pp. 204–217. IEEE Computer Society (2018)
13. Mavridou, A., Laszka, A.: Designing secure ethereum smart contracts: a finite state machine based approach. In: Meiklejohn, S., Sako, K. (eds.) FC 2018. LNCS, vol. 10957, pp. 523–540. Springer, Heidelberg (2018). https://doi.org/10.1007/978-3-662-58387-6_28
14. Mavridou, A., Laszka, A., Stachtiari, E., Dubey, A.: VeriSolid: correct-by-design smart contracts for ethereum. In: Goldberg, I., Moore, T. (eds.) FC 2019. LNCS, vol. 11598, pp. 446–465. Springer, Cham (2019). https://doi.org/10.1007/978-3-030-32101-7_27
15. Müller, P., Schwerhoff, M., Summers, A.J.: Viper: a verification infrastructure for permission-based reasoning. In: Jobstmann, B., Leino, K.R.M. (eds.) VMCAI 2016, LNCS, vol. 9583, pp. 41–62. Springer, Heidelberg (2016). https://doi.org/10.1007/978-3-662-49122-5_2

16. Namjoshi, K.S.: Certifying model checkers. In: Berry, G., Comon, H., Finkel, A. (eds.) CAV 2001, LNCS, vol. 2102, pp. 2–13. Springer, Heidelberg (2001). https://doi.org/10.1007/3-540-44585-4_2

17. Necula, G.C.: Proof-carrying code. In: POPL, pp. 106–119. ACM Press (1997)

18. Peled, D., Pnueli, A., Zuck, L.: From falsification to verification. In: Hariharan, R., Vinay, V., Mukund, M. (eds.) FSTTCS 2001, LNCS, vol. 2245, pp. 292–304. Springer, Heidelberg (2001). https://doi.org/10.1007/3-540-45294-X_25

19. Permenev, A., Dimitrov, D., Tsankov, P., Drachsler-Cohen, D., Vechev, M.: VerX: safety verification of smart contracts. In: IEEE Symposium on Security and Privacy (2020)

20. Sergey, I., Nagaraj, V., Johannsen, J., Kumar, A., Trunov, A., Hao, K.C.G.: Safer smart contract programming with Scilla. Proc. ACM Program. Lang. **3**(OOPSLA), 185:1–185:30 (2019)

21. SMTChecker and Formal Verification for Solidity. At https://docs.soliditylang.org/en/v0.8.23/smtchecker.html

22. Stephens, J., Ferles, K., Mariano, B., Lahiri, S., Dillig, I.: SmartPulse: automated checking of temporal properties in smart contracts. In: IEEE Symposium on Security and Privacy (2021)

23. Sun, H., Singh, K., Ramos-Dávila, J.P., Aldrich, J., DiVincenzo, J.: Gradual verification for smart contracts. In: PriSC (2024). https://arxiv.org/pdf/2311.13351.pdf

24. Zhong, J.E., et al.: The move prover. In: Lahiri, S.K., Wang, C. (eds.) CAV 2020, LNCS, vol. 12224, pp. 137–150. Springer, Cham (2020). https://doi.org/10.1007/978-3-030-53288-8_7

Author Index

The manufacturer's authorised representative in the EU is Springer
Nature Customer Service Centre GmbH, Europaplatz 3, 69115 Heidelberg,
Germany. If you have any concerns regarding our products, please
contact ProductSafety@springernature.com

Printed and bound by CPI Group (UK) Ltd, Croydon, CR0 4YY

27/04/2026

02097586-0003